Chemical Process Industries

The rapid growth and expansion of the chemical process industry during the past century have been accompanied by a simultaneous rise in human health problems as well as material and property losses because of fires, explosions, hazardous and toxic spills, equipment failures, other accidents, and business interruptions. Concern over the potential consequences of emissions of harmful chemicals (along with catastrophic accidents) has sparked interest at both the industrial and regulatory levels in obtaining a better understanding of the potential for environmental health risks in chemical and related industries. This practical book presents and examines the environmental and health risk assessment calculations as they apply to various chemical process industries.

Chemical Process Industries: Environmental and Health Risk Calculations can be used as a college text designed to provide new engineers and scientists some comprehension of the industries into which they may enter. It also serves as a useful reference for practitioners and will help them better understand the health risk aspects of various industrial operations. The chemical process industries employ mechanical, electrical, and civil engineers and a host of other scientists; these professions should also benefit from material in this book that applies to their fields of work.

Chemical Process Industries
Environmental and Health Risk Calculations

Louis Theodore
R. Ryan Dupont

CRC Press
Taylor & Francis Group
Boca Raton London New York

CRC Press is an imprint of the
Taylor & Francis Group, an **informa** business

Designed cover image: ©Shutterstock

First edition published 2023
by CRC Press
6000 Broken Sound Parkway NW, Suite 300, Boca Raton, FL 33487-2742

and by CRC Press
4 Park Square, Milton Park, Abingdon, Oxon, OX14 4RN

CRC Press is an imprint of Taylor & Francis Group, LLC

© 2023 Louis Theodore & R. Ryan Dupont

Reasonable efforts have been made to publish reliable data and information, but the author and publisher cannot assume responsibility for the validity of all materials or the consequences of their use. The authors and publishers have attempted to trace the copyright holders of all material reproduced in this publication and apologize to copyright holders if permission to publish in this form has not been obtained. If any copyright material has not been acknowledged please write and let us know so we may rectify in any future reprint.

Except as permitted under U.S. Copyright Law, no part of this book may be reprinted, reproduced, transmitted, or utilized in any form by any electronic, mechanical, or other means, now known or hereafter invented, including photocopying, microfilming, and recording, or in any information storage or retrieval system, without written permission from the publishers.

For permission to photocopy or use material electronically from this work, access www.copyright.com or contact the Copyright Clearance Center, Inc. (CCC), 222 Rosewood Drive, Danvers, MA 01923, 978-750-8400. For works that are not available on CCC please contact mpkbookspermissions@tandf.co.uk

Trademark Notice: Product or corporate names may be trademarks or registered trademarks and are used only for identification and explanation without intent to infringe.

Library of Congress Cataloging-in-Publication Data

Names: Theodore, Louis, author. | Dupont, R. Ryan, author.
Title: Chemical process industries: environmental and health risk calculations / Louis Theodore, R. Ryan Dupont.
Description: First edition. | Boca Raton: CRC Press, [2023] | Includes bibliographical references and index.
Identifiers: LCCN 2022034549 (print) | LCCN 2022034550 (ebook) | ISBN 9781032041858 (hardback) | ISBN 9781032254883 (paperback) | ISBN 9781003283454 (ebook)
Subjects: LCSH: Chemical industry--Environmental aspects. | Chemical processes--Environmental aspects. | Environmental health. | Pollution--Risk assessment.
Classification: LCC TD195.C45 T45 2023 (print) | LCC TD195.C45 (ebook) | DDC 363.738/4--dc23/eng/20221011
LC record available at https://lccn.loc.gov/2022034549
LC ebook record available at https://lccn.loc.gov/2022034550

ISBN: 978-1-032-04185-8 (hbk)
ISBN: 978-1-032-25488-3 (pbk)
ISBN: 978-1-003-28345-4 (ebk)

DOI: 10.1201/9781003283454

Typeset in Times New Roman
by KnowledgeWorks Global Ltd.

*To my newsletter, www.theodorenewsletter.com,
15,000 hits and going strong.*
(Louis Theodore)

*To my late parents Bob and Marie Dupont, thanks
for all you did and all you gave.*
(R. Ryan Dupont)

Look to your health; and if you have it, praise God, and value it next to a good conscience; for health is the second blessing that we mortals are capable of, a blessing that money cannot buy.

Izaaic Walton (1593–1683)

Contents

About the Authors xxi
Contributor xxiii
Preface xxv

PART I Introduction to Environmental and Health Risk

Chapter 1 Definitions/Glossary of Environmental and Health Risk Terms 3

    1.1 Introduction 3
    1.2 Terms and Definitions 3
    1.3 Illustrative Examples 7
        1.3.1 Illustrative Example 1 7
        1.3.2 Illustrative Example 2 7
        1.3.3 Illustrative Example 3 7
        1.3.4 Illustrative Example 4 7
    Problems 8
    References 8

Chapter 2 Introduction to Environmental and Health Risk 9

    2.1 Introduction 9
    2.2 Risk Variables and Categories 10
        2.2.1 Risk Variables 10
        2.2.2 Risk Categories 11
    2.3 Financial Risk 11
    2.4 Sports Risk 13
    2.5 Environmental and Health Risk Terms 14
    2.6 Risk Errors 16
    2.7 Illustrative Examples 18
        2.7.1 Illustrative Example 1 18
        2.7.2 Illustrative Example 2 18
        2.7.3 Illustrative Example 3 19
        2.7.4 Illustrative Example 4 19
    Problems 19
    References 20

Chapter 3 Environmental and Health Risk Analysis 21

    3.1 Introduction 21
    3.2 Health Risk Assessment/Analysis 23

		3.2.1	Health Problem Identification	24
		3.2.2	Toxicity and Dose-response Assessment	25
		3.2.3	Exposure Assessment	26
		3.2.4	Health Risk Characterization	27
	3.3	Hazard Risk Assessment/Analysis		28
		3.3.1	Hazard/event Problem Identification	30
		3.3.2	Hazard/event Probability	31
		3.3.3	Hazard/event Consequences	32
		3.3.4	Hazard Risk Characterization	32
	3.4	Risk Uncertainties/Limitations		33
		3.4.1	Health Risk	34
		3.4.2	Hazard Risk	35
	3.5	Environmental Regulations		37
		3.5.1	Air	38
		3.5.2	Water	38
		3.5.3	Solid Waste	38
		3.5.4	Pollution Prevention	39
	3.6	Illustrative Examples		39
		3.6.1	Illustrative Example 1	39
		3.6.2	Illustrative Example 2	40
		3.6.3	Illustrative Example 3	40
		3.6.4	Illustrative Example 4	40
	Problems			40
	References			41
Chapter 4	Introduction to Probability and Statistics			43
	4.1	Introduction		43
	4.2	Probability Definitions and Interpretations		43
	4.3	Basic Probability Theory		46
	4.4	Median, Mean, and Standard Deviation		49
	4.5	Random Variables		51
	4.6	Linear Regression		51
	4.7	Illustrative Examples		56
		4.7.1	Illustrative Example 1	56
		4.7.2	Illustrative Example 2	57
		4.7.3	Illustrative Example 3	57
		4.7.4	Illustrative Example 4	58
	Problems			59
	References			59
Chapter 5	Probability Distributions			61
	5.1	Introduction		61
	5.2	Discrete Probability Distributions		62
		5.2.1	Binomial Distribution	64
		5.2.2	Multinominal Distribution	65

		5.2.2.1	Permutations and Combinations	65
		5.2.2.2	Multinomial Theorem	69
	5.2.3	Hypergeometric Distribution		70
	5.2.4	Poisson Distribution		72
5.3	Continuous Probability Distributions			74
	5.3.1	Weibull Distribution		76
	5.3.2	Normal Distribution		82
	5.3.3	Exponential Distribution		87
	5.3.4	Log-normal Distribution		89
	5.3.5	Other Continuous Probability Distributions		92
5.4	Illustrative Examples			93
	5.4.1	Illustrative Example 1		93
	5.4.2	Illustrative Example 2		94
	5.4.3	Illustrative Example 3		94
	5.4.4	Illustrative Example 4		95
	5.4.5	Illustrative Example 5		96
	5.4.6	Illustrative Example 6		96
	5.4.7	Illustrative Example 7		97
Problems				97
References				98

PART II Chemical Process Industries

Chapter 6 Definitions/Glossary of Chemical Process Terms 101

Emma Parente

6.1	Introduction		101
6.2	Terms and Definitions		102
6.3	Illustrative Examples		120
	6.3.1	Illustrative Example 1	120
	6.3.2	Illustrative Example 2	121
	6.3.3	Illustrative Example 3	121
	6.3.4	Illustrative Example 4	121
	6.3.5	Illustrative Example 5	122
	6.3.6	Illustrative Example 6	122
	6.3.7	Illustrative Example 7	122
	6.3.8	Illustrative Example 8	122
Problems			123
References			123

Chapter 7 History 125

7.1	Introduction	125
7.2	Early History	125
7.3	The Role of Science	126
7.4	The Modern Chemical Process Industry	128

		7.5	The History of Engineering	129
		7.6	Sources of Information for the CPI	130
			7.6.1 Traditional Sources	130
			7.6.2 Engineering and Science Sources	131
			7.6.3 Internet Sources	131
			7.6.4 Personal Experience	132
		7.7	Illustrative Examples	132
			7.7.1 Illustrative Example 1	132
			7.7.2 Illustrative Example 2	132
			7.7.3 Illustrative Example 3	133
			7.7.4 Illustrative Example 4	133
	Problems			134
	References			134
Chapter 8	Chemical Process Equipment			135
	8.1	Introduction		135
	8.2	Chemical Reactors		135
		8.2.1 Reactor Definition		135
		8.2.2 Reactor Type		136
	8.3	Heat Exchangers		138
	8.4	Mass Transfer Equipment		140
		8.4.1 Distillation		140
		8.4.2 Adsorption		141
		8.4.3 Absorption		142
		8.4.4 Evaporation		142
		8.4.5 Extraction		143
		8.4.6 Drying		143
	8.5	Fluid Flow Equipment		143
		8.5.1 Pipes and Tubing		144
		8.5.2 Ducts		144
		8.5.3 Fittings		145
		8.5.4 Valves		146
		8.5.5 Fans and Blowers		146
		8.5.6 Pumps		147
		8.5.7 Compressors		147
		8.5.8 Stacks		148
	8.6	Ancillary Equipment		148
		8.6.1 Electricity		148
		8.6.2 Steam		149
		8.6.3 Water		149
			8.6.3.1 Cooling Water	149
			8.6.3.2 Potable and General Use Water	149
			8.6.3.3 Demineralized Water	150
		8.6.4 Refrigeration		150
		8.6.5 Compressed Air		150
		8.6.6 Inert Gas Supplies		150

Contents   xiii

      8.7   Material Transportation and Storage Equipment 151
            8.7.1   Gases 151
            8.7.2   Liquids 151
            8.7.3   Solids 152
      8.8   Instrumentation and Controls 152
            8.8.1   Feedback Loop Instrumentation
                    and Control Systems 153
            8.8.2   Automatic Trip Systems and Interlocks 154
      8.9   Process Diagrams 155
            8.9.1   Block Diagrams 156
            8.9.2   Graphic Flow Diagrams 157
            8.9.3   Process Flow Diagrams 157
            8.9.4   Process P&IDs 159
            8.9.5   Tree Diagrams 165
                    8.9.5.1   Fault Tree Analysis 165
                    8.9.5.2   Event Tree Analysis 165
            8.9.6   Preparing Flow Diagrams 165
      8.10 Illustrative Examples 168
            8.10.1 Illustrative Example 1 168
            8.10.2 Illustrative Example 2 169
            8.10.3 Illustrative Example 3 170
            8.10.4 Illustrative Example 4 172
            8.10.5 Illustrative Example 5 172
  Problems 173
  References 174

Chapter 9   Chemical Processes: Fundamentals and Principles 177

      9.1   Introduction 177
      9.2   The Chemical Process 177
      9.3   The Conservation Law 178
      9.4   Conservation of Mass, Energy, and Momentum 180
      9.5   Stoichiometry 182
      9.6   Limiting and Excess Reactants 184
      9.7   Optimum Process Design 185
      9.8   Problem Solving 187
            9.8.1   Generic Problem-solving
                    Techniques 187
            9.8.2   A Specific Problem-solving
                    Approach 188
            9.8.3   Some General Comments 189
      9.9   Illustrative Examples 190
            9.9.1   Illustrative Example 1 190
            9.9.2   Illustrative Example 2 191
            9.9.3   Illustrative Example 3 192
            9.9.4   Illustrative Example 4 192
            9.9.5   Illustrative Example 5 193

	9.9.6	Illustrative Example 6 193
	9.9.7	Illustrative Example 7 193
Problems 193
References 194

Chapter 10 Industry-Specific Processes 195

    10.1   Introduction 195
    10.2   The Early CPI 195
    10.3   The Shreve CPI 197
    10.4   The Theodore-Dupont CPI 198
    10.5   Illustrative Examples 198
            10.5.1   Illustrative Example 1 198
            10.5.2   Illustrative Example 2 198
            10.5.3   Illustrative Example 3 199
            10.5.4   Illustrative Example 4 199
    Problems 199
    Reference 199

Chapter 11 Emergency Planning and Response 201

    11.1   Introduction 201
    11.2   The Need for Emergency Response Planning 202
    11.3   The Planning Committee 203
    11.4   Hazard Surveys 207
    11.5   Planning for Emergencies 209
    11.6   Training of Personnel 212
    11.7   Notification of Public and Regulatory Officials 213
    11.8   Plan Implementation 215
    11.9   Illustrative Examples 216
            11.9.1   Illustrative Example 1 216
            11.9.2   Illustrative Example 2 216
            11.9.3   Illustrative Example 3 217
            11.9.4   Illustrative Example 4 217
            11.9.5   Illustrative Example 5 218
            11.9.6   Illustrative Example 6 218
            11.9.7   Illustrative Example 7 219
    Problems 220
    References 221

PART III Health Risk Calculations for Specific Chemical Process Industries

Chapter 12 Inorganic Chemicals 225

    12.1   Introduction 225
    12.2   General Comments 225

	12.3	Sulfuric Acid 227
	12.4	Hydrochloric Acid 227
	12.5	Ammonium Nitrate 228
	12.6	Sodium Chloride 229
	12.7	Cement 230
	12.8	Glass 231
	12.9	Illustrative Examples 232
		12.9.1 Illustrative Example 1 232
		12.9.2 Illustrative Example 2 233
		12.9.3 Illustrative Example 3 233
		12.9.4 Illustrative Example 4 234
	Problems 234
	References 235

Chapter 13 Organic Chemicals 237

	13.1	Introduction 237
	13.2	General Comments 237
	13.3	Phenol 238
	13.4	Ethyl Acetate 240
	13.5	Plastic 241
	13.6	Paper 242
	13.7	Rubber 243
	13.8	Illustrative Examples 243
		13.8.1 Illustrative Example 1 243
		13.8.2 Illustrative Example 2 244
		13.8.3 Illustrative Example 3 244
		13.8.4 Illustrative Example 4 245
	Problems 245
	References 246

Chapter 14 Petroleum Refining 247

	14.1	Introduction 247
	14.2	Drilling 248
	14.3	Refining/Processing 249
	14.4	Petrochemicals 252
	14.5	Transportation/Transmission 253
		14.5.1 Pipelines 254
		14.5.2 Ships 254
		14.5.3 Trains 254
	14.6	Illustrative Examples 255
		14.6.1 Illustrative Example 1 255
		14.6.2 Illustrative Example 2 255
		14.6.3 Illustrative Example 3 256
		14.6.4 Illustrative Example 4 256
	Problems 257
	References 257

Chapter 15 Energy and Power 259

15.1 Introduction 259
15.2 Fossil Fuels 259
    15.2.1 Coal 260
    15.2.2 Oil 260
    15.2.3 Natural Gas 260
    15.2.4 Oil Shale 261
15.3 Nuclear Energy 261
15.4 Solar Energy 262
15.5 Hydroelectric and Geothermal Energy 263
15.6 Power Generation 264
15.7 Air Conditioning and Refrigeration 265
    15.7.1 Air Conditioning 265
    15.7.2 Refrigeration 266
15.8 Illustrative Examples 267
    15.8.1 Illustrative Example 1 267
    15.8.2 Illustrative Example 2 267
    15.8.3 Illustrative Example 3 268
    15.8.4 Illustrative Example 4 269
    15.8.5 Illustrative Example 5 269
    15.8.6 Illustrative Example 6 270
Problems 270
References 271

Chapter 16 Pharmaceuticals 273

16.1 Introduction 273
16.2 General Comments (Barboza et al. 1977) 273
16.3 History 274
16.4 PhRMA 275
16.5 Research and Development 275
16.6 Process Descriptions 276
    16.6.1 Chemical Synthesis 277
    16.6.2 Fermentation 277
    16.6.3 Extraction 278
    16.6.4 Formulation 278
16.7 Other Operation and Facility Considerations 278
    16.7.1 Storage and Transfer 278
    16.7.2 Power and Steam Generation 279
    16.7.3 Waste Disposal 279
    16.7.4 Wastewater Treatment 279
    16.7.5 QA/QC 280
16.8 Illustrative Examples 280
    16.8.1 Illustrative Example 1 280
    16.8.2 Illustrative Example 2 280

Contents   xvii

            16.8.3   Illustrative Example 3 281
            16.8.4   Illustrative Example 4 282
   Problems 283
   References 284

Chapter 17   Food Products Industry 285

   17.1   Introduction 285
   17.2   History 285
   17.3   The Food and Drug Administration 286
   17.4   Food Processing and Preservation 287
            17.4.1   Canning 288
            17.4.2   Freezing 288
            17.4.3   Dehydration 288
            17.4.4   Miscellaneous Methods 289
   17.5   Refrigeration 289
   17.6   Food Additives 291
   17.7   Illustrative Examples 291
            17.7.1   Illustrative Example 1 292
            17.7.2   Illustrative Example 2 293
            17.7.3   Illustrative Example 3 294
            17.7.4   Illustrative Example 4 294
            17.7.5   Illustrative Example 5 294
   Problems 295
   References 295

Chapter 18   Nanotechnology 297

   18.1   Introduction 297
   18.2   Nanotechnology 297
            18.2.1   Nanomaterials 298
            18.2.2   Nanomaterial Production 299
                      18.2.2.1   High-Temperature Processes 300
                      18.2.2.2   Chemical Vapor Deposition (CVD) 300
                      18.2.2.3   Electrodeposition 300
                      18.2.2.4   Sol-Gel Synthesis 300
                      18.2.2.5   Mechanical Crushing Via Ball
                                     Milling 301
                      18.2.2.6   Naturally Occurring Materials 301
   18.3   Current Applications 301
   18.4   Environmental Implications 302
   18.5   Health Risk Assessment 304
   18.6   Hazard Risk Assessment 306
   18.7   Environmental Regulations 307
   18.8   Future Trends 310

18.9	Illustrative Examples	311
	18.9.1 Illustrative Example 1	311
	18.9.2 Illustrative Example 2	311
	18.9.3 Illustrative Example 3	312
	18.9.4 Illustrative Example 4	313
	18.9.5 Illustrative Example 5	313
Problems		318
References		319

Chapter 19 Military and Terrorism 321

19.1	Introduction	321
19.2	The US Military	321
19.3	Explosives	322
19.4	Terrorism	324
	19.4.1 International Terrorism	325
	19.4.2 The Need for Emergency Response Planning	326
	19.4.3 Anti-Terrorism Efforts	328
19.5	Current Risks and Prioritization for Risk Reduction	329
19.6	Illustrative Examples	330
	19.6.1 Illustrative Example 1	330
	19.6.2 Illustrative Example 2	331
	19.6.3 Illustrative Example 3	331
	19.6.4 Illustrative Example 4	332
	19.6.5 Illustrative Example 5	332
Problems		333
References		333

Chapter 20 Weather and Climate 335

20.1	Introduction	335
20.2	History	335
20.3	Climate	336
20.4	Meteorological Factors	337
	20.4.1 Plume Rise	338
	20.4.2 Effective Stack Height	339
20.5	Atmospheric Dispersion Modeling	340
20.6	The National Weather Service	341
20.7	Weather Observations	341
20.8	Weather Forecasting	342
20.9	Illustrative Examples	344
	20.9.1 Illustrative Example 1	344
	20.9.2 Illustrative Example 2	344
	20.9.3 Illustrative Example 3	345
	20.9.4 Illustrative Example 4	345
Problems		346
References		347

Contents   xix

Chapter 21 Architecture and Urban Planning 349

    21.1   Introduction 349
    21.2   History 350
    21.3   Current Debate on the Need for Sustainable Architecture 352
    21.4   Siting 353
    21.5   Design Considerations.. 353
    21.6   Materials Considerations 356
    21.7   Illustrative Examples 356
            21.7.1   Illustrative Example 1 357
            21.7.2   Illustrative Example 2 357
            21.7.3   Illustrative Example 3 357
            21.7.4   Illustrative Example 4 358
    Problems 358
    Reference 359

Chapter 22 Environmental Considerations 361

    22.1   Introduction 361
    22.2   Air Pollutants... 362
            22.2.1   Ozone and Carbon Monoxide 362
            22.2.2   Airborne Particulates 363
            22.2.3   Sulfur Dioxide and Acid Rain 364
            22.2.4   Hazardous Air Pollutants 365
            22.2.5   Indoor Air Pollutants 366
                    22.2.5.1   Radon 367
                    22.2.5.2   Secondhand Smoke 367
                    22.2.5.3   Other Combustion Products 368
                    22.2.5.4   Volatile Organic Compounds................... 369
                    22.2.5.5   Biologicals 369
    22.3   Water Pollutants... 370
            22.3.1   Drinking Water Supplies 370
                    22.3.1.1   Lead 371
                    22.3.1.2   Arsenic 372
                    22.3.1.3   Radionuclides 372
                    22.3.1.4   Microbiological Contaminants 373
                    22.3.1.5   Disinfection By-products 375
            22.3.2   Surface Water Pollutants 376
                    22.3.2.1   Point Sources 376
                    22.3.2.2   Nonpoint Sources 378
    22.4   Solid Waste 379
            22.4.1   Non-Hazardous Waste 379
            22.4.2   Hazardous Waste 380
    22.5   Toxic Substances 381
            22.5.1   New and Existing Chemicals Program 382
            22.5.2   Polychlorinated Biphenyls Program 384
            22.5.3   Asbestos Program 384

	22.5.4	Lead-Based Paint Program	384
	22.5.5	Formaldehyde	385
22.6	Illustrative Examples		385
	22.6.1	Illustrative Example 1	385
	22.6.2	Illustrative Example 2	386
	22.6.3	Illustrative Example 3	387
	22.6.4	Illustrative Example 4	387
Problems			388
References			389

Index 391

About the Authors

Louis Theodore, EngScD, was raised in Hell's Kitchen and received MChE and EngScD degrees from New York University and a BChE degree from The Cooper Union. During a 50-year tenure, Dr. Theodore was a successful educator (holding the rank of Full Professor of chemical engineering), Graduate Program Director (raising extensive financial support from local industries), researcher, professional innovator, and communicator in the engineering field. During this period, he was primarily responsible for his program achieving a No. 2 ranking by US News & World Report and was particularly successful in placing students in internships, jobs, and graduate schools across the United States.

Dr. Theodore is an internationally recognized lecturer who has provided more than 200 courses to industry, government, and technical associations and has served as an after-dinner or luncheon speaker on numerous occasions. He developed and served as principal moderator/lecturer for the US Environmental Protection Agency, for courses on hazardous waste incineration, air control equipment and health and hazard risk assessment, consulted for several industrial companies in the field of pollution prevention and environmental management, and served as a consultant/expert witness for the US EPA and US Department of Justice. Dr. Theodore is the author of more than 130 text/reference books (plus Basketball Coaching 101) ranging from pollution prevention to air pollution control to hazardous waste incineration to health risk assessment and to engineering and environmental ethics.

He is the recipient of the Air and Waste Management Association's prestigious Ripperton award that is "presented to an outstanding educator who through example, dedication and innovation has so inspired students to achieve excellence in their professional endeavors." He was also the recipient of the American Society for Engineering Education AT&T Foundation award for "excellence in the instruction of engineering students."

He currently serves as a part-time consultant to Theodore Tutorials. Dr. Theodore is also a member of Phi Lambda Upsilon, Sigma Xi, Tau Beta Pi, the American Chemical Society, the American Society of Engineering Education, and the Royal Hellenic Society and is a Fellow of the Air and Waste Management Association; he was also recently honored at Madison Square Garden for his contributions to basketball and the youth of America.

R. Ryan Dupont, PhD, has over 40 years of experience teaching and conducting applied and basic research in environmental engineering at the Utah Water Research Laboratory at Utah State University (USU). His main research areas have addressed soil and groundwater bioremediation, stormwater management via green infrastructure, and field remediation technology demonstration and treatment system performance verification and has lectured extensively in these and other environmental areas. He received his BS degree in civil engineering and MS and PhD degrees in environmental health engineering from the University of Kansas, Lawrence.

Dr. Dupont has been a Full Professor of Civil and Environmental Engineering at USU since 1995, served as the Head of the Environmental Engineering Program for 10 years, was instrumental in establishing an undergraduate degree in Environmental Engineering at USU, and has been responsible for attracting more than $7 million in extramural funding at the Water Research Lab since joining the faculty in 1982. He is currently a Co-Principal Investigator of a $35 million US AID project designed to improve water engineering practices in Egypt.

Dr. Dupont is a member of Sigma Xi, Tau Beta Pi, Chi Epsilon, the American Society of Civil Engineers, the American Society of Engineering Educators, the Water Environment Federation, Engineers without Borders, and the Air and Waste Management Association. Dr. Dupont was recognized as an Outstanding Young Engineering Educator by the American Society of Engineering Education in 1988 and was a 2015 recipient of the Richard I. Stessel Waste Management Award for "distinguished achievement as an educator in the field of waste management" from the Air and Waste Management Association. In 2021 he was awarded a Cazier Endowed Professorship from USU in recognition of lifetime achievement at the university.

Contributor

Emma Parente
Muhlenberg College
Allentown, PA, USA

Preface

The rapid growth and expansion of the chemical process industry during the past century have been accompanied by a simultaneous rise in enormous human health problems as well as material, and property losses because of fires, explosions, hazardous and toxic spills, equipment failures, other accidents, and business interruptions. Concern over the potential consequences of emissions of "harmful" chemicals (along with catastrophic accidents), particularly at the chemical process industry level, has sparked interest at both the industrial and regulatory levels in obtaining a better understanding of the subject of this book. The writing of this book was undertaken, in part, as a result of this growing concern.

Environmental and health risk comes into play in countless real-world process industrial applications and poses unique challenges for the environmental community. An integral part of this subject area is the role that elementary statistics and probability distributions play in health risk analysis. Therefore, the authors considered writing a book that highlights pragmatic material with problems (and their solutions) as they relate to this topic. This book will hopefully serve as a training tool for those individuals in academia and industry involved with environmental and health risk analysis. Although the literature is inundated with texts emphasizing traditional environmental and health topics, the goal of this book is to present this subject of environmental and health risk by employing a pragmatic approach.

This book is divided into three parts; Part I (Introduction to Environmental and Health Risk) serves as an introduction to both environmental and health risk and probability and statistics. This part basically presents principles required in the solution of technical topics covered in the remainder of the book; in particular, it treats the broad subject of health risk assessments. Part II (Chemical Process Industries) presents, lists, and examines the process industries (primarily chemical) in significant detail. The chapters in this part review a host of industries. The application and calculations of environmental and health risk assessment analyses for many of the processes in the previous part receive treatment in Part III (Chemical Process Industry Specific Health Risk Calculations) through the presentation and solution of numerous illustrative examples.

Part I

Introduction to Environmental and Health Risk

The first part of this book serves to introduce the reader to both environmental and health risk and basic statistics. Following a chapter concerned with definitions and a glossary of terms related to environmental and health risk, the remaining four chapters in Part I are split between environmental and health risk (Chapters 2 and 3) and probability and statistics (Chapters 4 and 5).

1 Definitions/Glossary of Environmental and Health Risk Terms

1.1 INTRODUCTION

The following list of Terms and Definitions is not a complete glossary of all terms that appear in the risk and risk-related fields. It should also be noted that many of the terms can mean different things to different people; this will become evident as one delves deeper into the literature (Theodore, Reynolds and Morris 1997; Shaefer and Theodore 2007; Theodore and Dupont 2012; Theodore and Theodore 2021). The definition of terms related to the chemical process industry receives treatment in Part II, Chapter 6.

1.2 TERMS AND DEFINITIONS

Acute (risk): Risk associated with short periods of time. For health risk, it usually represents short exposures to high concentrations of a hazardous agent.
Autoignition temperature: The lowest temperature at which a flammable gas in air will ignite without an ignition source.
Average rate of death (ROD): The average number of fatalities that can be expected per unit time (usually on an annual basis) from all possible risks and/or incidents.
C (ceiling): The term used to describe the maximum allowable exposure concentration of a hazardous agent related to industrial exposures to hazardous vapors.
Cancer: A tumor formed by mutated cells.
Carcinogen: A cancer-causing chemical.
CAS (Chemical Abstract Service) number: CAS numbers are used to identify chemicals and mixtures of chemicals.
Catastrophe: A major loss in terms of death, injuries, and damage.
Cause-consequence analysis: A method for determining the possible consequences or outcomes arising from a logical combination of input events or conditions that determine a cause.
Chronic (risk): Risks associated with long-term chemical exposure, usually at low concentrations.
Conditional probability: The probability of occurrence of an event given that a precursor event has occurred.
Confidence interval: A range of values of a variable with a specific probability that the true value of the variable lies within this range. The conventional confidence interval probability used by most engineers is the 95% confidence interval, defining the range of a variable in which its true values fall with 95% confidence.

Confidence limits: The upper and lower range of values of a variable defining its specific confidence interval.
Consequences: A measure of the expected effects of an incident outcome or cause.
CPQRA (chemical process quantitative risk analysis): It is analogous to a hazard risk assessment (HZRA) defined below.
Deflagration: The chemical reaction of a substance in which the reaction front advances into the unreacted substance present at less than sonic velocity.
Dermal: Applied to the skin.
Detonation: A release of energy caused by a rapid chemical reaction of a substance in which the reaction front advances into the unreacted substance present at greater than sonic velocity.
Domino effects: The triggering of secondary events; usually considered when a significant escalation of the original incident could result.
Dose: A weight (or volume) of a chemical agent, usually normalized to a unit body weight.
Epidemic: A disease prevalent and spreading rapidly among many individuals in a community at the same time.
Episodic release: A massive release of limited or short duration, usually associated with an accident.
Equipment reliability: The probability that, when operating under stated conditions, the equipment will perform its intended purpose for a specified period of time.
Event: An occurrence associated with an incident either as the cause or a contributing cause of the incident, or as a response to an initiating event.
Event sequence: A specific sequence of events composed of initiating events and intermediate events that may lead to a problem or an incident.
Event tree analysis (ETA): A graphical logic model that identifies and attempts to quantify possible outcomes following an initiating event.
Explosion: A release of energy that causes a pressure discontinuity or blast wave.
Exposure period: The duration of an exposure.
External event: A natural or man-made event; often an accident.
Failure frequency: The frequency (relative to time) of a failure.
Failure mode: A symptom, condition, or manner in which a failure occurs.
Failure probability: The probability that a failure will occur, usually for a given time interval.
Failure rate: The number of failures divided by the total elapsed time during which these failures occur.
Fatal accident rate (FAR): The estimated number of fatalities per 10^8 exposure hours (roughly 1,000 employee work lifetimes).
Fault tree: A method of representing the logical combinations of events that lead to a particular outcome (top event).
Fault tree analysis (FTA): A logic model that identifies and attempts to quantify possible causes of an event.
Federal Register: A daily government publication of laws and regulations promulgated by the US Federal Government.
Flammability limits: The range in which a gaseous compound in air will burst into flames or explode if ignited.
Frequency: Number of occurrences of an event per unit time.

Half-life: The time required for a chemical concentration or quantity to decrease by half its current value.
Hazard (problem): An event associated with an accident which has the potential to cause damage to people, property, or the environment.
Hazard and operability study (HAZOP): A technique to identify process hazards and potential operating problems using a series of guide words that key-in on process deviations.
Hazard risk assessment (HZRA): A technique associated with quantifying the risk of a hazard problem employing probability and consequence information.
Health problem: A problem normally associated with health arising from the continuous emission of a chemical into the environment.
Health risk assessment (HRA): A technique associated with quantifying the risk of a health problem employing toxicology and exposure information.
Human error: Actions by engineers, operators, managers, and so on that may contribute to or result in accidents.
Human error probability: The ratio between the number of human errors and the number of opportunities for human error.
Human factors: Factors attempting to match human capacities and limitations.
Human reliability: A measure of human errors.
Incident: An event.
Individual risk: The risk to an individual.
Ingestion: The intake of a chemical through the mouth.
Initiating event: The first event in an event sequence.
Instantaneous release: Emissions that occur over a very short duration.
Intermediate event: An event that propagates or mitigates the initiating event during an event sequence.
Isopleth: A plot of constant concentration, usually downwind from a release source.
Lethal concentration (LC): The concentration of a chemical that will kill a test animal usually based on a 1- to 4-hr exposure duration.
Lethal concentration 50 (LC_{50}): The concentration of a chemical that will kill 50% of animals, usually based on a 1- to 4-hr exposure duration.
Lethal dose (LD): The quantity of a chemical that will kill a test animal, usually normalized to a unit of body weight.
Lethal dose 50 (LD_{50}): The quantity of a chemical that will kill 50% of test animals, usually normalized to a unit of body weight.
LEL/LFL: The lower explosive/flammability limit of a chemical in air that will produce an explosion or flame if ignited.
Level of concern (LOC): The concentration of a chemical above which there may be adverse human health effects.
Likelihood: A measure of the expected probability or frequency of occurrence of an event.
Malignant tumor: A cancerous tumor.
Mutagen: A chemical capable of changing (mutating) a living cell.
Pandemic: An outbreak of a disease that occurs over a wide geographic area (as multiple countries or continents) and typically affects a significant proportion of the population.

PEL (permissible exposure limit): The permissible exposure of humans to a chemical continuously for 8 hr without any danger to health and safety as established by the Occupation Safety and Health Administration (OSHA).

Personal protection equipment (PPE): Material/equipment worn to protect a worker from exposure to hazardous agents.

Precision: The degree of "exactness" of repeated measures relative to the true or actual value.

ppm: The concentration of a chemical, parts per million, in air almost always presented on a volume basis, often designated as ppm_v as opposed to ppm_m on a mass basis for chemical concentrations in water.

ppb: The concentration of a chemical, part per billion, in air almost always on a volume basis; often designated as ppb_v as opposed to ppb_m on a mass basis for chemical concentrations in water.

Maximum individual risk: The highest individual risk in an exposed population.

Probability: An expression for the likelihood of occurrence of an event or an event sequence, usually over an interval of time.

Propagating factors: Influences that contribute to the sequence of events following the initiating event.

Protective system: Systems, such as pressure vessel relief valves, that function to prevent or mitigate the occurrence of an accident or incident.

Risk: A measure of economic loss or human injury in terms of both the incident likelihood and the magnitude of the loss or injury.

Risk analysis: The engineering evaluation of incident consequences, frequencies, and risk assessment results.

Risk assessment: The process by which risk estimates are made.

Risk contour: Lines on a risk graph that connect points of equal risk.

Risk estimation: A combination of the estimated consequences and likelihood of a risk.

Risk management: The application of management policies, procedures, and practices in analyzing, assessing, and controlling risk.

Risk perception: The perception of risk that is a function of age, race, sex, personal history and background, familiarity with the potential risk, dread factors, perceived benefits of the risk-causing action, marital status, residence, and so on.

Societal risk: A measure of risk to a group of individuals.

Source term: The estimation of the release rate of a hazardous agent from a source.

Time of failure: The time period associated with the inability to perform a duty or intended function.

TLV-C: The ceiling exposure limit representing the maximum concentration of a chemical in air that should never be exceeded.

TLV-STEL: The short-term exposure limit (maximum concentration in air) for a continuous 15-min averaged exposure duration.

TLV (threshold limit value): Established by the American Council of Governmental Industrial Hygienists, it is the concentration of a chemical in air that produces no adverse effects in exposed individuals.

TLV-TWA: The allowable time-weighted average concentration of a chemical in air for an 8-hr workday/40-hr workweek that produces no adverse effect on the exposed individual.

Toxic dose: The combination of concentration and exposure period for a toxic agent to produce a specific harmful effect.

UEL/UFL: The upper explosive/flammability limit of a chemical in air that will produce an explosion or flame if ignited.

Uncertainty: A measure, often quantitative, of the degree of doubt or lack of certainty associated with an estimate.

1.3 ILLUSTRATIVE EXAMPLES

Four illustrative examples complement the material presented above.

1.3.1 ILLUSTRATIVE EXAMPLE 1

Provide a qualitative definition of risk.

Solution: Risk can be defined as the product of two factors: (1) the probability of an undesirable event and (2) the measured consequences of the undesirable event. Measured consequences may be stated in terms of impacts on health, financial loss, injuries, deaths, or other variables.

1.3.2 ILLUSTRATIVE EXAMPLE 2

Define failure in the context of risk assessment.

Solution: Failure represents an inability to perform some required function, mostly related to process equipment operations and performance.

1.3.3 ILLUSTRATIVE EXAMPLE 3

Define reliability in the context of risk assessment.

Solution: Reliability is the probability that a system or one of its components will perform its intended function under certain conditions for a specified period. The reliability of a system and its probability of failure are complementary in the sense that the sum of these two probabilities is unity. The basic concepts and theorems of probability that find application in the estimation of risk and reliability are considered in Chapter 4.

1.3.4 ILLUSTRATIVE EXAMPLE 4

Define an accident in the context of risk assessment.

Solution: As noted earlier, an accident is an unexpected event that has undesirable consequences and can be quantitatively described through an HZRA. The causes of accidents must be identified in order to help prevent accidents from occurring in the future. Any situation or characteristic of a system, plant, or process that has the potential to cause damage to life, property, or the environment is considered a hazard. A hazard can also be defined as any characteristic that has the potential to cause an accident.

PROBLEMS
1.1 Describe the word epidemic in laymen's terms.
1.2 Describe the word pandemic in laymen's terms.
1.3 Describe what a virus is in technical terms.
1.4 Describe some of the characteristics of viruses in technical terms and explain their "life cycle."

REFERENCES
Shaefer, S., and Theodore, L. 2007. *Probability and Statistics Applications in Environmental Science*. Boca Raton, FL: CRC Press/Taylor & Francis Group.

Theodore, L., and Dupont, R.R. 2012. *Environmental Health Risk and Hazard Risk Assessment: Principles and Calculations*. Boca Raton, FL: CRC Press/Taylor & Francis Group.

Theodore, L., Reynolds, J., and Morris, K. 1997. *Dictionary of Concise Environmental Terms*. Amsterdam, The Netherlands: Gordon and Breach Science Publishers.

Theodore, M.K., and Theodore, L. 2021. *Introduction to Environmental Management*, 2nd Edition. Boca Raton, FL: CRC Press/Taylor & Francis Group.

2 Introduction to Environmental and Health Risk

2.1 INTRODUCTION

The rapid growth and expansion of the chemical process industry has been accompanied not only by a spontaneous rise in chemical emissions to the environment but also human, material, and property losses because of fires, explosions, hazardous and toxic spills, equipment failures, other accidents, and business interruptions. Concern over the potential consequences of these continuous emissions and catastrophic accidents, particularly at chemical, petrochemical, and utility plants, has sparked interest at both the industrial and regulatory levels in obtaining a better understanding of the main subject of an earlier book: *Environmental Health and Hazard Risk Assessment: Principles and Calculations* (Theodore and Dupont 2012). The writing of that "risk" book was undertaken, in part, as a result of this growing concern.

Risk of all types (health risk, hazard risk, individual risk, societal risk, sports risk, etc.) has surged to the forefront of numerous engineering and science areas of interest. Why? A good question. Some of the more obvious reasons include (not in the order of environmental importance) the following:

1. Increased environmental health and safety legislation
2. Accompanying regulations
3. Regulatory fines
4. Liability concerns
5. Environmental activists and their organizations
6. Public concerns
7. Skyrocketing health care costs
8. Skyrocketing workers' compensation costs
9. Codes of ethics

These factors, individually and in toto, have created a need for engineers and scientists to develop a proficiency in environmental risk and environmental risk-related topics, particularly as they apply to health risk. This need, in turn, gave rise to the driving force that led to the writing of this book.

Members of society are confronted with risks on a daily basis. Here is a sampling of some activities for which risk can potentially play a role:

1. Electrocution when turning on the TV
2. Using soap with chemical additives

3. Tripping down the stairs
4. Drinking Starbucks coffee
5. Indoor air pollution
6. Driving to work
7. Eating a hot dog for lunch
8. Living downwind of a refinery
9. Being struck by an automobile while returning from lunch

Risks abound. They are all around and society often has little to no control over many of them. Perhaps a careful analysis of risk is indeed in order.

Health problems and accidents can also occur in many ways other than from routine, daily, and "normal" activities. There may be a chemical spill, a round-the-clock emission from a power plant, an explosion, a terrorist attack, or a runaway reaction in a nuclear plant. There are also potential risks and accidents in the transport of people and materials such as trucks overturning, trains derailing, and ships capsizing. There are "acts of God" such as earthquakes, tsunamis, and tropical storms. It is painfully clear that health and hazard problems are a fact of life. The one common thread through all these situations is that these problems are rarely understood and, unfortunately, are frequently mismanaged.

The job of the practicing engineer and scientist is to measure or calculate the magnitude of health and environmental risk and often compare the magnitude of one risk to others that are similar in nature. Perhaps more difficult is the task of comparing the risk of one event with risks arising from events of a totally different nature.

2.2 RISK VARIABLES AND CATEGORIES

2.2.1 Risk Variables

Placing a risk in perspective entails translating myriad technical risk analyses into concepts of risk that both the technical community and the general public can understand. The most effective technique for presenting risks in perspective is to contrast risks to other, similar risks. There are several comparison variables that affect acceptance of risk. Ten such variables include the following:

1. Voluntary versus involuntary
2. Delayed versus immediate
3. Natural versus man-made
4. Controllable versus uncontrollable
5. Known versus unknown
6. Ordinary versus catastrophic
7. Chronic versus acute
8. Necessary versus luxury
9. Occasional versus continuous
10. Old versus new

Introduction to Environmental and Health Risk

The public generally accepts voluntarily assumed risk more easily than an involuntarily imposed risk. Similarly, a naturally occurring risk is more easily accepted than a man-made risk. The more similar risks are with regard to these variables, the more meaningful it is to compare those risks.

2.2.2 Risk Categories

There are dozens of risk categories. Topping the list, for purposes of this book, are *environmental and health risks*. Two other important risks include financial risk and sports risk. Both of these risks are briefly reviewed in the following two sections. Some other risks (in alphabetical order) include:

1. Aerospace
2. Architecture
3. Chemical
4. Construction
5. Education
6. Energy
7. Governance
8. Medical
9. Pharmaceutical
10. Travel
11. Urban planning

However, irrespective of the risk, it is fair to say that the calculation of environmental and health risks has now become mandatory in environmental assessment and analysis studies/applications. More and more technical individuals are now required in this field as risk analysis is often performed using the best available data and information.

There are, of course, many other risk categories that the engineer, scientist, bureaucrat, society, and so on are exposed to on a regular basis. Details of these "other" risks are available in the literature. As mentioned earlier, the next two sections explain financial and sports risk. The former topic is directly relevant to environmental risk and the latter is of interest to one of the authors (Theodore 2015).

2.3 FINANCIAL RISK

As noted previously, there are other risks, in addition to environmental ones, that the practicing engineer and applied scientist must be proficient in understanding. Perhaps the most important of these is financial risk. And, although this book is primarily concerned with environmental and health risk, the authors would be negligent if the topic of financial risk was not at least qualitatively addressed.

A company or individual hoping to increase profitability must carefully assess a range of investment opportunities and risks and select the most profitable options from those available. Increasing competitiveness also requires that efforts need to be made to reduce the costs of existing processes. In order to accomplish this, engineers and scientists should be fully aware of not only technical factors but also economic factors of their design options, particularly those that have the largest effect on financial risk and profitability.

In earlier years, engineers and scientists concentrated on the technical side of projects and left financial studies to the economists. In effect, those involved in making estimates of the capital and operating costs of engineering design alternatives have often left the overall economic analysis and investment decision-making to others. This approach is no longer acceptable.

Some technical personnel are not equipped to perform a financial or economic analysis. Furthermore, many already working for companies have never taken courses in this area. This short-sighted attitude is surprising for a group of individuals who normally go to great lengths to obtain all the available technical data they can prior to making an assessment of a project or study. The attitude is even more surprising when one notes that data are readily available to enable an engineer or scientist to assess the economic prospects of both his or her own company and those of his or her particular industry (Theodore and Ricci 2010).

The term *economic analysis* in real-world problems generally refers to calculations made to determine the conditions for realizing minimum cost or maximum financial return for a design or operation. The same general principles apply whether one is interested in the choice of competing alternatives for projects, in the design of plants so that the various components are economically integrated, or in the economical operation of existing plants. General considerations that form the framework on which sound decisions must be made are often simple. Sometimes their application to the problems encountered in the development of a commercial enterprise involves too many intangibles to allow exact analysis; in that case, judgment must be intuitive. Occasionally, such calculations may be made with a considerable degree of exactness.

Concern with maximum financial return implies that the criterion for judging projects involves risk and profit. While this is usually true, there are many important objectives, which, though ultimately aimed at increasing profit, cannot be immediately evaluated in quantitative terms. Perhaps the most significant is the recent increased concern with public health, environmental degradation, safety, and sustainability. Thus, there has been some tendency in recent years to regard management of commercial organizations as a profession with social obligations and responsibilities where considerations other than the profit motive may govern business decisions. However, these additional social objectives are, for the most part, often not inconsistent with the economic goal of satisfying human wants with the minimum risk. In fact, even in the operation of primarily nonprofit organizations, it is still important to determine the effect of various policies on both risk and long-term economic viability (Reynolds, Jeris and Theodore 2004).

If all industrial financial studies simply involved running costs, where a day-to-day expenditure of appropriate raw materials and labor would produce a product of immediate market value, risk predictions as to future demand and prices would be minimized. However, any future return over a period of time can best be evaluated by a host of different methods.

Introduction to Environmental and Health Risk

In order to accomplish this, appropriate data for the value of money, i (interest rate), and the lifetime, n, of the process are needed. To a certain extent, the values chosen are interdependent. A large n and a small i can give the same result as a small n and a large i.

The result is often evaluated in terms of a lump sum expressed as a present worth factor. The question now arises as to how the element of risk or uncertainty enters into these formulations. Several characteristics of financial and business risks are of interest in connection with any attempt to formulate methods for taking them into consideration.

First, financial risks are not governed purely by chance like the roll of the dice. What appears as a sound investment to engineers, scientists, and business executives familiar with the know-how and experience in a given company might represent a highly speculative venture for a concern engaged largely in a different type of business. Similarly, one investor in the common stock of a given company may not agree with another who does not see growth possibilities in the same stock. Thus, the situation exists where some ventures (investments) require a higher rate of return than others simply because such a rate is necessary to attract venture capital.

Second, aside from the chance of success or failure, a given company is limited in the amount of funds it can invest either from surplus or by borrowing. Thus, in offshore crude oil exploration, a large company (such as British Petroleum) that can finance the drilling of a number of oil wells can recover the costs of unsuccessful ventures from the profits of successful ones and, on the average, show attractive returns, even though four out of five wells drilled turn out to be "dry" holes. The position of the wildcatter or small operator is different in that an unlucky run of failures can put him or her out of business. Companies, like individuals, are limited in the absolute amount of capital they can afford to invest, and, as proposed ventures approach this limit, the rate of return required will increase, even though the financial risk remains unchanged. The utility of a large gain must therefore be balanced against the dis-utility of a smaller loss, which may mean disaster.

In modern business, which is often run by corporations, the entrepreneur is, for the most part, the common stockholder. It is true that the actual operations of the company are in the hands of business executives, and their salaries depend in large part on their ability to show profits. Often, however, their fortunes may not be intimately linked with those of the companies they manage if they own only modest amounts of stock in their companies. Furthermore, their salaries, as reported in the media in recent years – though astronomically high – typically do not represent a major expense in company operations. If the company they represent fails, they are often able to find opportunities for employment elsewhere. Similarly, the bondholder and preferred stockholder are protected to varying degrees from the risk of company failure. The holder of common stock, on the other hand, is subject to all the risks inherent in running the business. A proper procedure for evaluating new venture capital risk should, therefore, take these factors into consideration.

2.4 SPORTS RISK

Another important risk consideration that has become an integral part of modern day society is sports injury and its associated risks. What sports? Skiing, tennis, football, baseball, boxing, and so on, and, of course, basketball.

Injuries, like accidents, inevitably happen. It is just a matter of when, where, and their severity. For athletes, and in particular basketball (a topic that is dear to one of the authors) athletes, injuries can occur at any time during the lifetime of an athlete's participation period; they, for the most part, do not respect age. "Youth" are physically stronger but usually more reckless, while "seniors" are weaker but usually less reckless.

Most basketball injuries involve the knee, either through twisting or through the application of a lateral force. Surgery for such injuries has become much simpler with the invention of the arthroscope. Another common problem is a stress fracture: a weakening of the front of the shinbone from overuse, with pain and possible bone cracking as the result. Ligament tears are more common. Almost all these conditions heal with rest. Prevention of injuries depends primarily on good conditioning. The improper or illegal use of drugs and enhancing substances such as steroids for the temporary improvement of athletic performance in all competitions has been a frequent subject of inquiry but does not fall within the scope of this book.

Unfortunately, risk and risk factors associated with nearly every sport have received only superficial treatment in the literature. Data on comparative risks with various sports are available. Data on comparative sports risks with non-sports activities, for example, driving a car, BBQing, and so on are also available. But little is available on predicting risks from the probability of occurrence to the consequences associated with the injury. Hopefully this will change in the future.

Most readers are not familiar with health risk concerns with professional football. However, there is concern elsewhere. For example, a May 12, 2014, New York Times front page article was titled "Football Risks Sink In, Even in the Heart of Texas: Town Cuts 7th-Grade Tackle Program." The article was about Marshall, Texas where there has been a shift in perceptions about football that would have been hard to imagine when the school made a cameo in the book *Friday Night Lights* nearly 25 years ago. Amid widespread and growing concerns about the physical dangers of the sport, the school board in Marshall approved plans in February of 2014 to shut down the district's entry-level, tackle-football program for seventh graders in favor of flag football. Interestingly, there was little objection. Only time will tell if the decision is the beginning of the end of scholastic football in Texas.

2.5 ENVIRONMENTAL AND HEALTH RISK TERMS

Is environmental and health risk important to the practitioner? The reader can decide since all actions, objects, processes, gambling, and so on entail *risk*. It is no wonder that this four-letter word has become a hot button for practitioners.

The previous chapter contains a host of environmental terms and their accompanying definitions. Several of the terms contain the word risk. The so-called traditional definitions associated with these words or phrases are presented there. This section attempts to review not only the myriad of risk and risk-related terms but also some of the myriad of accompanying definitions of these terms that have been used in industry and have appeared in the literature. No attempt has been made to present this list in alphabetical order. Rather, this approach has attempted to provide the various terms in a logical, sequential order.

Introduction to Environmental and Health Risk

The four major risk terms include:

1. Risk
2. Risk assessment
3. Risk analysis
4. Risk management

"Risk" was defined earlier as a measure of economic loss, human injury, or human health effects and in terms of both the likelihood (probability) and the magnitude (consequences) associated with either a loss or injury. Although it is a quantifiable term, it has been misused by practitioners. "Risk assessment" involves the process of determining the events or problems that can produce a risk, the corresponding probabilities, and consequences, and finally, the characterization of the risk. "Risk analysis" employs the results of the aforementioned risk assessment and attempts to optimally use these results; in effect, it analyzes risk assessment information. Finally, "risk management" uses all the information provided by the risk assessment and risk analysis steps to reduce or eliminate the risk, select the optimum action(s), or evaluate the net benefits versus health/safety concerns of a specific action. Note that in line with its title, this book is primarily concerned with risk calculations from both a health and environmental perspective.

Of course, there are other risks. The definitions (for purposes of this section) for these other risks are as follows: *Individual risk* is defined as the risk to an individual; this can include a health problem or injury, the likelihood of occurrence, and the time period over which the problem might occur. The *maximum individual risk* is the aforementioned individual risk to a person exposed to the highest risk in an *exposed* population; this can be determined by calculating individual risks at every "location" and selecting the result for the maximum value. The *average individual risk* (in an exposed population) is the aforementioned individual risk averaged over the total population that is exposed to the risk in question. Alternatively, the average individual risk (in a total population) is the individual risk averaged over the entire population without regard to whether or not all the individuals in the population are actually exposed to the risk. Unfortunately, this particular average risk, whether applied to employees or the public, can be (at times) extremely misleading. These average risks have, on occasion, been expressed as exposed hours per worked hours; thus, the risk may be calculated for a given duration of time or averaged over the working day. *Societal risk* provides a measure of risk to a specific group of people, that is, it is based on the people affected by an event or scenario. *Time to respond risk* characterizes the time that a response occurs following a given event/scenario.

Risk communication is concerned with communicating the information generated from a risk assessment, risk analyses, and a risk management study. *Ecological risk* is a risk that describes the likelihood that adverse ecological effects resulting from an event/scenario will occur. *Total risk* is the term generally employed to describe the summation of the risk from all scenarios. *De minimus risk* has recently taken on significant importance in toxicology. It is defined as a risk judged to be too insignificant to be of societal concern or too small to be effectively applied to standard risk assessment studies. Financial risk is important enough to receive treatment in an earlier section. Finally, the new kid on the block is *unreasonable risk*, a term that has yet to be clearly defined. Perhaps it would be best to simply say that unreasonable risk is unreasonable in comparison to some other risk that is reasonable, i.e., when one compares two risks and selects one as more reasonable.

2.6 RISK ERRORS

No discussion of risk would be complete without some mention of errors, accuracy, and precision. The significance of conclusions based upon numerical results is necessarily determined by the reliability of the data and of the methods of calculation in which they are employed. It should be understood at the outset that most numerical calculations are by their very nature inexact. Errors are primarily due to one of three sources: inaccuracies in the original data, lack of precision in carrying out elementary operations, and inaccuracies introduced by approximate methods of solution.

Of particular significance in some applications are the errors due to roundoff and the inability of carrying, in a given calculation, more than a certain number of significant figures. Terms such as *absolute error*, *relative error*, and *truncation error* have a very real meaning, and an analysis parallel to another in question must frequently be carried out to establish the reliability of a given answer.

Accuracy is a term often employed in any discussion of risk. Accuracy is defined as "the condition or quality of being true, correct, or exact; freedom from error or defect; precision or exactness; correctness" (Dictionary.com 2021a). *Precision* is defined as "the state or quality of being precise ... mechanical or scientific exactness ... exact in measuring, recording, etc." (Dictionary.com 2021b). The accuracy is poor if, for example, the reported weight of a beaker is 35 g when the actual weight is 55 g. If the weight is reported as 35.3 g, the reading is more accurate. If the weight is reported as 35.29, the reading is more precise. Data or readings or calculations upon which risk may be based are subject to errors, and these errors may also be subject to error distributions which are discussed in Chapter 5.

Standard statistical techniques can be utilized to determine an estimate for an error in a specific piece of data where that data point is calculated from several other previously determined pieces of data. The error to be determined is known as the *estimated accumulated error*. Unfortunately, some earlier engineering textbooks give the impression that there should be a 100% match between data points and graphs drawn from theoretical equations. As discussed above, real-world data are limited in its accuracy by experimental hardware, measuring devices, and other factors; therefore, it is important that engineers and scientists become familiar with one or more methods for determining the magnitude of errors associated with data and calculations.

The subject of error in data recorded and the propagation of individual errors from various other data is a topic that requires careful attention and study. The focus of this presentation is to handle the relatively straightforward topic of the estimated

Introduction to Environmental and Health Risk

accumulated error for calculations. For any function Y, the estimated accumulated error, S_Y, is defined by Equation 2.1 as:

$$S_Y^2 = \sum_{i=1}^{N} a_i^2 S_i^2 \qquad (2.1)$$

where

$$a_i = \left(\frac{\delta Y}{\delta u_i}\right) \qquad (2.2)$$

where u_i = the independent variable in the defining equation for Y; N = the number of independent variables; and S_i = the estimate of the error for each independent variable u_i. The reader is referred to the literature (Theodore 2014; Byrd and Cothern 2000) for additional details and sample calculations.

What about *uncertainty*? Qualitatively, uncertainty may be viewed as having two components: variability and lack of knowledge. Uncertainty, whether applied to toxicological values, probability, consequences, risks, and so on may be described qualitatively or quantitatively. Qualitatively, descriptions include large, huge, monstrous, tiny, very small, and so on. Quantitative terms describing the uncertainty associated with a value x are normally in the form $x \pm w_{zx}$ where w_{zx} provides a measure of the uncertainty (i.e., standard deviation, 95% confidence limit, etc.) (Theodore 2014).

A substantial amount of information on uncertainty and uncertainty analysis is available. Useful references abound, but, in general, there are three main sources of uncertainty that have been earmarked by practicing engineers and scientists:

1. Model uncertainty
2. Data uncertainty
3. General quality uncertainty

Model uncertainty reflects the weaknesses, deficiencies, and inadequacies present in any model and may be viewed as a measure of the displacement of the model from reality. *Data uncertainty* results from incomplete data measurement, estimation, inference, or supposed expert opinion. *General quality uncertainties* arise because the practitioner often cannot identify every health problem or other incident. Naturally, the risk engineer's objective is to be certain that the major contributors to the risk are identified, addressed, and quantified. Uncertainty here arises from not knowing the individual risk contributions from those risk problems that have not been considered; one, therefore, may not be able to accurately predict the overall (or combined) risk. Byrd and Cothern (2000) have expanded this three-part uncertainty categorization in the following manner:

1. Subjective judgment
2. Linguistic imprecision

3. Statistical variation
4. Sampling error
5. Inherent randomness
6. Mathematical modeling
7. Causality
8. Lack of data or information
9. Problem formulation

Sensitivity and *importance* are also issues in the utilization of risk results. As noted earlier, uncertainty analysis is used to estimate the effect of data and model uncertainties on the risk estimate. Sensitivity analysis estimates the effect on calculated outcomes of varying inputs to the models individually or in combination. Importance analysis quantifies and ranks risk estimate contributions from subsystems or components of the complete system (e.g., individual incidents, groups of incidents, sections of a process, etc.).

To summarize, different assumptions can change any quantitative risk characterization by several orders of magnitude. The uncertainty that arises is related to how well (and often, consistently) input data are obtained, generated, or measured, and the degree to which judgment is involved in developing risk scenarios and selecting input data. Simply put, uncertainty arises from how data/evidence was both measured and interpreted. Despite these limitations and uncertainties, risk characterizations provide the practitioner with some analysis and assessment capabilities required for overall risk management/reduction efforts.

2.7 ILLUSTRATIVE EXAMPLES

Four illustrative examples complement the material presented above.

2.7.1 Illustrative Example 1

Compare annual versus lifetime risks.

Solution: A time frame must be included with a risk estimate for the numbers to be meaningful. For both health and hazard risks, annual or lifetime risks are commonly used. Direct evidence is usually expressed annually because the information is often collected and summarized annually. However, predictive information is commonly expressed as a lifetime probability, for example, when expressing cancer risk or a terrorist-related risk.

2.7.2 Illustrative Example 2

Describe what a cancer risk number of 10^{-6} probability means.

Solution: A cancer risk number usually represents the probability of developing cancer over an individual's lifetime. A risk of 10^{-6} indicates an individual has a 1 in

Introduction to Environmental and Health Risk

1,000,000 chance of developing cancer throughout a lifetime of exposure (assumed to be 70 years). One generally can also assume an upper 95% confidence limit on the maximum likelihood estimate. Since the predicted risk is an upper bound, the actual risk is unlikely to be higher but may be lower than the predicted risk.

2.7.3 ILLUSTRATIVE EXAMPLE 3

Qualitatively describe the following terms:

1. Substantial risk
2. Unreasonable risk
3. Insignificant risk

Solution: Each individual generally has a different concept of the meaning of the term *substantial* as it applies to health and hazard risk. One might consider substantial to be a risk higher than one might be exposed to at their home from normal activities of life, i.e., slipping in the shower, falling up or downstairs, tripping over kids or animal toys, etc.

Unreasonable risk is an unacceptable risk, a risk level that is/significantly (usually 1–2 orders of magnitude) above the US Environmental Protection Agency (EPA) quantitative risk standard for lifetime probabilities of adverse effects.

The EPA has generally described *insignificant risk* quantitatively in terms of lifetime probabilities below which the risk is assumed low enough to be ignored.

2.7.4 ILLUSTRATIVE EXAMPLE 4

Qualitatively describe the following terms:

1. De minimus risk
2. Acceptable risk

Solution: *De minimus risk* is analogous to insignificant risk and has become a legal term that is decided upon on a case-by-case basis.

Acceptable risk generally suggests that the risk is either insignificant or perhaps zero.

PROBLEMS

2.1 Describe the meaning of a cancer risk of 0.1×10^{-6}.
2.2 Describe the difference between an epidemic and pandemic in laymen's terms.
2.3 Provide a technical definition of epidemiology.
2.4 Briefly describe the causation factors that may arise between chemical exposure and injury or death in humans.

REFERENCES

Byrd, D., and Cothern, C. 2000. *Introduction to Risk Analysis*. Rockville, MD: Government Institutes.

Dictionary.com. 2021a. *Unabridged Based on The Random House Unabridged Dictionary*, © Random House, Inc., New York, NY. https://www.dictionary.com/browse/accuracy (accessed August 8, 2021).

Dictionary.com. 2021b. *Unabridged Based on The Random House Unabridged Dictionary*, © Random House, Inc., New York, NY. https://www.dictionary.com/browse/precision, precise (accessed August 8, 2021).

Reynolds, J., Jeris, J.S., and Theodore, L. 2004. *Handbook of Chemical and Environmental Engineering Calculations*. Hoboken, NJ: John Wiley & Sons.

Theodore, L. 2015. *Basketball Coaching 101*. East Williston, NY: Theodore Tutorial.

Theodore, L. 2014. *Chemical Engineering: The Essential Reference*. New York, NY: McGraw-Hill.

Theodore, L., and Dupont, R.R. 2012. *Environmental Health Risk and Hazard Risk Assessment: Principles and Calculations*. Boca Raton, FL: CRC Press/Taylor & Francis Group.

Theodore, L., and Ricci, F. 2010. *Mass Transfer Operations for the Practicing Engineer*. Hoboken, NJ: John Wiley & Sons.

3 Environmental and Health Risk Analysis

3.1 INTRODUCTION

People face all kinds of risks everyday, some voluntarily and others involuntarily. Therefore, risk plays a very important role in today's world. Earlier studies on cancer and the recent COVID-19 pandemic have caused a turning point in the world of risk because they opened the eyes of risk scientists and health professionals to the world of health risk assessments (HRAs).

The usual objective of HRA and the accompanying calculations is to evaluate the potential for adverse health effects from the release of chemicals into the environment. Unfortunately, the environment is very complex since there is a vast array of potential receptors present. The task of testing and evaluating each of the enormous number of chemicals on the market for their impact on human populations and ecosystems becomes extremely difficult. To further complicate the problem, health is a concept that has come to mean different things to different people. Some have defined it as: "… a state of complete physical, mental and social well-being and not merely the absence of disease or infirmary." Many other definitions and concepts have been proposed and appear in the literature.

Since 1970 the field of HRA has received widespread attention within both the scientific and regulatory communities. It has also attracted the attention of the public. Properly conducted risk assessments and risk assessment calculations have received fairly broad acceptance, in part because they put into perspective the terms toxic, health, hazard, and risk. *Toxicity* is an inherent property of all substances. It states that all chemical and physical agents can produce adverse health effects at some dose or under some specific exposure conditions. In contrast, exposure to a chemical that has the capacity to produce a particular type of adverse effect represents a health "hazard." *Risk* (in a general sense), however, is the probability or likelihood that an adverse outcome will occur in a person or a group that is exposed to a particular concentration or dose of the hazardous agent. *Health risk* is a function of exposure and dose. Consequently, *HRA* is defined as the process or procedure used to estimate the likelihood that humans or ecological systems will be adversely affected by a chemical or physical agent under a specific set of conditions.

The term *risk assessment* is not only used to describe the likelihood of an adverse response to a chemical or physical agent, but it has also been used to describe the likelihood of any unwanted event. These include risks such as: explosions or injuries in the workplace; natural catastrophes; injury or death due to various voluntary activities such as skiing, sky diving, flying, and bungee jumping; diseases; death due to natural causes; and many others (Paustenbach 1989).

Risk assessment of accidents serves a dual purpose. It estimates the probability that an accident will occur and also assesses the severity of the consequences of an accident. Consequences may include damage to the surrounding environment, financial loss, injury, or loss of life. This chapter is also concerned with introducing the reader to the methods used to identify these hazards and the causes and consequences of accidents. Risk assessment of accidents (or hazard risk assessment, HZRA) provides an effective way to help ensure that a mishap either does not occur or reduces the likelihood of severe consequences as a result of the accident. The results of the HZRA allow concerned parties to take precautions to prevent an accident before it happens.

There are other classes of environmental health risks that do not pertain to chemicals but are an integral part of HZRA. For example, health problems can arise immediately/soon after a hazard, such as a hurricane or earthquake, that can leave local inhabitants without potable water for an extended period of time.

Environmental risk assessment may be broadly defined as a scientific enterprise in which facts and assumptions are used to estimate the potential for adverse effects. *Risk management*, as the term is used by the US Environmental Protection Agency (EPA) and other regulatory agencies, refers to a decision-making process which involves such considerations as risk assessment, technological feasibility, economic information about costs and benefits, statutory requirements, public concerns, and other factors. Risk communication is the exchange of information about risk.

Risk assessment may also be defined as the characterization of potential adverse effects to humans or to an ecosystem resulting from environmental hazards. Risk assessment supports risk management, the set of choices centering on whether and how much to control future exposure to the suspected. Risk managers face the necessity of making difficult decisions involving uncertain science, potentially grave consequences to health or the environment, and large economic effects on industry and consumers. What risk assessment provides is an orderly, explicit, and consistent way to deal with scientific issues in evaluating whether a health problem or a hazard exists. This evaluation typically involves large *uncertainties*, because the available scientific data are limited, and the mechanisms for adverse health impacts or environmental damage are only imperfectly understood.

Risk assessment and *risk management* are two different processes, but they are intertwined. Risk assessment and risk management give a framework not only for setting regulatory priorities but also for making decisions that cut across different environmental areas. Risk management refers to a decision-making process that involves such considerations as risk assessment, technology feasibility, economic information about costs and benefits, statutory requirements, public concerns, and other factors. Therefore, risk assessment supports risk management in that the choices on whether and how much to control future exposure to a suspected problem may be determined during the risk management process (Burke, Singh and Theodore 2000).

From a risk management standpoint, whether dealing with a site-specific situation or a national standard, the deciding question is ultimately, "What degree of risk is acceptable?" In general, this does not mean a "zero risk" standard, but rather a concept of *negligible risk*. At what point is there really no significant health or environmental risk, and at what point is there an adequate safety margin to protect public health and the environment? In addition, some environmental statutes require consideration of benefits together with risks in making risk management decisions.

Thus, it should be noted that health risk addresses risks that arise from health and health-related problems. Chemicals are generally the culprit. Both the effect on and exposure to a receptor (in this case, generally a human) ultimately determine the risk to the individual for the health problem of concern. The risk can be described in either qualitative or quantitative terms, and there are various terms that may be used, e.g., 10 individuals will become sick, or 1×10^{-6} (one in a million) will die, or something as simple as "it is a major problem."

Another category of environmental risk is the aforementioned hazard risk. This class of risk is employed to describe risks associated with hazards or hazard-related problems, for example, accidents, negative events, and catastrophes. Unlike most health problems, these usually occur over a short period of time, say a few seconds or minutes. Both the probability and the consequence associated with the accident/event ultimately determine the hazard risk. Once again, the risk can be described in either qualitative or quantitative terms, and there are various terms that may be used.

As noted earlier, once a risk bas been calculated, one needs to gauge the estimated consequences (or opportunities if examining financial/economic scenarios) and evaluate and prioritize options for risk management or mitigation. These potentially strategic evaluations are usually fraught with uncertainties at numerous levels. Thus, the risk assessment process is normally followed by alternatives analyses; these options are usually based on decision-making procedures that are beyond the scope of this book. However, it is fair to say that there may be a full range of outcomes and consequences to various scenarios. It should also be noted that risk assessment is a dynamic process that can very definitely be a function of time. Much of this material is addressed later in the chapter.

3.2 HEALTH RISK ASSESSMENT/ANALYSIS

HRAs provide an orderly, explicit, and consistent way to deal with issues in evaluating whether a health problem exists and what the magnitude of the problem may be. This evaluation typically involves large uncertainties (to be discussed in a later section) because the available scientific data are limited and the mechanisms for adverse health impacts or environmental damage are only imperfectly understood.

When one examines risk, how does one decide how safe is "safe," or how clean is "clean?" To begin with, one has to look at both inputs of the risk equations, that is, both the toxicity of a pollutant and the extent of exposure. Information is required for both the current and the potential exposure, considering all possible exposure pathways. In addition to human health risks, one needs to look at the potential ecological or other environmental effects.

In recent years, several guidelines and handbooks have been published to help explain approaches for conducting HRAs. As discussed by a special National Academy of Sciences Committee which convened in 1983, most human or environmental health hazards can be evaluated by dividing the analysis into four parts: health problem identification, dose-response assessment or toxicity assessment, exposure assessment, and risk characterization (see Figure 3.1). The risk assessment might stop with the first step,

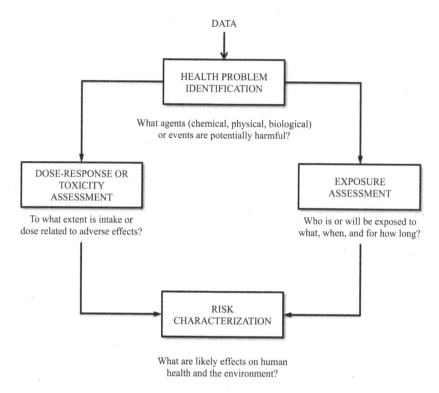

FIGURE 3.1 Health risk evaluation process.

health problem identification, if no adverse effect is identified or if an agency elects to take regulatory actions without further analysis (Burke, Singh and Theodore 2000). Regarding identification, a *health problem* is defined as a toxic agent or a set of conditions that has the potential to cause adverse effects to human health or the environment. Health problem identification involves an evaluation of various forms of information in order to identify the different problems potentially caused by the toxic agent. Dose-response or toxicity assessment is also required in an overall assessment; responses and effects can vary widely since all chemicals and contaminants vary in their capacity to cause adverse effects. This step frequently requires that assumptions be made to extrapolate experimental results from animal tests to expected effects on exposed humans. *Exposure assessment* is the determination of the magnitude, frequency, duration, and routes of exposure of toxic agents to human populations and ecosystems. Finally, in *health risk characterization* toxicology and exposure data/information are combined to obtain a qualitative or quantitative expression of risk. An expanded presentation on each of the four HRA steps is provided in the next four sections.

3.2.1 HEALTH PROBLEM IDENTIFICATION

Health problem identification is defined as the process of determining whether human exposure to a chemical at some dose could cause an increase in the incidence of an adverse health condition (cancer, birth defect, etc.), or whether exposure to non-humans, such as fish, birds, and other forms of wildlife, could cause adverse ecological effects. In other words, does exposure to a chemical have the potential to cause harm? It involves characterizing the quality, nature, and strength of the evidence of causation. It may not give a yes or no answer; however, it is intended to provide an assessment on which to base a decision as to whether a health problem has been identified. This identification characterizes the problem in terms of the agent and dose of the agent of concern. Since there are few hazardous chemicals or hazardous agents for which definitive exposure data in humans exist, the identification of health hazards is often characterized by the effects of health problems on laboratory test animals or other species and test systems (Paustenbach 1989).

There are numerous methods available to identify the potential for chemicals to cause both adverse health conditions and significant effects on the environment. These can include, but are not limited to toxicology, epidemiology, molecular and atomic structural analysis, safety data sheets (SDSs), standardized mortality ratios (observed deaths or expected deaths), engineering approaches to problem solving, analysis of the fate of chemicals in the environment, and evaluations of carcinogenic versus noncarcinogenic health hazards.

3.2.2 Toxicity and Dose-response Assessment

Dose-response assessment is the process of characterizing the relationship between the dose of an agent administered or received and the incidence of an adverse health effect in exposed populations. This process considers such important factors as intensity of exposure, age, pattern of exposure, and other variables that might modify the response, such as sex and lifestyle. In effect, it involves the evaluation of the effects expected from various quantity/concentration levels of particular chemical in the environment. Dose and response are therefore fundamental concepts that provide a relationship between the dosage of a toxic agent and the biological response. The magnitude of the biological response depends on the concentration of the contaminant/physical agent at the site of action, while the concentration of the contaminant at the active site depends on the dose. Thus, the dose and the response are causally related. Toxicity data exhibit a dose-response relationship if a mathematical model can be formulated to describe the response of the receptor and/or test organism in terms of the dose administered. The relation often takes the form of a percentage or number of receptors responding in a given manner either to a dose or to a specified range of concentrations over a given period of time. A dose-response assessment usually requires extrapolation from high to low doses and from animal to humans or one laboratory animal species to a wildlife species. A dose-response assessment should also describe and justify the methods of extrapolation used to predict incidence, and it should characterize the statistical and biological uncertainties in these methods. When possible, the uncertainties should be described numerically rather than quantitatively (see the section "Risk Uncertainties/Limitations" later in this chapter).

Why is toxicology so important? As noted earlier, it is the dose that makes the poison. A low-level dose may cause no effect. Yet, a larger dose may lead to either an adverse health effect or even death. This dose variation is also a function of the chemical of concern. Furthermore, the manner in which the dose impacts a chemical's absorption, distribution, metabolism in the human body, and ultimate excretion from the body can vary with both the chemical and the dose.

Once again, it is important to differentiate between the terms chronic and acute as they relate to toxicity. *Chronic toxicity* is caused by long-term or repeated exposure to low doses of the chemical, and the intensity is usually less than with acute exposures. *Acute toxicity* is caused by large doses of a chemical over short time periods and is often characterized by the effects of health problems on laboratory test animals or other species and test systems (Paustenbach 1989).

3.2.3 Exposure Assessment

As noted, a critical component of environmental HRA is *exposure assessment*. It is defined as the determination of the concentration of chemicals in time and space at the location of receptors and/or target populations. This description must therefore also include an identification of all major pathways for movement and transformation of a toxic material from a source to receptors. Ideally, concentrations should be identified as a function of time and location and should include all major transformation processes. The principal pathways generally considered in exposure assessments are atmospheric transport and surface and groundwater transport. Since atmospheric dispersion has received the bulk of treatment in the literature, a good part of the material to follow will address this topic.

The exposure assessment process consists of two basic methods for determining the concentration of a chemical to which receptor target populations are exposed:

1. The first is the direct measurement of the intensity, frequency, and duration of human or animal exposure to a pollutant currently present in the environment. This is a common practice in occupational settings.
2. In some situations, however, either concentrations are too low to be detected against the background, or direct measurement is too costly or difficult to implement. Under these circumstances, the second method is employed. It involves the use of mathematical models to estimate hypothetical exposures that might arise from the release of new chemicals into the environment.

In its most complete form, an exposure assessment should describe the magnitude, duration, timing, and route of exposure of the hazardous agent, along with the size, nature, and classes of the human, animal, aquatic, or wildlife populations exposed, and the uncertainties in all estimates. The exposure assessment can often be used to identify feasible prospective control options and predict the effects of available treatment technologies for controlling or limiting exposure (Paustenbach 1989). However, the estimation of the likelihood of exposure to a chemical remains a difficult task. Attention in the past focused on too many overly conservative assumptions. This in turn resulted in an overestimation of the actual exposure risk posed to vulnerable receptors.

Environmental and Health Risk Analysis

Obviously, without exposure(s), there are no risks. To experience adverse effects, one must first come into contact with the toxic agent(s). Exposures to chemicals can occur via inhalation of air (breathing), intake into the body via ingestion of water and food, or adsorption through the skin. These intake processes are followed by chemical distribution through the body via the bloodstream. After being absorbed and distributed, the chemical(s) may be metabolized and excreted, either as the parent compound or as their metabolites. The principal excretory organs are the kidney, liver, and lungs.

As noted earlier, the main pathways of exposure considered in human exposure assessments are via atmospheric, surface, and groundwater transport. However, the ingestion of toxic materials that have passed through the aquatic and terrestrial food chains, and dermal absorption, are two other pathways of potentially significant human exposure.

The physical and chemical properties of the chemicals under study will dictate the primary route(s) by which exposure will occur. Naturally, the chemical under study should be analyzed for the primary route(s) of human exposure. There are instances where humans may be exposed to a compound via more than one route, for example, by inhalation and oral ingestion. Which is the most significant route of exposure? Assuming approximately equal exposure by both routes, it is recommended that the chemical exposure assessment should focus on the route posing the greater risk. For those situations where one route of exposure predominates over another, the dominant route should be considered. Once an exposure assessment determines the quantity of a chemical with which human populations may come in contact, the information can be combined with toxicity data to estimate potential health risks (Theodore, Reynolds and Morris 1996).

The reader should once again note that two general types of potential health risk from chemical exposures exist. These are classified as follows:

1. *Chronic*: Risk related to continuous exposures over long periods of time, generally several months to a year. Concentrations of emitted chemicals are usually relatively low. This subject area falls in the general domain of HRA, and it is this subject that is addressed in this section. Thus, in contrast to the acute (short-term) exposures that predominate in HZRAs, chronic (long-term) exposures are the major concern in HRAs.
2. *Acute*: Risk related to exposures that occur for a relatively short period of time, generally from minutes to 1 to 2 days. Concentrations of emitted chemicals are usually high relative to their no-effect levels. In addition to inhalation, airborne substances might directly contact the skin, or liquids and sludges may be splashed on the skin or into the eyes, leading to adverse health effects in acute risk settings. This subject area falls, in a general sense, in the domain of HZRA and is addressed in a later section.

3.2.4 HEALTH RISK CHARACTERIZATION

Health risk characterization is the process of estimating the incidence of a health effect under the various conditions of human or animal exposure described in an exposure assessment. It is performed by combining the exposure assessment with dose-response information. From a receptor's perspective, the risk from exposure to any chemical also depends on the potency associated with the effects and the duration of the exposure. The summary effects of the uncertainties in the preceding steps should also be included in this analysis.

The quantitative estimate of the risk is of principal interest to the regulatory agency or risk manager making a decision. The risk manager must consider the results of the risk characterization when evaluating the economics, societal aspects, and various benefits of the assessment. Factors such as societal pressure, technical uncertainties, and severity of the potential impacts influence how the decision-makers respond to the risk assessment. As one might suppose, there is room for improvement in this step of the risk assessment process (Paustenbach 1989; Masters 1991).

A risk estimate indicates the likelihood of occurrence of the different types of health or environmental effects in exposed populations. Risk assessment should include both human health and environmental evaluations (e.g., impacts on ecosystems). Ecological impacts include actual and potential effects on plants and animals (other than domesticated species). The number produced from the risk characterization, often representing the probability of adverse health effects being caused, must be carefully interpreted.

3.3 HAZARD RISK ASSESSMENT/ANALYSIS

There are several steps in evaluating the risk of an accident (see Figure 3.2). A more detailed figure is presented in Figure 3.3. If the system in question is a chemical plant, the specific steps to be followed in the risk evaluation process are listed here:

1. A brief description of the equipment and chemicals used in the plant is needed.
2. Any hazard in the system has to be identified. Hazards that may occur in a chemical plant include:
   a. Fire
   b. Toxic vapor releases
   c. Slippage
   d. Corrosion
   e. Explosions
   f. Rupture of pressurized vessels
   g. Runaway reactions
3. The event (or series of events) that will initiate an accident has (have) to be identified. An event could be a failure to follow correct safety procedures, improperly repaired equipment, or failure of a safety mechanism.
4. The probability that the accident will occur has to be determined. For example, if a chemical plant has a 10-year life span, what is the probability that the temperature in a reactor will exceed the specified temperature range over that lifetime? The probability can be ranked qualitatively from low to high. A low probability means that it is unlikely for the event to occur in the lifetime of the plant. A medium probability suggests that there is a

Environmental and Health Risk Analysis

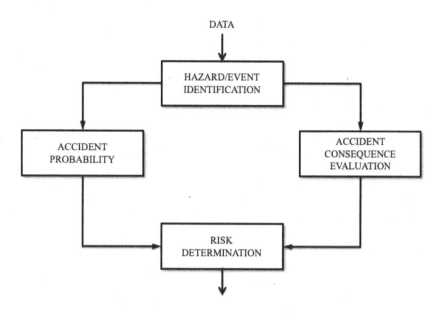

FIGURE 3.2 Simplified HZRA flowchart.

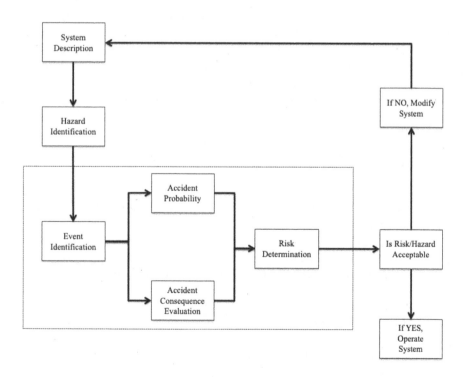

FIGURE 3.3 HZRA flowchart for a chemical plant.

possibility that the event will occur. A high probability means that the event will likely occur during the lifetime of the plant.
5. The severity of the consequences of the accident must be determined. This will be described in detail later in this section.
6. If the probability of the accident and the severity of its consequences are low, then the risk is usually deemed acceptable and the plant should be allowed to operate. If the probability of occurrence is too high or the damage to the surroundings is too great, then the risk is usually unacceptable and the system needs to be modified to minimize these effects.

As indicated in Figure 3.3, the heart of the HZRA approach is enclosed in the dashed box comprising Steps 3 through 6. The algorithm allows for re-evaluation of the process if the risk is deemed unacceptable (the process is repeated after system modification starting with Step 1.

Once again, it is important to note that an accident generally results from a sequence of events. Each individual event, therefore, represents an opportunity to reduce the frequency, consequence, and/or risk associated with the accident culminating from the individual events.

3.3.1 Hazard/event Problem Identification

Hazard or event identification provides information on situations or chemicals and their releases, which can potentially harm the environment, life, or property. Information at is required to identify chemical hazards includes chemical identities; quantities and location of the chemicals in question; and chemical properties such as boiling points, ignition temperatures, and toxicity to humans (Santoleri, Reynolds, and Theodore 2000). Obviously, the key word in this section is "identify," and the identification subject can be assisted by reviewing the following topics:

1. Process equipment
2. Classification of accidents
3. Fires, explosions, and hazardous spills

The next section addresses hazard/event evaluation techniques and covers a number of methods used to identify some of the hazards common to the chemical process and manufacturing industries.

Generally, the hazard in question will take the form of either a chemical release or a "disaster arising from a blast/fire," and some of the hazard/event evaluation techniques for these sorts of accidents include (Theodore and Dupont 2013):

1. System checklists
2. Safety review/safety audit
3. "What if" analysis
4. Preliminary hazard analysis (PHA)
5. Hazard and operability (HAZOP) studies

Environmental and Health Risk Analysis

3.3.2 HAZARD/EVENT PROBABILITY

There are a host of reasons why accidents occur in industry. The primary causes are mechanical failure, operational error (human error), process upsets, and design errors. There are three steps that normally lead to an accident:

1. Initiation
2. Propagation
3. Termination

The chemical industry today is involved in a broad spectrum of manufacturing processes that range from biological preparation to the manufacturing of plastics and explosives. Although the basic plans and designs for these processes may be similar, each individual plant will have its own unique set of potential hazards.

Deviations from normal conditions, as manifested by the following circumstances, must be well understood if accidents are to be prevented:

1. Abnormal (excursion) temperatures
2. Abnormal (excursion) pressures
3. Material flow stoppage
4. Equipment leaks or spills
5. Failure of equipment

Chemical processing under "extreme conditions" of high temperatures and pressure requires a more thorough analysis and extra safeguards. Explosions at high initial temperatures and pressures are much more severe than explosions at lower temperature and pressure conditions. Therefore, chemical processes under extreme conditions also require specialized equipment design and fabrication. Other factors that should be considered when evaluating a chemical process are the rate and order of the reaction, stability of the reaction, and the potential human health problems caused by the raw materials used.

Under the above circumstances, there is a need for high standards in equipment design, operation, and maintenance. Regardless of adequate safeguards and control of highly technological processes, accidents do and will continue to occur. It is therefore important to examine the causes of such accidents associated with specific pieces of equipment, supporting systems, and material being handled. As noted earlier, the sequence of events resulting in an accident can generally be traced back to one or a combination of the following causes:

1. Equipment failure
2. Control system failure
3. Utilities and ancillary equipment outage
4. Human error
5. Fire exposure/explosions
6. Natural causes
7. Plant layout

3.3.3 Hazard/event Consequences

Consequences of accidents can be classified qualitatively by the degree of severity. Factors that help to determine the degree of severity for chemicals are the concentration at which the hazard is released, the relative toxicity of the hazard, and, in the case of a chemical release, the length of time that a person is exposed to the hazardous agent. From a qualitative perspective, the worst-case consequence of a scenario is defined as a conservatively high estimate of the severity of the accident identified (AIChE 1992). On this basis, one can qualitatively rank the consequences of accidents into *low*, *medium*, and *high* degrees of severity (US EPA 1987a). A low degree of severity means that the hazard consequence is nearly negligible, and the injury to persons, property, or the environment may be observed only after an extended period of time. The degree of severity is considered to be medium when the accident is serious but not catastrophic. An example of this could include a case where there is a release of a low concentration of a chemical that is considered to be highly toxic. Another example of a medium degree of severity could be a highly concentrated release of a less toxic chemical large enough to cause injury or death to persons and damage to the environment unless immediate action is taken. There is a high degree of severity when the accident is catastrophic or the concentrations and toxicity of a chemical hazard are large enough to cause injury or death to many individuals, and there is long-term damage to the surrounding environment.

Potential consequences of other specific hazard/accident conditions

1. Flying shrapnel
2. Rocketing tank parts
3. Fireballs created by mechanically atomized drops of burning liquid and vapor
4. Secondary fires and explosions caused by flaming tank contents
5. Release of toxic or corrosive substances to the surroundings

Additional details regarding identification, probability, and consequences of accidents and risks are available in the literature (American Insurance Association 1979; US EPA 1987a; AIChE 1992; Santoleri, Reynolds, and Theodore 2000; Theodore and Dupont 2012).

3.3.4 Hazard Risk Characterization

Risk characterization often involves a judgment of the probability and severity of consequences based on the history of previous incidents, local experience, and the best available current technological information. It provides an estimation of:

1. The likelihood (probability) of an accidental release based on the history of current conditions and controls at a facility, consideration of any unusual environmental conditions (e.g., areas on floodplains), or the possibility of simultaneous emergency incidents (e.g., flooding or fire hazards resulting in the release of hazardous materials)
2. The severity of consequences of human injury that may occur (acute, delayed, and/or chronic health effects), the number of possible injuries and deaths, and the associated high-risk groups

Environmental and Health Risk Analysis

3. The severity of consequences on critical facilities (e.g., hospitals, fire stations, police departments, and communication centers)
4. The severity of consequences of damage to property (temporary, repairable, and permanent)
5. The severity of consequences of damage to the environment

The risk characterization process also attempts to attach meaning to a risk that has been calculated, including such factors as economic, social, and technological, plus selecting a course of action concerning the risk. Finally, interpreting and communicating the calculated risk value(s) to the public can be accomplished in a number of ways including the following:

1. Comparing the calculated risk with other known risks
2. Providing a perspective on the frequency or occurrence(s) of the risk
3. Explaining the sensitivity of the risk results and calculations to changes in input model data and scenario assumptions

The risk management process generally involves selecting a course of action that best addresses the risk in question and can include the following:

1. A cost-benefit analysis of risk
2. Measuring public perception
3. Determining acceptable levels of risk

Because so many hazards exist in everyday life, risk assessment must be used as a tool for evaluating those that are the most pressing or most hazardous. Over time, one may find that some activities are more hazardous than once perceived (i.e., smoking cigarettes or manufacturing polychlorinated biphenyls [PCBs]). Once the evidence is evaluated, these practices may be either stopped or limited. An assessment on an unknown chemical hazard or potentially unsafe practice attempts to predict what the consequences might be without waiting for final proof of an adverse impact.

Risk characterization estimates the risk associated with a process under investigation. The result of this characterization is a number that represents the probability of adverse effects from that process and/or from a substance released from that process. For instance, a risk characterization for all effects from a nuclear power plant might be expressed as one additional cancer case per one million people.

Once a risk characterization is made, the meaning of that risk must be evaluated. Public environmental and health agencies generally only consider risk greater than 10 in 1 million (1×10^{-5}) or 1 in 1 million (1×10^{-6}) to be significant risks warranting action.

3.4 RISK UNCERTAINTIES/LIMITATIONS

The general subject of uncertainties/limitations is discussed in this section. The approach will examine the topic by briefly reviewing health risk and hazard risk concerns separately with their uncertainties/limitations highlighted. However, before proceeding to these two topics, it is necessary to once again define and differentiate between *accuracy* and *precision*, and *variability* and *uncertainty*. The accuracy of a measurement or a calculation refers to that value relative to the correct value while precision relates to the repeatability of that measurement or calculation and expressed as the confidence interval around the mean of that value. Variability refers to the variation in the values that arise due to a calculation or measurement while uncertainty, the concern at hand, refers to variations that occur due to the operations or process itself. For example, in measuring the diameter of a heat exchanger tube in a heat exchanger, accuracy refers to the value of the mean measurement compared to the true value of the tube diameter, precision refers to the range of values that arises from repeated measurements, while variability occurs from repeated measurements of the tube's diameter and uncertainty arises from the measurement of many tubes (in the population) of the heat exchanger.

3.4.1 HEALTH RISK

Although great controversy can surround results of risk assessments, especially quantitative risk assessments, they are useful in particular applications. They can help establish priorities for regulatory action or intervention of any type. A consistent risk assessment performed across a range of substances can create a relative estimate of their health risks to humans. The limits of risk assessment can also be tested when government agencies (faced with the absence of other types of data and the need for action) must rely on risk assessment methods to establish health-based standards or guidelines to prevent human exposure to hazardous substances. Because of risk assessment shortcomings and the desire for greater specificity in measuring exposure, increasing interest is being shown in understanding pathologic changes at the molecular level with the hope that these investigations will lead to toxicological and epidemiological analyses of greater accuracy and sensitivity than are currently available (US EPA 1987b; Shields and Hanes 1991). In a general sense, problems in this area arise because of:

1. Uncertainty associated with available data
2. Uncertainty associated with governing equations
3. Concerns associated with assumed information
4. Concerns associated with limited and/or constrained governing equations
5. Concerns associated with overall analysis quality

In the risk characterization steps of a human HRA, conclusions about health and dose-response are integrated with those from the exposure assessment step. In addition, confidence about these conclusions, including information about the uncertainties associated with each aspect of the assessment in the final risk summary, should be highlighted. In the previous assessment steps and in the risk characterization, the risk assessor should also distinguish between variability and uncertainty.

Variability arises from true heterogeneity in characteristics such as dose-response differences within a population or differences in contaminant levels in the environment. The values of some variables used in an assessment often change with time and space or across the population whose exposure is being estimated. Assessments should address the resulting variability in doses received by members of the target population. Individual exposure, dose, and risk can vary widely in a large population.

The central tendency and high-end individual risk descriptors are intended to capture the variability in exposure, lifestyles, and other factors that lead to a distribution of risk across a population. Uncertainty, on the other hand, represents lack of knowledge about factors such as adverse effects or contaminant levels, which may be reduced or increased with additional study. Generally, risk assessments involve several categories of uncertainty, and each merits consideration. Measurement uncertainty refers to the usual error that accompanies scientific measurements and standard statistical techniques applied to analytical quality control data can often be used to express measurement uncertainty. A substantial amount of uncertainty is often inherent in environmental sampling, and assessments should also address these uncertainties. Similarly, there are uncertainties associated with the use of scientific models, for example, dose-response models, and models of environmental fate and transport. Evaluation of model uncertainty should consider the scientific basis for the model and its available empirical validation.

It should be noted that there is no completely satisfactory way to generate accurate risk data since it is an inexact science fraught with uncertainties. At the very least, risk characterization should be checked against experience for reasonableness since the size and quality of the data employed does not permit an accurate quantitative estimate with a high degree of confidence. Careful documentation of all four parts of a risk assessment should also be maintained to prevent the practitioner from falling into traps that can influence the final results or pass the risk via a cross-media process onto another location or vulnerable population. The authors believes that the EPA and OSHA have compounded the human HRA uncertainty problem by some of their ambiguous and conflicting rules and regulations.

Finally, it should also be noted that less information is available on the similarities or differences in the degree of response of experimental animals as compared to humans to varying doses of a chemical. In these tests, the animals are, out of necessity, administered high doses of the chemical whereas humans are usually exposed to much lower levels. This makes it necessary to extrapolate from results seen at high doses to the results expected at low doses in humans. The validity of these extrapolations is, in most cases, not amenable to experimental verification. Thus, while the test species may serve as an approximate measure of the potential of a chemical to cause toxic effects in humans, attempts to quantify human risk on the basis of such studies remain subject to considerable scientific uncertainty. This uncertainty is particularly critical when, for example, attempts are made to predict carcinogenic responses in humans using data from tests in rats and mice.

3.4.2 Hazard Risk

Estimating the magnitude of risks that cannot be measured accurately or directly often requires employing assumptions that cannot be verified or tested experimentally. Obviously, knowledge about the present and the future is never completely accurate. Inadequate knowledge is usually the largest cause of uncertainty. The inadequacy of knowledge means that the full extent of the uncertainty is also unknown. Uncertainty due to variability occurs when a single number (as often employed in risk analysis) is used to describe something that truly has multiple or variable values.

Variability is often ignored by using values based on the mean of all the values occurring within a group. Information on sources of uncertainties and limitations of input data are available.

Uncertainties and limitations in system description data could include the following:

1. Process description or drawings that are incorrect or out of date (Burke, Singh, and Theodore 2000).
2. Procedures do not represent actual operations.
3. Site area maps and population data that be incorrect or out of date.
4. Weather data from the nearest available site may be inappropriate due to its distance from the site or dissimilarity of microclimatic conditions.

Hazard identification data could have uncertainties and/or limitations because:

1. Recognition of major hazards may be incomplete.
2. Screening techniques employed for the selection of hazards for further evaluation may omit important cases

Frequently techniques may have sources of uncertainty or limitation due to:

1. *Uncertainties in modeling due to*:
    a. Extrapolation of historical data to larger-scale operations that may overlook hazards introduced by scaling-up to larger equipment
    b. Limitation of fault tree theory that requires system simplification
    c. Incompleteness in fault and event tree analysis (Shaefer and Theodore 2007)
2. *Uncertainties in data that may be caused because*:
    a. Data may be inaccurate, incomplete, or inappropriate
    b. Data from related activities might not be directly applicable
    c. Data generated by expert judgment may be inaccurate
    d. Characterization of the general population is improper or incomplete

Consequence techniques may have sources of uncertainty or limitations due to calculational burdens (even with computers). For example, they may arise from a number of dispersion modeling variables for chemical accidents, including:

1. *Uncertainties in physical modeling due to*:
    a. Inappropriate model selection
    b. Incorrect or inadequate physical basis for a model
    c. Inadequate validation
    d. Inaccurate model parameters
2. *Uncertainties in physical model data due to*:
    a. Uncertain input data (composition, temperature, pressure)
    b. Uncertain source terms for dispersion and other models

3. *Uncertainties in effects modeling due to*:
   a. Animal data that may be inappropriate for humans (especially for toxicity)
   b. Omission of mitigating effects
   c. Lack of epidemiological data on humans of the same sex, age, education, etc.

Risk estimation may have sources of uncertainty or limitation due to:

1. *Assumptions of symmetry such as*:
   a. Uniform wind roses that rarely occur
   b. Uniform ignition sources that may be incorrect
   c. Single point source for all incidents that may be inaccurate
2. *Assumptions to reduce the complexity of the analysis such as*:
   a. A single condition of wind speed and stability that may be too restrictive
   b. A limited number of ignition cases that can reduce accuracy
   c. General problems with the quality of data

The reader should note that since many risk analyses have been conducted on the basis of fatal effects, there are also uncertainties on precisely what constitutes a fatal dose of thermal radiation, blast effect, a toxic chemical, and so on.

3.5 ENVIRONMENTAL REGULATIONS

Environmental regulations are not simply a collection of laws on environmental topics. They are an organized system of statutes, regulations, and guidelines that minimize, prevent, and punish those responsible for the consequences of damage to the environment. This system requires each individual, whether an engineer, physicist, chemist, attorney, or consumer, to be familiar with its concepts and case-specific interpretations. Environmental regulations deal with the problems of human activities and the environment, and the uncertainties of law associated with them.

The National Environmental Policy Act (NEPA), enacted on January 1, 1970, is considered a political anomaly by some. NEPA is not based on specific legislation; instead, it is concerned in a general manner with environmental and quality of life issues. The Nixon Administration at the time became preoccupied not only with trying to pass more extensive environmental legislation but also implementing the laws. Nixon's White House Commission on Executive Reorganization proposed in the Reorganizational Plan 3 of 1970 that a single independent agency be established separate from the Council for Environmental Quality (CEQ). The plan was sent to Congress by President Nixon on July 9, 1970, and this new EPA began operation on December 2, 1970. EPA was officially born.

In many ways, EPA is the most far-reaching regulatory agency in the federal government because its authority is very broad. It is charged with protecting the nation's land, air, and water systems. Under a mandate of national environmental laws, it continues to strive to formulate and implement actions that lead to a compatible balance between human activities and the ability of natural systems to support and nurture life.

3.5.1 Air

Air management issues involve several different areas related to air pollutants and their control and are regulated at the federal level through the Clean Air Act. Atmospheric dispersion of pollutants can be mathematically modeled (generally in equation form) to predict where pollutants emitted from a particular source, such as a utility stack, will reach the ground and at what concentration. Pollution control equipment can be added to various sources to reduce the quantity of pollutants before they are emitted into the air. Acid rain, the greenhouse effect, and global warming are all indicators of adverse effects to the air, land, and sea, which result from excessive quantities of pollutants being released into the air.

One topic that few people are aware of is the issue of indoor air quality. Inadequate ventilation systems in homes and businesses directly affect the health of the building occupants. For example, the episode of Legionnaires' disease, which occurred in Philadelphia in the 1970s, was related to microorganisms that grew in the cooling water of the air conditioning system. Noise pollution, although not traditionally an air pollution issue, is included in this topic. The effects of noise pollution are seldom noticed until hearing is impaired. Although hearing impairment is a commonly known result of noise pollution, few people realize that stress is also a significant result of excessive noise exposure. The human body enacts its innate physiologic defensive mechanisms under conditions of loud noise, and the fight to control these physical instincts can cause significant stress on the individual.

3.5.2 Water

Pollutants entering rivers, lakes, and oceans originate from a wide variety of sources, including stormwater runoff, industrial discharges, and accidental spills and are regulated at the federal level through the Clean Water Act, while drinking water quality is regulated by the Safe Drinking Water Act. It is important to understand how these substances disperse in their environment in order to determine how to control them. Studies on pollutants in municipal and industrial wastewater treatment systems, industrial use systems, drinking water supplies, and other water systems are on-going. Often, wastewater from industrial plants must be pretreated before it can be discharged into a municipal treatment system and are controlled by federally mandated municipal pretreatment programs.

3.5.3 Solid Waste

Solid waste management through the Resource Conservation and Recovery Act (RCRA) addresses treatment and disposal methods for municipal, commercial, industrial, and medical wastes. Programs to reduce and dispose of municipal and industrial waste include reuse, reduction, recycling, and composting, in addition to incineration and landfilling. Potentially infectious waste generated in medical facilities must be specially packaged, handled, stored, transported, treated, and disposed of to ensure the safety of both the waste handlers and the general public. Incineration (Santoleri, Reynolds, and Theodore 2000) has been a typical treatment method for hazardous waste for many years. The

Environmental and Health Risk Analysis

Comprehensive Environmental Response, Compensation and Liability Act (CERCLA), also known as Superfund, was enacted in 1984 to identify and remedy uncontrolled hazardous waste sites. It also attempted to place the burden of cleanup on the generator rather than on local municipalities, states, or the federal government. Asbestos, metals, and underground storage tanks, for example, either contain or are inherently hazardous materials that require special handling and disposal. Further, it is important to realize that both small and large generators of hazardous waste are regulated.

3.5.4 POLLUTION PREVENTION

Pollution prevention covers domestic, commercial, municipal, and industrial means of reducing pollution and is supported at the federal level through technology transfer, guidance, and municipal and industrial collaboration and support through the 1990 Pollution Prevention Act. Pollution prevention covers both energy and water conservation as well as material use efficiency and can be accomplished through: (1) proper residential and commercial building design; (2) proper heating, cooling, and ventilation systems; (3) energy conservation; (4) reduction of water consumption; and (5) attempts to reuse or reduce materials before they become wastes. Domestic and industrial solutions to environmental problems can be addressed by considering ways to make homes and workplaces more energy efficient as well as ways to reduce the quantities of wastes generated within them.

3.6 ILLUSTRATIVE EXAMPLES

Four illustrative examples complement the material presented above.

3.6.1 ILLUSTRATIVE EXAMPLE 1

What are the two concepts that generally arise in a discussion associated with uncertainty.

Solution: Generally, uncertainty consists of two parts: variability and inadequate knowledge. Uncertainty due to variability occurs when a single number is used to describe something that truly has multiple or variable values. Variability is often ignored by using values based on the mean of all the values occurring within a group. A second type of variability is when a single value exists but constantly changes over time.

Despite the importance of variability of data, inadequate knowledge is usually the largest cause of uncertainty. Three common sources of inadequate knowledge include:

1. Parameter uncertainty or lack of knowledge of accurate parameter values due to measurement errors, random errors, systematic errors, etc.
2. Model uncertainty due to errors arising from incorrect concepts of reality and the use of incorrect models for describing chemical releases, transport, and exposure
3. Decision rule uncertainty or lack of knowledge regarding how best to interpret modeling outcomes and resulting consequences

3.6.2 ILLUSTRATIVE EXAMPLE 2

Discuss health problems with delayed effects and their impact on risk estimates.

Solution: The cause of a health problem may not be obvious if the health problem effect is delayed. This applies to many chronic diseases. The direct proof of causality is important for this class of risk. The delayed effect obviously increases the uncertainties associated with any risk estimate.

3.6.3 ILLUSTRATIVE EXAMPLE 3

Can a health risk characterization provide information on exactly what to do about a specific problem?

Solution: No. Risk characterization is often imprecise in that it draws upon available information about a problem, applies scientific principles, and then provides guidance on potential risk. It does help identify hazards. How that information is used to decide what steps, if any, to take to reduce the hazard is not part of the risk characterization process, but is entailed in risk management activities.

3.6.4 ILLUSTRATIVE EXAMPLE 4

List some of the types of information a HZRA can provide.

Solution: A HZRA can provide the following information:

1. Identification and description of hazards and accident events that could lead to undesirable consequences
2. A qualitative estimate of the likelihood and consequence of each accident event sequence
3. A relative ranking of the risk of each hazard and accident event sequence
4. Some suggested approaches to risk reduction

For the chemical process industry, these results are normally provided to plant management and engineering or research groups, as appropriate, so that overall plant and process safety can be improved, and so that both on- and off-site risks can be minimized.

PROBLEMS

3.1 The word "what" appears in numerous problem identification procedures. List some questions/comments that are related to this term.
3.2 List and briefly discuss human errors that can occur in a chemical processing plant that can lead to accidents and uncontrolled releases of hazardous materials.
3.3 Discuss the problems in valuing life in the context of cost-benefit analysis in risk characterization.
3.4 Discuss EPA's relationship and role in addressing risk and environmental pollution at the state and local levels.

REFERENCES

American Institute of Chemical Engineers 1992. *Guidelines for Hazard Evaluation Procedures.* New York, NY: Center for Chemical Process Safety.

American Insurance Association 1979. *Hazard Survey of the Chemical and Allied Industries.* Technical Survey No. 3. New York, NY: Engineering and Safety Service.

Burke, G., Singh, B., and Theodore, L. 2000. *Handbook of Environmental Management and Technology*, 2nd Edition. Hoboken, NJ: John Wiley & Sons.

Masters, G. 1991. *Introduction to Environmental Engineering and Science.* Upper Saddle River, NJ: Prentice Hall.

Paustenbach, D. 1989. *The Risk Assessment of Environmental and Human Health Hazards: A Textbook of Case Studies.* Hoboken, NJ: John Wiley & Sons.

Santoleri, J., Reynolds, J., and Theodore, L. 2000. *Introduction to Hazardous Waste Incineration*, 2nd Edition. Hoboken, NJ: John Wiley & Sons.

Shaefer, S., and Theodore, L. 2007. *Probability and Statistics Applications in Environmental Science.* Boca Raton, FL: CRC Press/Taylor & Francis Group.

Shields, P., and Hanes, N. 1991. Molecular Epidemiology and the Genetics of Environmental Cancer. *Journal of the American Medical Association* 266: 681–687.

Theodore, L., and Dupont, R.R. 2012. *Environmental Health Risk and Hazard Risk Assessment: Principles and Calculations.* Boca Raton, FL: CRC Press/Taylor & Francis Group.

Theodore, L., Reynolds, J., and Morris, K. 1996. *Health, Safety and Accident Prevention: Industrial Applications.* Theodore Tutorials (originally published by USEPA, RTP, NC), East Williston, NY, 1996.

US Environmental Protection Agency 1987a. *Technical Guidance for Hazard Analysis, Emergency Planning for Extremely Hazardous Substances.* EPA-OSWER-88-0001. Washington, DC: Office of Solid Waste and Emergency Response.

US Environmental Protection Agency 1987b. *Unfinished Business: A Comparative Assessment of Environmental Problems, Overview Report.* EPA-230-2-87-025a. Washington, DC: Office of Policy, Planning and Evaluation.

4 Introduction to Probability and Statistics

4.1 INTRODUCTION

This chapter addresses the key fundamentals and principles associated with both *probability* and *statistics*. Webster defines probability as "the ratio of the number of outcomes in an exhaustive set of equally likely outcomes that produce a given event to the total number of possible outcomes" (Merriam-Webster 2021a). Most applied scientists, including engineers, would claim that probability is concerned with describing the phenomena of chance and randomness. Statistics is defined by Webster (Merriam-Webster 2021b) as follows: "a branch of mathematics dealing with the collection, analysis, interpretation, and presentation of masses of numerical data." There are obviously many other definitions.

Generally, *statistics* can be simply defined as the branch of science which deals with collecting, arranging, and using numerical facts or data arising from natural phenomena or experiments. Thus, a *statistic* is an item of information deduced from the application of statistical methods.

The most frequently encountered statistical problems are those which involve one or more of the following features:

1. Reduction of data
2. Relationships between two or more variables
3. Estimates and tests of significance
4. Reliability of inferences depending on one or more variables
5. Analysis of the fluctuations in a measurable quantity which can arise naturally during the course of a particular operation (Carnahan, Luther and Wilkes 1969)

No attempt is made in this chapter to introduce and apply packaged computer programs that are presently available; the emphasis is to provide the reader with an understanding of the fundamental principles so that he or she learns how statistical methods can be used in environmental and health risk analysis. The bulk of the material presented in this chapter is drawn from Theodore and Taylor (1996).

4.2 PROBABILITY DEFINITIONS AND INTERPRETATIONS

Probabilities are nonnegative numbers associated with the outcomes of the so-called random experiments. A random experiment is an experiment whose outcome is uncertain. Examples include throwing a pair of dice, tossing a coin, counting the number of defects in a sample from a lot of manufactured items, or observing the time to failure of a tube in a heat exchanger, a seal in a pump, or a bus section in an electrostatic precipitator. The set of possible outcomes of a random experiment is called the sample space and is usually designated by S. The probability of event A, $P(A)$, is the sum of the probabilities assigned to the outcomes constituting the subset A of the sample space S.

Consider, for example, tossing a coin twice. The sample space of heads (H) and tails (T) can be described as:

$$S = \{HH, HT, TH, TT\} \tag{4.1}$$

If probability ¼ is assigned to each element of S, and A is the event of at least one head, then:

$$A = \{HH, HT, TH\} \tag{4.2}$$

The sum of the probabilities assigned to the elements of A is ¾. Therefore, $P(A) = ¾$. The description of the sample space is not unique. The sample space S in the case of tossing a coin twice could be described in terms of the number of heads obtained. Then,

$$S = \{0, 1, 2\} \tag{4.3}$$

Suppose probabilities ¼, ½, ¼ are assigned to the outcomes 0, 1, and 2, respectively. Then, A, the event of at least one head, would have for its probability:

$$P(A) = P\{1, 2\} = ¾ \tag{4.4}$$

How probabilities are assigned to the elements of the sample space depends on the desired interpretation of the probability of an event. Thus, $P(A)$ can be interpreted as the *theoretical relative frequency*, that is, a number about which the relative frequency of event A tends to cluster as n, the number of times the random experiment is performed, increases indefinitely. This is the objective interpretation of probability. Under this interpretation, saying $P(A)$ is ¾ in the example given above means that if a coin is tossed twice n times, the proportion of times one or more heads occur clusters about ¾ as n increases indefinitely.

As another example, consider a single valve that can stick in an open (O) or closed (C) position. The sample space can be described as follows:

$$S = \{O, C\} \tag{4.5}$$

Suppose that the valve sticks twice as often in the open position as it does in the closed position. Under the theoretical relative frequency interpretation, the probability assigned to element O in S would be 2/3, twice the probability assigned to element C. If two such valves are observed, the sample space S can be described as:

$$S = \{OO, OC, CO, CC\} \tag{4.6}$$

Assuming that the two valves operate independently, a reasonable assignment of probabilities to the elements of S as listed could be 4/9, 2/9, 2/9, and 1/9. The reason

Introduction to Probability and Statistics

for this assignment will become clear after consideration of the concept of independence. If A is the event of at least one valve sticking in the closed position, then:

$$S = \{OC, CO, CC\} \tag{4.7}$$

The sum of the probabilities assigned to the elements of A is 5/9. Therefore, $P(A) = 5/9$.

Probability $P(A)$ can also be interpreted subjectively as a measure of degree of belief, on a scale from 0 to 1, that the event A occurs. This interpretation is frequently used in ordinary conversation. For example, if someone says, "The probability that I will go to the racetrack today is 90%," then 90% is a measure of the person's belief that they will go to the racetrack. This interpretation is also used when, in the absence of concrete data needed to estimate an unknown probability, on the basis of observed relative frequency, the personal opinion of an expert is sought. For example, an expert might be asked to estimate the probability that the seals in a newly designed pump will leak at high pressures. The estimate would be based on the expert's familiarity with the history of pumps of similar design.

Various combinations of the occurrence of any two events A and B can be indicated in set notations as follows:

$\bar{A}$ – A does not occur
$\bar{B}$ – B does not occur
$A + B$ – A occurs or B occurs in the mutually inclusive sense to indicate the occurrence of A, B, or both A and B
AB – A occurs and B occurs
$A\bar{B}$ – A occurs and B does not occur

Venn diagrams (Figure 4.1) provide a pictorial representation of these events. In set terminology, $\bar{A}$ is called the *complement* of A, that is, the set of elements in S that

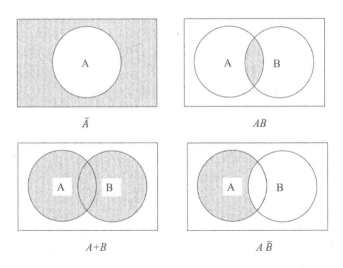

FIGURE 4.1 Venn diagrams.

are not in A. An alternate notation for the complement of A is A^c or A'. AB is called the *intersection* of A and B, that is, the set of elements in both A and B. An alternate notation in the literature for the intersection of A and B is $A \cap B$. $A + B$ is called the *union* of A and B, that is, the set of elements A, B, or both A and B. An alternate notation for the union of A and B is $A \cup B$.

When events A and B have no elements in common they are said to be *mutually exclusive*. A set having no elements is called the *null set* and is usually designated by ϕ. Thus, if events A and B are mutually exclusive, then $AB = \phi$. Note that the union of A and B consists of three mutually exclusive events: $A\bar{B}$, AB, $\bar{A}B$.

4.3 BASIC PROBABILITY THEORY

The mathematical properties of $P(A)$, the probability of event A, are deduced from the following postulates governing the assignment of probabilities to the elements of a sample space, S.

1. $P(S) = 1$
2. $P(A) \geq 0$ for any event A
3. If $A_1, \ldots, A_n \ldots$ are mutually exclusive, then $P(A_1 + A_2 + \cdots + A_n + \cdots) = P(A_1) + P(A_2) + \cdots + P(A_n) + \cdots$

In the case of a discrete sample space (i.e., a sample space consisting of a finite number or countable infinitude of elements), these postulates require that the numbers assigned as probabilities to the elements of S be nonnegative and have a sum equal to 1. These requirements do not result in the complete specification of the numbers assigned as probabilities. The approved interpretation of probability must also be considered. The mathematical properties of the probability of any event are the same regardless of how this probability is interpreted. These properties are formulated in theorems logically deduced from the postulates given earlier without the need for appeal to interpretation. Three basic theorems are:

$$\text{Theorem 1: } P(\bar{A}) = 1 - P(A) \tag{4.8}$$

$$\text{Theorem 2: } 0 \leq P(A) \leq 1 \tag{4.9}$$

$$\text{Theorem 3: } P(A+B) = P(A) + P(B) - P(AB) \tag{4.10}$$

Theorem 1 states that the probability that A does not occur is 1 minus the probability that A occurs. Theorem 2 states that the probability of any event lies between 0 and 1. Theorem 3, the addition theorem, provides an alternative way of calculating the probability of the union of two events as the sum of their probabilities minus the probability of their intersection. The addition theorem can be extended to three or more events. In the case of three events A, B, and C, the addition theorem becomes:

$$P(A+B+C) = P(A) + P(B) + P(C) - P(AB) - P(AC) - P(BC) + P(ABC) \tag{4.11}$$

For four events A, B, C, and D, the addition theorem becomes:

$$P(A+B+C+D) = P(A)+P(B)+P(C)+P(D)-P(AB)$$
$$-P(AC)-P(AD)-P(BC)-P(BD)-P(CD)+P(ABC)$$
$$+P(ABD)+P(BCD)+P(ACD)-P(ABCD) \quad (4.12)$$

To illustrate the application of the three basic theorems (Equations 4.8 through 4.10), consider what happens when one draws a card at random from a deck of 52 cards. The sample space S may be described in terms of 52 elements, each corresponding to one of the cards in the deck. Assuming that each of the 52 possible outcomes would occur with equal relative frequency in the long run leads to the assignment of equal probability, 1/52, to each of the elements of S. Let A be the event of drawing an ace and B the event of drawing a club. Thus, A is a subset consisting of four elements, each of which has been assigned probability 1/52, and $P(A)$ is the sum of these probabilities 4/52. Similarly, the following probabilities are obtained:

$$P(B) = 13/52, P(AB) = 1/52$$

Application of Theorem 1 gives:

$$P(\overline{A}) = 1 - P(A) = 48/52; P(\overline{B}) = 1 - P(B) = 39/52$$

Application of the addition theorem gives:

$$P(A+B) = P(A)+P(B)-P(AB) = 4/52 + 13/52 - 1/52 = 16/52$$

$P(A + B)$, the probability of drawing an ace or a club, can be calculated without using the addition theorem by calling $A + B$, the union of A and B, a set consisting of 16 elements. (The number, 16, is obtained by adding the number of aces, 4, to the number of clubs, 13, and subtracting the card that is counted twice—once as an ace and once as a club.) Since each of the 16 elements in $A + B$ has been assigned a probability of 1/52, $P(A + B)$ is the sum of the probabilities assigned, namely, 16/52.

The *conditional probability* of event B given A is denoted by $P(B|A)$ and is defined as:

$$P(B|A) = \frac{P(AB)}{P(A)} \quad (4.13)$$

where $P(B|A)$ can be interpreted as the proportion of A occurrences that also feature the occurrence of B.

For example, consider the random experiment of drawing two cards in succession from a deck of 52 cards. Suppose the cards are drawn *without* replacement (i.e., the first drawn is not replaced before the second card is drawn). Let A denote the event that the first card is an ace, and B, the event that the second card drawn is an ace. The sample space S can be described as a set of 52 times 51 pairs of cards. Assuming that each of these (52)(51) pairs has the same theoretical relative frequency, assign probability 1/(52)(51) to each pair. The number of pairs featuring an ace as the first and second card is (4)(3). Therefore,

$$P(AB) = \frac{(4)(3)}{(52)(51)}$$

The number of pairs featuring an ace as the first card and one of the other 51 cards as the second is (4)(51). Therefore,

$$P(A) = \frac{(4)(51)}{(52)(51)}$$

Applying the definition of conditional probability, Equation 4.13, yields:

$$P(B|A) = \frac{P(AB)}{P(A)} = \frac{(4)(3)/(52)(51)}{(4)(51)/(52)(51)} = \frac{3}{51} = 0.0588$$

as the conditional probability that the second card is an ace, given that the first is an ace. The same result could have been obtained by computing $P(B)$ on a new sample space consisting of 51 cards, three of which are aces. This illustrates the two methods for calculating a conditional probability. The first method calculates the conditional probability in terms of probabilities computed on the original sample space by means of the definition in Equation 4.13. The second method uses the given event to construct a new sample space on which the conditional probability is computed.

Conditional probability can also be used to formulate a definition for the independence of two events A and B. Event B is defined as independent of event A if and only if:

$$P(B|A) = P(B) \qquad (4.14)$$

Similarly, event A is defined to be independent if and only if:

$$P(A|B) = P(A) \qquad (4.15)$$

From the definition of conditional probability in Equation 4.13, one can deduce the logically equivalent definition that event A and event B are independent if and only if:

$$P(AB) = P(A) + P(B) \qquad (4.16)$$

To illustrate the concept of independence, consider again the random experiment of drawing two cards in succession from a deck of 52 cards. This time suppose that the cards are drawn *with* replacement (i.e., the first card is replaced in the deck before the second card is drawn). As before, let A denote the event that the first card is an ace, and B, the event that the second card drawn is also an ace. Then:

$$P(B|A) = P(B) = 4/52$$

and since $P(B|A) = P(B)$, B and A are independent events.

Introduction to Probability and Statistics

From the definition of $P(B|A)$ and $P(A|B)$, one can deduce from Equation 4.13 the multiplication theorem,

$$P(AB) = P(A)P(B|A) \tag{4.17}$$

$$P(AB) = P(B)P(A|B) \tag{4.18}$$

The multiplication theorem provides an alternate method for calculating the probability of the intersection of two events.

The multiplication theorem can be extended to the case of three or more events. For three events A, B, C, the multiplication theorem states:

$$P(ABC) = P(A)P(B|A)P(C|AB) \tag{4.19}$$

For four events A, B, C, and D, the multiplication theorem states:

$$P(ABCD) = P(A)P(B|A)P(C|AB)P(D|ABC) \tag{4.20}$$

Consider now n mutually exclusive events $A_1, A_2, ..., A_n$, whose union is the sample space S. Let B be any given event. Then, Bayes' theorem states:

$$P(A_i|B) = \frac{P(A_i)P(B|A_i)}{\sum_{i=1}^{n} P(A_i)P(B|A_i)}; i = 1, ..., \tag{4.21}$$

where $P(A_1)$, $P(A_2)$,..., $P(A_n)$ = *prior probabilities* of $A_1, A_2, ..., A_n$; and $P(A_1|B)$, $P(A_2|B)$ = *posterior probabilities* of $A_1, A_2, ..., A_n$.

Bayes' theorem provides the mechanism for revising prior probabilities, that is, for converting them into posterior probabilities on the basis of the observed occurrence of some given event.

4.4 MEDIAN, MEAN, AND STANDARD DEVIATION

One basic way of summarizing data is by the computation of a central value. The most commonly used central value statistic is the arithmetic average, or the *mean*. This statistic is particularly useful when applied to a set of data having a fairly symmetrical distribution. The mean is an efficient statistic in that it summarizes all the data in the set and because each piece of data is taken into account in its computation. The formula for computing the mean is:

$$\bar{X} = \frac{X_1 + X_2 + X_3 + \cdots + X_n}{n} = \frac{\sum_{i=1}^{n} X_i}{n} \tag{4.22}$$

where $\bar{X}$ = arithmetic mean; X_i = an individual measurement; n = total number of measurements; and $X_1, X_2, X_3, ...$ = the measurements 1, 2, 3, respectively.

The arithmetic mean is not a perfect measure of the true central value of a given data set. It can overemphasize the importance of one or two extreme data points.

Many measurements of a normally distributed data set will have an arithmetic mean that closely approximates the true central value.

When a distribution of data is asymmetrical, it is sometimes desirable to compute a different measure of central value. The second measure, known as the *median*, is simply the middle value of a distribution, or the quantity above which half the data lie and below which the other half lie. If n data points are listed in their order of magnitude, the median is the $[(n + 1)/2]$th value. If the number of data is even, then the numerical value of the median is the value midway between the two data nearest to the middle. The median, being a positional value, is less influenced by extreme values in a distribution than is the mean value. However, the median alone is usually not a good measure of central tendency. To obtain the median, the data provided must be first arranged in numerical order with the middle value determined from the reordered data set.

Another measure of central tendency used in specialized applications is the *geometric mean*, $\overline{X_G}$. The geometric mean can be calculated using the following equation:

$$\overline{X_G} = \sqrt[n]{(X_1)(X_2)...(X_2)} \qquad (4.23)$$

The most common measure of dispersion, or variability, of sets of data is the *standard deviation*, σ. Its defining formula is given by the expression:

$$\sigma = \sqrt{\frac{\Sigma(X_i - \overline{X})^2}{n-1}} \qquad (4.24)$$

where σ = the standard deviation (always positive); and X_i is the value of the ith data point.

The expression $(X_i - \overline{X})$ indicates that the deviation of each piece of data from the mean is taken into account by the standard deviation. *Variance* is denoted as σ^2.

As noted before, statistics is often concerned with the extraction of information from observed data. A distinction between a sample and the population (the totality of measurements from which the sample was drawn) needs to be made in order to draw an inference about a population from the information contained in a sample. To assist in this distinction, a characteristic of a *population* is normally symbolized with a Greek letter, for example, μ = 36.7 g, σ^2 = 100 g, and σ = 10 g. A characteristic of a *sample* is normally symbolized with an alphabet letter, for example, $\overline{X}$ = 24.3 g, s^2 = 81 g, and s = 9 g.

Chebyshev's theorem provides a rather unique and interesting interpretation of the sample standard deviation as a measure of the spread (dispersion) of sample observations about their mean (Theodore and Taylor 1996). The theorem states that at least $(1 - 1/k^2)$, $k > 1$, of the sample observations lie in the interval $(\overline{X} - ks, \overline{X} + ks)$. For $k = 2$, for example, this means that at least 75% of the sample observations lie in the interval $(\overline{X} - 2s, \overline{X} + 2s)$. The smaller the value of s, the greater the concentration of observations in the vicinity of $\overline{X}$.

In the case of a random sample of observations on a continuous random variable assumed to have a so-called normal pdf (see also Chapter 5), the graph of which is a

Introduction to Probability and Statistics

bell-shaped curve, the following statements give a more precise interpretation of the sample standard deviation s as a measure of spread or dispersion of observations of the variable (Theodore and Taylor 1996; Shaefer and Theodore 2007).

1. $\bar{X} \pm s$ includes approximately 68% of the sample observations.
2. $\bar{X} \pm 2s$ includes approximately 95% of the sample observations.
3. $\bar{X} \pm 3s$ includes approximately 99.7% of the sample observations.

The source of these percentages is the normal probability distribution, which is discussed later in more detail in Chapter 5.

4.5 RANDOM VARIABLES

A random variable is a real-valued function defined over the sample space S of a random experiment. The domain of the function is S, and the real numbers associated with the various possible outcomes of the random experiment constitute the range of the function. If the range of the random variable consists of a finite number or countable infinitude of values, the random variable is classified as *discrete*. If the range consists of a non-countable infinitude of values, the random variable is classified as *continuous*. A set has a countable infinitude of values if they can be put into one-to-one correspondence with positive integers. The positive even integers, for example, consist of a countable infinitude of numbers. The even integer $2n$ corresponds to the positive integers n for $n = 1, 2, 3, \ldots$. The real numbers in the interval $(0, 1)$ constitute a non-countable infinitude of values.

Defining a random variable on a sample space S amounts to coding the outcomes in real numbers. Consider, for example, the random experiment involving the selection of an item at random from a manufactured lot. Associate $X = 0$ with the drawing of a non-defective item and $X = 1$ with the drawing of a defective item. Then, X is a random variable with range $(0, 1)$ and is therefore discrete.

Let X denote the number of the throw on which the first failure of a switch occurs. Then, X is also a discrete random variable with range $\{1, 2, 3, \ldots, n, \ldots\}$. Note that the range of X consists of a countable infinitude of values and that X is therefore discrete. Alternatively, suppose that X denotes the time to failure of a bus section in an electrostatic precipitator. Then, X is a continuous random variable whose range consists of the real numbers greater than zero and is classified as continuous.

The probability distribution of a random variable concerns the distribution of probability over the range of the random variable. The distribution of probability is specified by the *probability density function (pdf)*. Chapter 5 is concerned with the general properties of the pdfs for the case of discrete and continuous random variables. Special pdfs find extensive application in environmental and health risk analysis.

4.6 LINEAR REGRESSION

The practicing engineer and scientist often encounters applications (e.g., describing future energy demands) that require the development of a mathematical relationship between data for two or more variables. For example, if Y (a dependent variable, e.g., health risk) is a function of or depends on t (an independent variable, e.g., time), that is:

$$Y = f(t) \tag{4.25}$$

one may be required to express this (Y, t) data in equation form. This process is referred to as *regression analysis*, and the regression method most often employed is the method of *least squares*.

An important step in this procedure, which is often omitted, is to prepare a plot of Y versus t. The result, referred to as a scatter diagram, could take on any form. Three such plots are provided in Figure 4.2. The first plot Figure 4.2a suggests a linear relationship between Y and t, that is:

$$Y = a_0 + a_1 t \tag{4.26}$$

where a_0 and a_1 are coefficients determined from regression analysis.

The second graph Figure 4.2b appears to be best represented by a second-order (or parabolic) relationship, that is:

$$Y = a_0 + a_1 t + a_2 t^2 \tag{4.27}$$

where a_0, a_1 and a_2 are coefficients determined from regression analysis.

The third plot suggests a linear model that applies over two different ranges; that is, it should represent the data where:

$$Y = a_0 + a_1 t; t_0 < t < t_M \tag{4.28}$$

and

$$Y = a_0' + a_1' t; t_M < t < t_L \tag{4.29}$$

where a_0, a_1, a_0', and a_1' are coefficients determined from regression analysis, and t_0 = the initial time point for the data set; t_M = the mid-range time point for the data set that ends one regression line and begins the other; and t_L = the last time point for the data set. This multi-equation model finds application in representing adsorption equilibria, multi-particle size distributions (Theodore 2007), quantum energy relationships, etc. In any event, a scatter diagram and individual judgment can suggest an appropriate model at an early stage in the analysis.

Some additional regression models often employed by technical individuals are as follows:

$$Y = a_0 + a_1 X + a_2 X^2 + a_3 X^3; \text{Cubic (third-order model)} \tag{4.30}$$

$$Y = a_0 + a_1 X + a_2 X^2 + a_3 X^3 + a_4 X^4; \text{Quadratic (fourth-order model)} \tag{4.31}$$

$$Y = 1/(a_0 + a_1 X); \text{Hyperbolic} \tag{4.32}$$

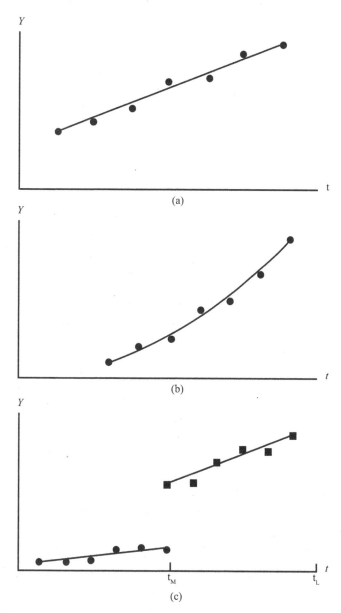

FIGURE 4.2 Scatter diagrams: (a) linear relationship; (b) parabolic relationship; and (c) dual-linear relationship.

$$Y = aX^b; \text{Exponential} \tag{4.33}$$

$$Y = ab^X; \text{Modified Exponential} \tag{4.34}$$

Procedures to evaluate the regression coefficients a, a_0, a_1, a_2, etc. are provided as follows. The reader should note that the analysis is based on the method of least

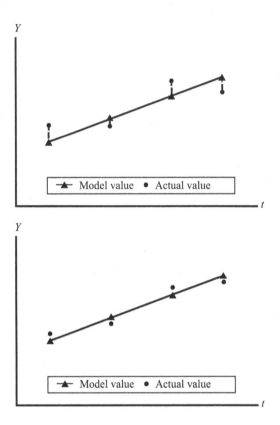

FIGURE 4.3 Error difference: actual versus predicted values.

squares. This technique provides numerical values for the regression coefficients a_i such that the sum of the square of the difference (error) between the actual Y and that predicted by the equation or model Ye is minimized. This shown in Figure 4.3.

In Figure 4.3 the dots (experimental values of Y) and triangles (equation or model value of Y, i.e., Y_e) represent the data and model values, respectively. On examining the two plots in Figure 4.3, one can conclude that the error $(Y - Y_e)$ squared and summed for the four points is less for the lower figure. Also note that a dashed line represents the error. The line that ultimately produces a minimum of the sum of the individual errors squared, that is, has the smallest possible value, is the appropriate regression model (based on the method of least squares). The proof is left as an exercise.

The calculation of the linear least squares model coefficients for Equation 4.35 can be calculated using Equations 4.36 and 4.37, where X and Y are measured data pairs, and n is the number of data pairs used to generate the best fit linear regression:

$$Y = a_0 + a_1 X \tag{4.35}$$

$$a_0 = Y_{intercept} = \frac{\sum X^2 \sum Y - \sum X \sum XY}{n \sum X^2 - (\sum X)^2} \tag{4.36}$$

$$a_1 = \text{Slope} = \frac{n\sum XY - \sum X \sum Y}{n\sum X^2 - (\sum X)^2} \tag{4.37}$$

The cubic model takes the form shown in Equation 4.30. For n pairs (more than four) of (Y, t) values, the constants a_0, a_1, a_2, and a_3 can be obtained by the method of least squares so that $\sum(Y - Y_e)^2$ again has the smallest possible value (i.e., is minimized). The coefficients a_0, a_1, a_2, and a_3 are the solution of the following system of four linear equations:

$$a_0 n + a_1 \sum X + a_2 \sum X^2 + a_3 \sum X^3 = \sum Y \tag{4.38}$$

$$a_0 \sum X + a_1 \sum X^2 + a_2 \sum X^3 + a_3 \sum X^4 = \sum XY \tag{4.39}$$

$$a_0 \sum X^2 + a_1 \sum X^3 + a_2 \sum X^4 + a_3 \sum X^5 = \sum X^2 Y \tag{4.40}$$

$$a_0 \sum X^3 + a_1 \sum X^4 + a_2 \sum X^5 + a_3 \sum X^6 = \sum X^3 Y \tag{4.41}$$

Because there are four equations and four unknowns, this set of equations can be solved for a_0, a_1, a_2, and a_3. This development can be extended to other regression equations (e.g., exponential, hyperbolic, higher-order models, etc.).

The correlation coefficient provides information on how well the model, or line of regression, fits the data. It is denoted by R and is given by:

$$R = \frac{\sum XY - \dfrac{\sum X \sum Y}{n}}{\sqrt{\left[\sum X^2 - \dfrac{(\sum X)^2}{n}\right]\left[\sum Y^2 - \dfrac{(\sum Y)^2}{n}\right]}} \tag{4.42}$$

The correlation coefficient satisfies the following six properties (Theodore and Taylor 1996; Shaefer and Theodore 2007):

1. If all points of a scatter diagram lie on a line, then $r = +1$ or -1. In addition, $r^2 = 1$. The square of the correlation coefficient is defined as the coefficient of determination.
2. If no linear relationship exists between the Xs and Ys, then $r = 0$. Furthermore, $r^2 = 0$. It can be concluded that r is always between -1 and $+1$, and r^2 is always between 0 and 1.
3. Values of r close to -1 or $+1$ are indicative of a strong linear relationship.
4. Values of r close to 0 are indicative of a weak linear relationship.
5. The correlation coefficient is positive or negative depending on whether the linear relationship has a positive or negative slope. Thus, positive values of r indicate that Y increases as t increases; negative values indicate that Y decreases as t increases.

6. If $r = 0$, it only indicates the lack of a linear correlation; Y and t might be strongly correlated by some nonlinear relation, as discussed earlier. Thus, r can only measure the strength of linear correlations; if the data are nonlinear, one should attempt to linearize before computing r.

It should be noted that the correlation coefficient only provides information on how well the model fits the data. It is emphasized that r provides *no* information on how good the model is or, in other words, whether this is the correct or best model to describe the functional relationship of the data.

This topic is addressed in another statistical procedure, analysis of variance (ANOVA), a subject that is beyond the scope of this text (Shaefer and Theodore 2007).

Problems involving more than two independent variables can be treated in a manner similar to that for two independent variables. Consider the equation:

$$Z = a_0 + a_1 X + a_2 Y \qquad (4.43)$$

which is called a *linear equation in the variables X, Y, and Z*. In a three-dimensional rectangular coordinate system this equation represents a plane, and the actual sample points $(X_1, Y_1, Z_1), (X_2, Y_2, Z_2), \ldots, (X_N, Y_N, Z_N)$ may "scatter" not too far from this plane, which is called an *approximating plane*.

By extension of the method of least squares discussed above, a *least squares plane* approximating the data can be developed. If we are estimating Z from given values of X and Y, this would be called a *regression plane of Z on X and Y*. The normal equations corresponding to the least squares plane are given by:

$$\sum Z = a_0 n + a_1 \sum X + a_2 \sum Y \qquad (4.44)$$

$$\sum XZ = a_0 \sum X + a_1 \sum X^2 + a_2 \sum XY \qquad (4.45)$$

$$\sum YZ = a_0 \sum Y + a_1 \sum XY + a_2 \sum Y^2 \qquad (4.46)$$

and can be remembered as being obtained from Equation 4.43 by multiplying by 1, X, and Y, respectively, and then summing. These three linear equations can be solved simultaneously for a_0, a_1, and a_2.

4.7 ILLUSTRATIVE EXAMPLES

Four illustrative examples complement the material presented above.

4.7.1 ILLUSTRATIVE EXAMPLE 1

Consider a case of a carton of 100 vaccine units from a pharmaceutical firm from which a sample of two units are drawn *without* replacement. If the carton contains five contaminated units, what is the probability that the sampled units are both contaminated?

Introduction to Probability and Statistics

Solution: Let A denote the event that the first unit drawn is contaminated, and B, the event that the second sampled unit is also contaminated. Then, the probability that the sample contains exactly two contaminated units is P(AB). By application of the multiplication theorem, one obtains:

$$P(AB) = P(A)P(B \mid A) = (5/100)(4/99) = 0.002 = 0.2\%$$

4.7.2 Illustrative Example 2

A continuous random variable X has a probability density function (pdf) of $X/2$ for $0 \leq X \leq 2$. What is the cumulative distribution function (cdf)?

Solution: By definition, for continuous variables,

$$F(X) = \int f(X)dX = \int \frac{X}{2} dX = X^2/4; 0 \leq X \leq 2$$

Applying the definition of cdf leads to:

$$F(X) = 0; X < 0$$

$$F(X) = X^2/4; 0 \leq X \leq 2$$

$$F(X) = 1; X \geq 2$$

4.7.3 Illustrative Example 3

Large low-pressure utility storage tanks are often among the most fragile items of plant equipment in use. They are usually designed to withstand a gauge pressure of only 8 in of water (0.3 psi); they will burst at about three times this pressure, and also if a vacuum is exceeded within the tank by more than a small amount. It is not surprising, therefore, that these tanks are often damaged.

A random variable X denoting the useful life in years of a storage tank handling explosives in a chemical plant has the probability density function (pdf):

$$f(X) = \frac{X^3}{20.25}; 0 < X < 3$$

$$f(X) = 0; \text{elsewhere}$$

Find the cdf of X.

Solution: As in the previous example, the cdf of X is given by:

$$F(X) = \int f(X)dX = \int \frac{X^3}{20.25} dX = X^4/81; 0 < X < 3$$

$$F(X) = 0; X \leq 0$$

$$F(X) = 1; X \geq 3$$

4.7.4 ILLUSTRATIVE EXAMPLE 4

The following concentration (C) time (t) data from a reactor in a plant are provided:

C, g/L = Y	t, hr = X
0.581	0
0.565	1
0.556	2
0.544	3
0.530	4
0.512	5

a. Generate a linear equation that describes the data given above.
b. Calculate the concentration at $t = 3.5$ hr and $t = 7$ hr.

Solution: (a) A linear regression of the data presented in the problem statement can be generated using Equations 4.35 for the value of the $Y_{intercept}$ and Equation 4.36 for the value of the regression line slope. In reviewing Equations 4.35 and 4.36 the following values can be generated from the data provided in the problem statement

$$n = 6$$

$$\Sigma X = 15$$

$$\Sigma Y = 3.288$$

$$\Sigma X^2 = 55$$

$$\Sigma Y^2 = 1.8049$$

$$\Sigma XY = 7.989$$

The coefficients of the linear regression model for these data can then be calculated as:

$$a_0 = Y_{intercept} = \frac{\Sigma X^2 \Sigma Y - \Sigma X \Sigma XY}{n \Sigma X^2 - (\Sigma X)^2} = \frac{(55)(3.288) - (15)(7.989)}{6(55) - (15)^2} = \frac{61.005}{105} = 0.581$$

$$a_1 = Slope = \frac{n \Sigma XY - \Sigma X \Sigma Y}{n \Sigma X^2 - (\Sigma X)^2} = \frac{(6)(7.989) - (15)(3.288)}{6(55) - (15)^2} = \frac{-1.386}{105} = -0.0132$$

The line of best fit for the linear regression for this data set is then:

$$Y = 0.581 - 0.0132X$$

Introduction to Probability and Statistics

(b) Using the equation for the line of best fit for these data, the concentration at 3.5 hr is:

$$Y_{3.5} = 0.581 - 0.0132(3.5) = 0.581 - 0.0462 = 0.5348 \text{ mg/L}$$

and for 7 hr is:

$$Y_7 = 0.581 - 0.0132(7) = 0.581 - 0.0924 = 0.4886 \text{ mg/L}$$

PROBLEMS

4.1 Comment on the difference between an equation describing an exponential curve and an equation describing a "modified" exponential curve.

4.2 Comment on freehand curve fitting method for describing relationships between independent variables.

4.3 Comment on the two concentration calculations requested in Illustrative example 4.7.4.

4.4 The following data were collected during a calibration of a chemiluminescent NO_x analyzer in a chemical plant.

Concentration of NO_x, ppm = x	Instrument response, volts = y
0.05	1.20
0.10	2.15
0.20	3.90
0.3	6.20
0.45	9.80

Obtain a linear regression equation that describes the relationship between x and y.

REFERENCES

Carnahan, H., Luther, H., and Wilkes, J. 1969. *Applied Numerical Methods.* Hoboken, NJ: John Wiley & Sons.

Merriam-Webster 2021a. *Probability.* Merriam-Webster.com. https://www.merriam-webster.com/dictionary/probability (accessed August 14, 2021).

Merriam-Webster 2021b *Statistics.* Merriam-Webster.com. https://www.merriam-webster.com/dictionary/statistics (accessed August 14, 2021).

Shaefer, S., and Theodore, L. 2007. *Probability and Statistics Applications in Environmental Science.* Boca Raton, FL: CRC Press/Taylor & Francis Group.

Theodore, L. 2007. *Air Pollution Control Equipment Calculations.* Hoboken, NJ: John Wiley & Sons.

Theodore, L., and Taylor, F. 1996. *Probability and Statistics.* Theodore Tutorials, originally published by U.S. EPA, RTP, NC. East Williston, NY.

5 Probability Distributions

5.1 INTRODUCTION

The probability distribution of a random variable concerns the distribution of probability over the range of the random variable. The distribution of probability is specified by the *probability density function* (pdf). This chapter is devoted to providing general properties of the pdf for the case of *discrete* and *continuous* random variables. These pdfs find extensive application in environmental and health risk analysis.

One way to express a likelihood quantitatively is to use a numerical value, termed the probability, to express its likelihood of occurrence. The statement that there is a 2% chance of a pandemic occurring is obviously more precise and less vague than saying the chance is very low. The probability can be expressed as a fractional number, for example, 0.37, or a percent number from 0% to 100%, for example, 37%. Naturally, the sum of fractional probabilities for all possible states of occurrence must be 1.0.

The probability variation is another factor that needs to be considered. This includes not only the variations of the reported single-valued data but also probability variations with time, for example, the chance of contracting cancer or the annual probability of an earthquake occurring of a given magnitude or the probability variation with time of a NASA spacecraft failing immediately after liftoff. It is for this reason that this chapter addresses the general subject of probability distributions, particularly as they apply to both health problems and hazards.

Before proceeding to probability distributions, it behooves the reader to grasp the concept of the pdf. In mathematics, a *function* is defined as a relationship between a quantity that depends on another quantity or quantities. *Probability distributions* are an integral part of the general subject of statistics. There are two distributions of concern that arise in environmental and health risk analysis studies:

1. *Probability density function* (*pdf*). A pdf is a distribution of probabilities of the values of a dependent variable as a function of a value of an independent variable (often time). In the context of environmental and health risk assessment, the pdf represents the probability that a given health or hazard problem will occur at or before a specified time. This describes the relative values of the probability (or likelihood of the occurrence) of all possible values of the independent variable.
2. *Cumulative distribution function* (*cdf*). A cdf is the cumulative sum of all probabilities of a dependent variable *less than or equal* to a specific value of an independent variable (often time). In environmental and health risk assessment, a cdf can provide information on ascending (or increasing) values of incident probabilities at increasing values of operating time since a previous component or system failure.

This chapter initially introduces discrete pdfs. Continuous pdfs are reviewed next. Four illustrative examples are provided to demonstrate concepts presented throughout the chapter. Much of the material presented in this chapter has been adopted from the literature (Theodore and Taylor 1996; Shaefer and Theodore 2007).

5.2 DISCRETE PROBABILITY DISTRIBUTIONS

The pdf of a discrete random variable X is specified by $f(x)$ where $f(x)$ has the following essential properties:

1. $f(x) = P(X = x)$ = probability assigned to the outcome corresponding to the number x in the range of X
2. $f(x) \geq 0$
3. $\Sigma_x f(x) = 1$

Property 1 indicates that the pdf of a discrete random variable generates probability by substitution. Properties 2 and 3 restrict the values of $f(x)$ to nonnegative real numbers whose sum is 1.

Consider, for example, a case of 100 vaccine units containing five contaminated vaccine units. Suppose that a vaccine unit selected at random is to be classified as contaminated or non-contaminated. Let X note the outcome, with $X = 0$ associated with the drawing of a non-contaminated vaccine unit, and $X = 1$ associated with the drawing of a contaminated one. Then X is a discrete random variable with a pdf specified by:

$$f(x) = 0.05; x = 1$$
$$f(x) = 0.95; x = 0$$

For another example of the pdf of a discrete random variable, let X denote the number of throws on which the first failure of an electrical switch occurs. Suppose that the probability that a switch fails on any throw is 0.001 and that successive throws are independent with respect to "failure." If the switch fails for the first time on throw x, it must have been successful on each of the preceding $x - 1$ trials. Therefore, the pdf of X is given by:

$$f(x) = (0.999)^{(x-1)}(0.001); x = 1, 2, 3, \ldots, n, \ldots$$

Note that the range of X consists of a countable infinitude of values. Verification of the earlier Property 3 for pdfs of discrete random variables can be accomplished by noting that $\Sigma_x f(x)$ is a geometric series with the first term equal to 0.001, a common ratio equal to 0.999, and therefore convergent to $(0.001)/(1 - 0.999)$, which is 1, that is, unity.

Another function used to describe the probability distribution of a discrete random variable X is the *cdf*. If $f(x)$ specifies the pdf of a random variable X, then $F(x)$ is used to specify the cdf. For *both* discrete and continuous random variables, the cdf of X is defined by:

$$F(x) = P(X \leq x); -\infty < x < \infty \tag{5.1}$$

Probability Distributions

Note that the cdf is defined for all real numbers, not just the values assumed by the random variable.

To illustrate the derivation of the cdf from the pdf, consider the case of a discrete random variable X whose pdf is specified by:

$$f(x) = 0.2 @ x = 2;\ = 0.3 @ x = 5;\ = 0.5 @ x = 7$$

Applying the definition of a cdf in Equation 5.1, one obtains for the cdf of X:

$$F(x) = 0 @ x < 2;\ = 0.2 @ 2 \le x < 5;\ = 0.5 @ 5 \le x < 7;\ = 1 @ x \ge 7$$

It is helpful to think of $F(x)$ as an accumulator of probability as x increases through all real numbers. In the case of a discrete random variable, the cdf is a step function increasing by finite jumps at the values of x in the range of X. In the earlier example, these probability jumps occur at the values 2, 5, and 7. The magnitude of each jump is equal to the probability assigned to the value where the jump occurs. This is depicted in Figure 5.1.

The following properties of the cdf of a discrete random variable X can be deduced directly from the definition of $F(x)$ in Equation 5.1 as follows:

1. $F(b) - F(a) = P(a < X \le b)$
2. $F(+\infty) = 1$
3. $F(-\infty) = 0$
4. $F(x)$ is a nondecreasing function of x

As will be noted in the next section, these properties also apply to continuous random variables.

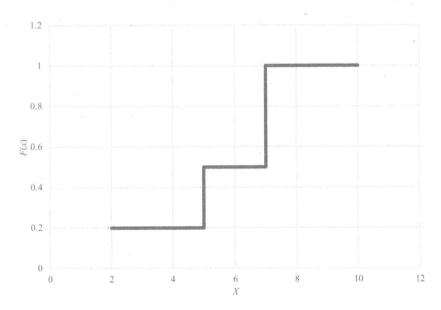

FIGURE 5.1 Graph of the cdf of a discrete random variable X.

5.2.1 BINOMIAL DISTRIBUTION

Consider n independent performances of a random experiment with mutually exclusive outcomes that can be classified as success or failure. The words success and failure are to be regarded as labels for two mutually exclusive categories of outcomes of the random experiment. Yet, they may not necessarily have the ordinary connotation of success or failure. Assume that p, the probability of success on any performance of the random experiment, is constant. Let q be the probability of failure, so that:

$$q = 1 - p \tag{5.2}$$

The probability distribution of X, the number of successes in n performances of the random experiment, is the binomial distribution, with a pdf specified by:

$$f(x) = \frac{n!}{x!(n-x)!} p^x q^{n-x}; x = 0, 1, \ldots, n \tag{5.3}$$

where $f(x)$ is the probability of x successes in n performances. One can show that the expected value of a random variable X is np and its variance is npq. For example, if the probability of a defective thermometer is 0.1, one can determine: (1) the mean and (2) the standard deviation for the distribution of defective thermometers in a total of 400 units. For this case, the expected value or mean is (400) (0.1) = 40, that is, one can therefore *expect* 40 thermometers to be defective. The variance is npq = (400) (0.1) (0.9) = 36; thus, the standard deviation = $\sqrt{36}$ = 6.

As a simple example of the binomial distribution, consider the probability distribution of the number of defects in a sample of five items drawn *with* replacement from a lot of 1,000 items, 50 of which are defective. Associate success with drawing a defective item from the lot. Then the result of each drawing can be classified as success (defective item) or failure (non-defective item). The sample of five items is drawn with replacement (i.e., each item in the sample is returned before the next is drawn from the lot; therefore, the probability of success remains constant at 0.05). Substituting in Equation 5.3 the values $n = 5$, $p = 0.05$, and $q = 0.95$ yield:

$$f(x) = \frac{5!}{x!(5-x)!}(0.05)^x (0.95)^{5-x}; x = 0, 1, 2, 3, 4, 5$$

as the pdf for X, the number of defects in the sample. The probability that the sample of five items contains exactly three defects is therefore given by:

$$f(X = 3) = \frac{5!}{3!(5-3)!}(0.05)^3 (0.95)^{5-3} = \frac{(5)(4)(3)(2)(1)}{(3)(2)(1)(2)(1)}(0.000125)(0.95)^2$$
$$= (0.00125)(0.9025) = 0.0011 = 0.11\%$$

Note that the term:

$$\frac{n!}{x!(n-x)!}$$

in Equation 5.3 is often denoted as:

$$\binom{n}{x}$$

where:

$$\binom{n}{x} = \frac{(n)(n-1)(n-2)\ldots(n-x+1)}{(x)(x-1)(x-2)\ldots 3,2,1} \qquad (5.4)$$

and

$$\binom{n}{0} = 1, \binom{n}{1} = n$$

Also note that

$$\binom{n}{x} = \frac{n}{(n-x)} \qquad (5.5)$$

The binomial distribution can be used to calculate the reliability of a redundant system. A redundant system consisting of n identical components is a system that fails only if more than r components fail. Familiar examples include single-usage equipment such as missile engines, short-life batteries, flash bulbs, which are required to operate for one time period and are not reused.

Once again, associate success with the failure of a component. Assume that the n components are independent with respect to failure and that the reliability of each is $1 - p$. Then X, the number of failures, has the binomial pdf in Equation 5.3, and the reliability of the redundant system is:

$$P(x \leq r) = \sum_{x=0}^{r} \frac{n!}{x!(n-x)!} p^x q^{n-x} \qquad (5.6)$$

5.2.2 MULTINOMINAL DISTRIBUTION

Any discussion on the multinomial distribution first requires an introduction to the general probability subject of *permutations* and *combinations*.

5.2.2.1 Permutations and Combinations

The problem of calculating probabilities of objects or events in a finite group defined as the sample space in which equal probabilities are assigned to the elements in the sample space requires counting the elements that make up the events. The counting of such events is often greatly simplified by employing the rules for permutations and combinations (Theodore and Taylor 1996; Shaefer and Theodore 2007).

TABLE 5.1
Subsets of Permutations and Combinations

Permutations (with regard to order)	Combinations (without regard to order)
Without replacement	Without replacement
With replacement	With replacement

Permutations and combinations deal with the grouping and arrangement of objects or events. By definition, each different ordering in a given manner or arrangement with regard to the order of all or part of the objects is called a *permutation*. Alternately, each of the sets that can be made by using all or part of a given collection of objects without regard to the order of the objects in the set is called a *combination*. Although permutations or combinations can be obtained with replacement or without replacement, most analyses of permutations and combinations are based on sampling that is performed without replacement, that is, each object or element can be used only once. For each of the two *with/without* pairs (with/without regard to order and with/without replacement), four subsets of two may be drawn. These four are provided in Table 5.1.

Each of the four paired subsets in Table 5.1 is considered in the following text with accompanying examples based on the letters A, B, and C. To personalize this, the reader could consider the options (games of chance) one of the authors faces while on a 1-day visit to a casino. The only three options normally considered are dice (often referred to as craps), blackjack (occasionally referred to as 21), and pari-mutuel (horses, trotters, dogs, and jai alai) simulcasting betting. All three of these may be played during a visit, although playing two or only one is also an option. In addition, the order may vary, and the option may be repeated. Some possibilities include the following:

- Dice, blackjack, and then simulcast wagering
- Blackjack, wagering, and dice
- Wagering, dice, and wagering
- Wagering and dice (the author's usual sequence)
- Blackjack, blackjack (following a break), and dice

In order to simplify the examples that follow, dice, blackjack, and wagering are referred to as *objects* represented by the letters A, B, and C, respectively.

1. Consider a scenario that involves three separate objects, A, B, and C. The arrangement of these objects is called a *permutation*. There are six different

orders or permutations of these three objects possible, as follows, while noting that ABC ≠ CBA:

$$ABC \to BAC \to CAB$$
$$ACB \to BCA \to CBA$$

Thus, BCA would represent blackjack, wagering, and dice.

The number of different permutations of n objects is always equal to $n!$, where $n!$ is normally referred to as factorial n. Factorial n or $n!$ is defined as the product of the n objects taken n at a time and denoted as $P(n, n)$ or $P(n|n)$. Thus,

$$P(n,n) = P(n|n) = n! \tag{5.7}$$

With three objects, the number of permutations is $3! = 3 \times 2 \times 1 = 6$. Note that $0!$ is 1. The number of different permutations of n objects taken r at a time is given by:

$$P(n,r) = P(n|r) = \frac{n!}{(n-r)!} \tag{5.8}$$

(Note: The permutation term P also appears in the literature as nP_r.) For the three objects A, B, and C, taken two at a time ($n = 3$ and $r = 2$). Thus,

$$P(3,2) = \frac{3!}{(3-2)!} = \frac{(3)(2)(1)}{1} = 6$$

These possible different orders, noting once again that AB ≠ BC, are:

$$AB \to BC \to AC$$
$$BA \to CB \to CA$$

Consider now a scenario involving n objects in which these can be divided into j sets with the objects within each set being alike. If $r_1, r_2, ..., r_j$ represent the number of objects within each of the respective sets, with $n = r_1 + r_2 + \cdots + r_j$, then the number of permutations of the n objects is given by:

$$P(n; r_1, r_2, ..., r_j) = \frac{n!}{r_1! r_2! ... r_j!} \tag{5.9}$$

This represents the number of permutations of n objects of which r_1 are alike, r_2 are alike, and so on. Consider, for example, 2As, 1B, and 1C; the number of permutations of these four objects is:

$$P(4; 2, 1, 1) = \frac{4!}{2!1!1!} = \frac{(4)(3)(2)(1)}{(2)(1)(1)(1)} = 12$$

2. Consider the arrangement of the same three objects in (1), but obtain the number of permutations (with regard to order) with replacement, *PR*. There are 27 different permutations possible.

AAA	BBB	CCC
AAB	BBA	CCB
AAC	BBC	CCA
ABA	BAB	CBC
ABB	BAA	CBB
ACA	BCB	CAC
ACC	BCC	CAA
ABC	BAC	CBA
ACB	BCA	CAB

For this scenario,

$$PR(n,n) = PR(n \mid n) = (n)^n \quad (5.10)$$

so that

$$PR(3,3) = (3)^3 = 27$$

For *n* objects taken *r* at a time,

$$PR(n,r) = PR(n \mid r) = (n)^r \quad (5.11)$$

3. The number of different ways in which one can select *r* objects from a set of *n* without regard to order (i.e., the order does not count) and without replacement is defined as the number of *combinations*, *C*, of the *n* objects taken *r* at a time. The number of combinations of *n* objects taken *r* at a time is given by:

$$C(n,r) = C(n \mid r) = \frac{P(n,r)}{r!} = \frac{n!}{(n-r)!\,r!} \quad (5.12)$$

(*Note.* The combination term *C* also appears in the literature as C_r^n.)

4. The arrangement of *n* objects, taken *r* at a time without regard to order and with replacement, is denoted by *CR* and given by:

$$CR(n,r) = CR(n \mid r) = C(n,r) + (n)^{r-1} \quad (5.13)$$

with

$$CR(n,n) = CR(n \mid n) = C(n,n) + (n)^{n-1}; \; r = n \quad (5.14)$$

Probability Distributions

TABLE 5.2
Describing Equations for Permutations and Combinations

	Without Replacement	With Replacement	Type
With Regard to Order	$P(n,r) = P(n\mid r) = \dfrac{n!}{(n-r)!}$ (Equation 5.8)	$PR(n,r) = PR(n\mid r) = (n)^r$ (Equation 5.11)	Permutation
Without Regard to Order	$C(n,r) = C(n\mid r) = \dfrac{n!}{(n-r)!}$ (Equation 5.12)	$CR(n,r) = CR(n\mid r) = C(n,r) + (n)^{r-1}$ (Equation 5.13)	Combinations

Table 5.1 can now be rewritten to include the describing equation for each of the earlier four subsets. This is presented in Table 5.2.

5.2.2.2 Multinomial Theorem

Suppose a random variable can not only assume two values (as in the binomial distribution) but can also fall into any one of n different classes with respective probabilities $p_1, p_2, \ldots, p_n$. Then, for a total of n observations, the probability that $x_1, x_2, \ldots, x_n$ observations will fall into classes $1, 2, \ldots, n$, respectively, is given by the frequency function defined as the *multinomial distribution*:

$$f(x_1, x_2 \ldots, x_n) = \frac{n!}{x_1! x_2! \ldots x_n!} (p_1)^{x_1} (p_2)^{x_2} \ldots (p_n)^{x_n} \tag{5.15}$$

With respect to any variable x_i, the multinomial distribution has a mean and variance of:

$$\mu = np \tag{5.16}$$

and

$$\sigma^2 = np(1-p) \tag{5.17}$$

When applied to any variable i, that is, x_i,

$$\mu_i = np_i \tag{5.18}$$

and

$$\sigma_i^2 = np_i(1-p_i) \tag{5.19}$$

Also note that (once again)

$$n_1 + n_2 + \cdots + n_n = n \tag{5.20}$$

and

$$p_1 + p_2 + \cdots + p_n = 1 \qquad (5.21)$$

In addition, if $n = 2$, one obtains the binomial distribution discussed above.

Consider the following example. If one die is thrown 12 times, one can calculate the probability of having the numbers 1, 2, 3, 4, 5, and 6 appear twice after the 12 throws using the multinomial distribution equation. Applying Equation 5.15 leads to:

$$P(x_1 = n_1, x_2 = n_2, x_3 = n_3, x_4 = n_4, x_5 = n_5, x_6 = n_6)$$

$$= \frac{n!}{n_1! n_2! n_3! n_4! n_5! n_6!} (p_1)^{x_1} (p_2)^{x_2} (p_3)^{x_3} (p_4)^{x_4} (p_5)^{x_5} (p_6)^{x_6} \qquad (5.22)$$

Substituting yields:

$$P(x_1 = 2, x_2 = 2, x_3 = 2, x_4 = 2, x_5 = 2, x_6 = 2)$$

$$= \frac{12!}{2!2!2!2!2!2!} \left(\frac{1}{6}\right)^2 \left(\frac{1}{6}\right)^2 \left(\frac{1}{6}\right)^2 \left(\frac{1}{6}\right)^2 \left(\frac{1}{6}\right)^2 \left(\frac{1}{6}\right)^2 = 0.00344 = 0.344\%$$

Also consider the following fluid flow example. There are five fans, four pumps, and three compressors in the shop of a refinery. If one of these prime movers is selected from the shop and then returned for later use, calculate the probability that out of six prime movers selected, three are fans, two are pumps, and one is a compressor. The probabilities for this application are as follows:

P(fan) = 5/12
P(pump) = 4/12
P(compressor) = 3/12

Applying Equation 5.15 with subscripts 1, 2, and 3 referring to fans, pumps, and compressors, respectively, leads to:

$$P(x_1 = 3, x_2 = 2, x_3 = 1) = \frac{6!}{3!2!1!} \left(\frac{5}{12}\right)^3 \left(\frac{4}{12}\right)^2 \left(\frac{3}{12}\right)^1 = (60)(0.0723)(0.111)(0.25)$$

$$= 0.122 = 12.2\%$$

5.2.3 Hypergeometric Distribution

The hypergeometric distribution is used to describe a situation in which a random sample of r items is drawn without replacement from a set of n items. *Without replacement* means that an item is not returned to the set after it is drawn. Recall that the binomial distribution is frequently applicable in cases where the item is drawn with *replacement*.

Probability Distributions

Suppose that it is once again possible to classify each of the n items as a success or failure. Again, the words success and failure do not have the usual connotation. They are merely labels for two mutually exclusive categories into which n items have been classified. Thus, each element of the population may be dichotomized as belonging to one of two disjointed classes.

Let a be the number of items in the category labeled success. Then $n - a$ is the number of items in the category labeled failure. Let X denote the number of successes in a random sample of r items drawn without replacement from the set of n items. Then the random variable X has a hypergeometric distribution, whose pdf is specified as follows:

$$f(x) = \frac{\frac{a!}{x!(a-x)!} \frac{(n-a)!}{(r-x)!(n-a-r+x)!}}{\frac{n!}{r!(n-r)!}}, x = 0, 1, \ldots, \min(a, r) \quad (5.23)$$

The term $f(x)$ is the probability of x successes in a random sample of r items drawn without replacement from a set of n items, where a are classified as successes and $n - a$ as failures. The term $\min(a, r)$ represents the smaller of the two numbers a and r, that is, $\min(a, r) = a$ if $a < r$ and $\min(a, r) = r$ if $r \le a$.

Consider the following example. A sample of five transistors is drawn at random without replacement from a lot of 1,000 units, 50 of which are defective. What is the probability that the sample of five transistors contains exactly three defective units? The number of items in the set from which the sample is drawn is the number of transistors in the lot. Therefore, $n = 1,000$. Associate success with drawing a defective transistor and failure with drawing a non-defective one. Determine a, the number of successes in the set of n items. Because 50 of the transistors in the lot are defective, $a = 50$. Also note that the sample is drawn without replacement and that the size of the sample, $r = 5$.

Substituting the values of n, r, and a in the hypergeometric pdf provided in Equation 5.23 gives

$$f(x) = \frac{\frac{50!}{x!(50-x)!} \frac{(1,000-50)!}{(5-x)!(1,000-50-5+x)!}}{\frac{1,000!}{5!(1,000-5)!}}$$

$$= \frac{\frac{50!}{x!(50-x)!} \frac{(950)!}{(5-x)!(945+x)!}}{\frac{1,000!}{5!(995)!}}, x = 0,1,2,3,4,5$$

Also, substitute the appropriate value of X above to obtain the required probability.

$$P(\text{sample contains exactly three defective units}) = P(X = 3)$$

Therefore,

$$P(X=3) = \frac{\dfrac{50!}{3!(50-3)!} \dfrac{(950)!}{(5-3)!(945+3)!}}{\dfrac{1,000!}{5!(995)!}} = \frac{\dfrac{50!}{3!(47)!} \dfrac{(950)!}{(2)!(948)!}}{\dfrac{1,000!}{5!(995)!}}$$

$$= \frac{\dfrac{(50)(49)(48)(47)!}{3!(47)!} \dfrac{(950)(949)(948)!}{(2)!(948)!}}{\dfrac{(1,000)(999)(998)(997)(996)(995)!}{5!(995)!}}$$

$$= \frac{\dfrac{(50)(49)(48)}{(3)(2)(1)} \dfrac{(950)(949)}{(2)(1)}}{\dfrac{(1,000)(999)(998)(997)(996)}{(5)(4)(3)(2)(1)}} = \frac{\dfrac{(50)(49)(8)}{(1)} \dfrac{(475)(949)}{(1)}}{\dfrac{(100)(333)(998)(997)(249)}{(1)}}$$

$$= \frac{8.83519 \times 10^9}{8.2509 \times 10^{12}} = 0.0011 = 0.11\%$$

Are there differences between the hypergeometric and binomial distributions? The hypergeometric distribution is obviously a special case of the binomial distribution when applied to finite populations. In particular, the hypergeometric distribution approaches the binomial distribution when the population size approaches infinity. Note, however, that others have claimed that the binomial distribution is a special case of a hypergeometric distribution.

Finally, the hypergeometric distribution is applicable in situations similar to those when the binomial distribution is used, except that samples are taken from a *small* population. Examples arise in sampling from small numbers of chemical, medical, and environmental samples, as well as from manufacturing lots.

5.2.4 Poisson Distribution

The pdf of the Poisson (named after Simeon Poisson) distribution can be derived by taking the limit of the binomial pdf as $n \to \infty$, $P \to 0$, and $nP = \mu$ remains constant. The Poisson pdf is given by:

$$f(x) = \frac{e^{-\mu}\mu^x}{x!}, x = 0,1,2,\ldots \tag{5.24}$$

The term $f(x)$ is the probability of x occurrences of an event that occurs on the average μ times per unit of space or time. Both the mean and the variance of a random variable X having a Poisson distribution are μ (Theodore and Taylor 1996; Shaefer and Theodore 2007).

In the 155 years that the Kentucky Derby has been run, only 68 horses that entered the race have been undefeated. What is the probability that two undefeated horses will be entered in the next "Run for the Roses?" Assume the Poisson distribution may be applied. For this exercise,

$$\mu = \frac{68}{155} = 0.439; n = 2$$

$$P(X=2) = P(2) = \frac{e^{-0.439}(0.439)^2}{2!} = \frac{0.645(0.193)}{(2)(1)} = \frac{0.1245}{2} = 0.062 = 6.2\%$$

In effect, the race will be run with two undefeated horses, approximately six times every 100 years!

Now, consider the following example. If λ is the failure rate (per unit time) of each component of a system, then λt is the average number of failures for a given unit of time. The probability of x failures in the specified unit of time is obtained by substituting $\mu = \lambda t$ in Equation 5.24 to obtain:

$$f(x) = \frac{e^{-\lambda t}(\lambda t)^x}{x!}, x = 0,1,2,... \qquad (5.25)$$

Suppose, for example, that in a certain country the average number of airplane crashes per year is 2.5. What is the probability of four or more crashes occurring during the next year? Substituting $\lambda = 2.5$ and $t = 1$ in Equation 5.25 yields:

$$f(x) = \frac{e^{-2.5}(2.5)^x}{x!}, x = 0,1,2,...$$

as the pdf of X, the number of airplane crashes in 1 year. The probability of four or more airplane crashes next year is then:

$$P(X \geq 4) = 1 - \sum_{x=0}^{3} \frac{e^{-2.5}(2.5)^x}{x!} = 1 - (0.0821 + 0.205 + 0.257 + 0.214) = 1 - 0.76 = 0.24$$
$$= 24\%$$

This is obviously not an acceptable risk!

The Poisson distribution plays a role in operations research, particularly in the analysis of *queueing systems* where a queueing system is defined as any service facility to which customers or jobs arrive, receive service, and then depart. Note that the word customer (in a general sense) can refer to:

1. A telephone call arriving at a telephone exchange
2. An order for a component stocked in a warehouse
3. A broken component brought to a repair shop, or
4. A packet of digital data arriving at some node in a complex computer network

The *service time S* is defined as the amount of time required to service the customer. The length of a telephone call and the time to repair a broken component are examples of service times. In most typical applications, both the customer arrivals and their service times are random variables. Queueing systems are classified according to three factors:

1. The *input process*, which denotes the probability distribution of the customer arrivals
2. The *service distribution*, which denotes the probability distribution of the service time, and
3. The *queueing discipline*, which refers to the order of service

Rosenkrantz (2009) provides additional information and illustrative examples.

5.3 CONTINUOUS PROBABILITY DISTRIBUTIONS

The pdf of a continuous random variable X has the following properties:

1. $\int_a^b f(x)dx = P(a < X < b)$
2. $f(x) \geq 0$
3. $\int_{-\infty}^{+\infty} f(x)dx = 1$

Property 1 indicates that the pdf of a continuous random variable generates probability by the integration of the pdf over the interval whose probability is required. When this interval contracts to a single value, the integral over the interval becomes zero. Therefore, the probability associated with any particular value of a continuous random variable is zero. Consequently, if X is continuous,

$$P(a \leq X \leq b) = P(a < X \leq b) = P(a < X < b) = P(a \leq X < b)$$

Property 2 restricts the values of $f(x)$ to nonnegative numbers. Property 3 follows from the fact that:

$$P(-\infty < X < \infty) = 1$$

As an example of the pdf of a continuous random variable, consider the pdf of the time X, in hours, between successive failures of an air conditioning system. Suppose the pdf of X is specified by:

$$f(x) = 0.01e^{-0.01x} \ @ \ x > 0; = 0 \text{ everywhere else}$$

Inspection of the graph in Figure 5.2 indicates that intervals in the lower part of the range of X are assigned greater probabilities than intervals of the same length in the upper part of the range of X because the areas over the former are greater than the areas over the latter. The expression $P(a \leq X < b)$ can be interpreted geometrically as the area under the pdf curve over the interval (a, b). Integration of the pdf over the interval yields the probability assigned to the interval. For example, the probability that the time

Probability Distributions

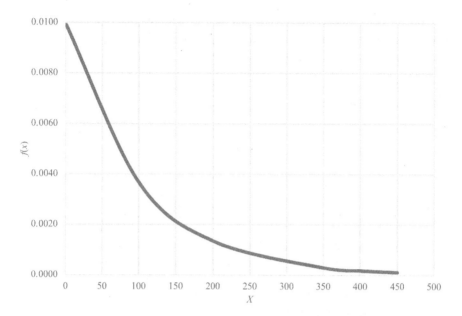

FIGURE 5.2 The pdf of time, in hours, between successive failures of an air conditioning system.

in hours between successive failures of the aforementioned air conditioning system is greater than 6 but less than 10 is:

$$P(6 \leq X < 10) = \int_{6}^{10} 0.01 e^{-0.01x} dx = 0.04 = 4\%$$

Another function used to describe the probability distribution of a continuous random variable X is the cdf. If $f(x)$ specifies the pdf of a random variable X, then $F(x)$ is used to specify the cdf. For *both* discrete and continuous random variables, the cdf of X is defined by:

$$F(x) = P(X \leq x); -\infty < x < \infty \quad (5.1)$$

Note once again that the cdf is defined for all real numbers, not just the values assumed by the random variable.

To illustrate the derivation of the cdf from the pdf, consider the case of a continuous random variable. The cdf is a continuous function. Suppose, for example, that X is a continuous random variable with pdf specified by:

$$f(x) = 2x @ 0 \leq x < 1; = 0 \text{ elsewhere} \quad (5.26)$$

Applying the definition of the cdf in Equation 5.1, one obtains:

$$F(x) = 0 @ x < 0; = \int_{0}^{x} 2x\, dx = x^2 @ 0 \leq x < 1; = 1 @ x \geq 1 \quad (5.27)$$

The pdf of a continuous random variable can also be obtained by differentiating its cdf and setting the pdf equal to zero where the derivative of the cdf does not exist. For example, differentiating the cdf obtained in Equation 5.27 yields the pdf in Equation 5.26. In this case, the derivative of the cdf does not exist for $x = 1$.

The following properties of the cdf of a continuous random variable X can be deduced directly from the definition of $F(x)$ in Equation 5.1:

1. $F(b) - F(a) = P(a < X \leq b)$
2. $F(+\infty) = 1$
3. $F(-\infty) = 0$
4. $F(x)$ is a nondecreasing function of x

Note that these properties are identical to those of the discrete random variables discussed in Section 5.2.

These continuous probability distributions generally are based on time (which is continuous) as the independent variable. Which distribution has received the greatest attention from a risk analysis/assessment perspective? This is a tough question to answer, but the author would rank them in the following order: Weibull (most attention), normal, exponential, and log-normal (least attention). As noted above other than the log-normal distribution, time is almost always the independent variable.

5.3.1 WEIBULL DISTRIBUTION

Simply put, the Weibull distribution describes the *failure rate* as a function of time. It has served the technical community for over 50 years; the chemical industry, refineries, the Pentagon, NASA, and so on have been the beneficiaries. Thus, it is a key continuous distribution to consider for environmental and health risk assessment. This section not only reviews the traditional material in the literature on the Weibull distribution, but also includes illustrative examples concerned with reliability relations plus series and parallel systems. Recent efforts of the authors to improve on Weibull's distribution, are also detailed.

Unlike the exponential distribution, the failure rate of equipment frequently exhibits three stages as: a break-in (Bl) stage with a declining failure rate, a useful life stage characterized by a fairly constant failure rate, and a wear-out (WO) period characterized by an increasing failure rate. Many industrial parts and components follow this path. A failure rate curve exhibiting these three phases (see Figure 5.3) is called a *bathtub curve*.

Weibull introduced the distribution, which bears his name principally on empirical grounds, to represent certain life-test data. The Weibull distribution provides a mathematical model of all three stages of the bathtub curve. An assumption about the *failure rate*, $Z(t)$, that reflects all three stages of the bathtub stage is:

$$Z(t) = \alpha \beta t^{\beta-1}; t > 0 \qquad (5.28)$$

where α and β are constants, referred to as the shape parameters of curve-fitting parameters. For $\beta < 1$ the failure rate $Z(t)$ decreases with time. For $\beta = 1$ the failure

Probability Distributions

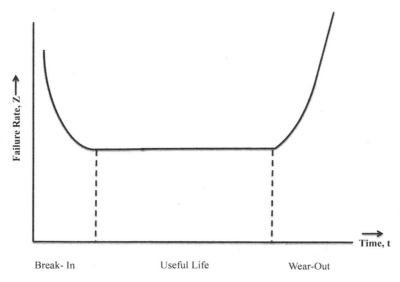

FIGURE 5.3 Bathtub curve described by the Weibull distribution.

rate is constant and equal to α. For $\beta > 1$ the failure rate increases with time. One can then show that the pdf of T, time to failure, is:

$$f(t) = \alpha\beta t^{\beta-1} \exp\left(-\int_0^t \alpha\beta t^{\beta-1} dt\right) = \alpha\beta t^{\beta-1} \exp(-\alpha t^\beta); t > 0; \alpha > 0, \beta > 0 \quad (5.29)$$

Equation 5.29 defines the pdf of the Weibull distribution. The variety of assumptions about failure rate and the probability distribution of time to failure that can be accommodated by the Weibull distribution make it especially attractive in describing failure time distributions in industrial and process plant applications.

To illustrate probability calculations involving the exponential and Weibull distributions introduced in conjunction with the bathtub curve of failure rate, consider first the case of a transistor having a constant rate of failure of 0.01 per 1,000 hr. To find the probability that the transistor will operate for at least 25,000 hr, first note that:

$$Z(t) = 0.01$$

Using Equations 5.28 and 5.29 ultimately yields:

$$f(t) = 0.01 \exp\left(-\int_0^t 0.01 dt\right) = 0.01 e^{-0.01t}; t > 0$$

as the pdf of T, the time to failure of the transistor. Because t is measured in thousands of hours, the probability that the transistor will operate for at least 25,000 hr is given by:

$$P(T > 25) = \int_{25}^{\infty} 0.01 e^{-0.01t} dt = -e^{-\infty} + e^{-0.01(25)} = 0 + 0.78 = 78\%$$

The reader should note that this example reduces to (because of the constant rate specification) a calculation of an exponential distribution that will be discussed below.

Now suppose it is desired to determine the 10,000-hr reliability of a circuit of five such transistors connected in series. The 10,000-hr reliability of one transistor is the probability that it will last at least 10,000 hr. This probability can be obtained by integrating the pdf of T, time to failure, which gives:

$$P(T > 10) = \int_{10}^{\infty} 0.01 e^{-0.01t} dt = -e^{-\infty} + e^{-0.01(10)} = 0 + 0.90 = 90\%$$

This result also represents the *reliability*, R, that the circuit will last at least 10,000 hr, i.e., $R(10) = 0.90$.

The 10,000-hr reliability of a circuit of five transistors connected in series is obtained by applying the formula for the reliability of series system to obtain:

$$R_s = [R(10)]^5 = (0.9)^5 = 0.59 = 59\%$$

As another example of this type of probability calculation, consider a component whose time to failure, T, in hours, has a Weibull pdf with parameters $\alpha = 0.01$ and $\beta = 0.50$ in Equation 5.29. (Note that this involves a *nonconstant* rate application that applies over the entire time domain.) This gives:

$$f(t) = (0.01)(0.50)t^{(0.5-1)}e^{(0.01t^{0.5})} = (0.005)t^{(-0.5)}e^{(0.01t^{0.5})}; t > 0$$

as the Weibull pdf of the failure time of the component under consideration. The probability that the component will operate at least 8,100 hr is then given by:

$$P(T > 8,100) = \int_{8,100}^{\infty} (0.005)t^{(-0.5)}e^{(0.01)t^{0.5}} dt = e^{-(0.01)t^{0.5}}\Big|_{8,100}^{\infty}$$

$$= 0 + e^{(0.01)(8,100)^{0.5}} = 0.41 = 41\%$$

Estimates of the parameters α and β in Equation 5.29 can be obtained by using a graphical procedure described by Bury (1975). This procedure is based on the fact that:

$$\ln\left[\ln\frac{1}{1-F(t)}\right] = \ln(\alpha) + \beta \ln(t) \tag{5.30}$$

where $F(t) = 1 - \exp(-\alpha t^\beta)$, $t > 0$; and $F(t) = 0$, $t < 0$ defines the cdf of the Weibull distribution. In Equation 5.30, the expression

$$\ln\left[\ln\frac{1}{1-F(t)}\right]$$

serves as a linear function of $\ln(t)$ with slope β and intercept $\ln(\alpha)$. The graphical procedure for estimating α and β on the basis of a sample of n observed values of t, time to failure, first involves the ordering of the observations from smallest ($i = 1$) to

Probability Distributions

largest ($i = n$). The value of the ith observation varies from sample to sample. It can be shown that the average value of $F(t)$ for t equal to the value of the ith observation on T is $i/(n + 1)$. One may then plot

$$\ln\left[\ln\frac{1}{1-\dfrac{i}{n+1}}\right]$$

against the natural logarithm of the ith observation $i = 1$ to $i = n$. Under the assumption that T has a Weibull distribution, the plotted points lie on a straight line whose slope is β and whose intercept is $\ln(\alpha)$. Special Weibull probability paper allows plotting the ith observation against $i/(n + 1)$ to achieve the same result.

To illustrate this procedure, suppose that a sample of ten observations on the time to failure of an electric component yields the observations in the first column of Table 5.3. Fitting a straight line to the derived values in Columns 3 and 4 yields a line (Figure 5.4) with slope = 1.4 and intercept = −6.5. Therefore, the estimated value of α is 0.0015 and the estimated value of β is 1.4.

A variety of conditional failure distributions, including wear-out patterns, can be described by the Weibull distribution. Therefore, this distribution has been frequently recommended instead of the exponential distribution as an appropriate failure distribution model. Empirically satisfactory fits have been obtained from failure data of electron tubes, relays, ball bearings, metal fatigue, and even human mortality.

In recent years, the Weibull distribution has come under fire, due primarily to the efforts of both of the authors of this book. The last 6 years have provided an opportunity for Dupont and Theodore (2011), along with Dupont and others (Dupont,

TABLE 5.3
Data for Estimation of Weibull Parameters

Time to Failure, t (days)	Order of Failure, t	$\ln(t)$	$\ln\left[\ln\dfrac{1}{1-\dfrac{i}{n+1}}\right]$
18	1	2.89	−2.36
36	2	3.58	−1.62
40	3	3.69	−1.16
53	4	3.97	−0.81
71	5	4.26	−0.51
90	6	4.50	−0.23
106	7	4.66	0.02
127	8	4.84	0.27
149	9	5.00	0.54
165	10	5.11	0.88

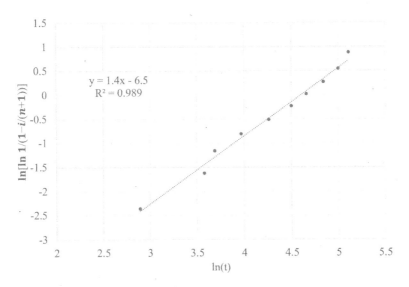

FIGURE 5.4 Curve fitting results to determine Weibull distribution parameters α and β.

McKenna and Theodore 2013; Dupont, Ricci and Theodore 2014, 2015), to carefully analyze the merits and limitations of the Weibull distribution. This has led to the development of the Dupont, Ricci and Theodore (DRaT) models, details of which are provided below.

As noted above, the general two-coefficient Weibull model is represented by an equation that can be applied to three failure rate periods representing three failure mode stages. As such, the model consists of six coefficients, two for each of the three failure mode stages. Dupont, Ricci, and Theodore (2015) viewed this six-coefficient relationship as both cumbersome *and* unnecessary. After some deliberation, it was decided to explore a new, simpler approach to represent failure behavior, specifically via the failure-time relationship presented in Figure 5.5 (as opposed to the original failure rate-time relationship). After even more deliberation and analysis, these authors settled on four models that are based on combinations of either a power function (P) or an exponential function (E) for the BI period, with either a P or E function for the WO period, the sum of which results in a curve as shown in Figure 5.5. There are four combinations of the power and exponential functions that result: BI(E) + WO(E), BI(P) + WO(P), BI(E) + WO(P), and BI(P) + WO(E). The corresponding equations resulting from these combinations were defined as DRaT II, DRaT III, DRaT IV, and DRaT V models, respectively (Dupont, Ricci and Theodore 2015).

The failure-time functional relationship for the BI and WO periods is represented by $g(t)$ and $h(t)$, respectively, in the development that follows, with the sum of these two functions resulting in the DRaT models estimating the number of failures as a function of time, $f(t)$, i.e.,

$$f(t) = g(t) + h(t) \qquad (5.31)$$

Probability Distributions

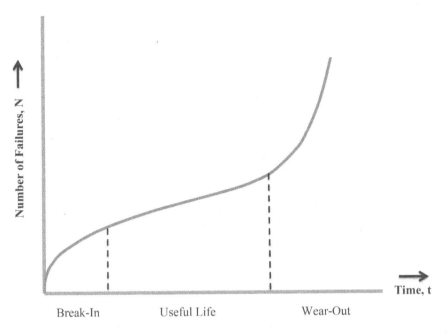

FIGURE 5.5 Failure-time relationship for the Weibull distribution.

The general form of the DRaT model can be written as follows if an exponential function is assumed to describe both the BI and WO period failure-time relationships. This is labeled the DRaT II model (Equation 5.32):

$$f(t)_{II} = BI(E) + WO(E) = A_1\left(1 - e^{-A_2 t}\right) + B_1\left(e^{B_2 t} - 1\right) \qquad (5.32)$$

Another form of the DRaT model can be written as follows if a power function is assumed to describe the BI and an exponential function is assumed to describe the WO period failure-time relationships. This is labeled the DRaT V model (Figure 5.6):

$$f(t)_V = BI(P) + WO(E) = A_1 t^{A_2} + B_1\left(e^{B_2 t} - 1\right) \qquad (5.33)$$

Specific details of the other forms of the DRaT model are provided in the literature (Dupont, Ricci and Theodore 2015).

Three improvements in the DRaT models relative to the Weibull model become immediately apparent.

1. The DRaT models contains four (not six) coefficients to be estimated.
2. The DRaT models are continuous over the entire time range as opposed to the Weibull model that is evaluated separately over three compartmentalized failure stages.
3. The requirement of a constant failure rate period in the Weibull model has been replaced by a more realistic failure rate in the DRaT models that can continue to increase slightly with time during the supposed *constant* failure rate period.

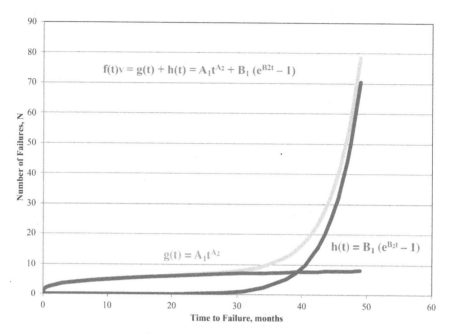

FIGURE 5.6 DRaT model representation for the combined failure pattern when BI(P) and WO(P) are assumed in the DRaT V model.

The reader should note once again that the failure rate for most applications is a *calculated* quantity obtained from the number of failures (N) versus time (t) data. The failure rate at a specified time is approximately equal to the slope of N versus t at the time point in question. Theodore and Ricci (2011) provide six different numerical differentiation procedures to calculate this derivative, and the reader is directed to that reference for more details.

5.3.2 Normal Distribution

When T, time to failure, has a normal distribution with mean μ and variance σ, its pdf is given by:

$$f(t) = \frac{1}{\sqrt{2\pi}\sigma} \exp\left[-\frac{1}{2}\left(\frac{t-\mu}{\sigma}\right)^2\right]; -\infty < t < \infty \quad (5.34)$$

The graph of $f(t)$ is the familiar bell-shaped curve in Figure 5.7 (Theodore and Taylor 1996; Shaefer and Theodore 2007).

There are two points at which the curve in Figure 5.7 changes shape in the normal curve. The curve changes at the first point, from being concave with respect to the horizontal axis, to being convex at the first point. At the second point, it again changes to concavity. These are known as points of inflection. Interestingly, the points of inflection are located at an interval σ from the mean, that is, at $\bar{x} \pm \sigma$, or 68% of the area under the curve symmetrically displaced from the mean.

Probability Distributions

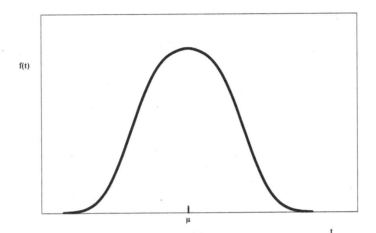

FIGURE 5.7 Normal pdf of time to failure.

If a variable T is normally distributed with mean μ and standard deviation σ, then the random variable $(T-\mu)/\sigma$ is also normally distributed with a mean of 0 and standard deviation of 1. The term $(T-\mu)/\sigma$ is called a *standard normal variable* that is represented by Z, not to be confused with the failure rate $Z(t)$.

Table 5.4 is a tabulation of areas under a standard normal curve to the right of z_o for *nonnegative* values of z_o. Probabilities about a standard normal variable Z can be determined from this table. For example,

$$P(Z > 1.54) = 0.062$$

is obtained directly from the table as the area to the right of 1.54. The symmetry of the standard normal curve about zero implies that the area to the right of zero is 0.5 and the area to the left of zero is 0.5. Consequently, one can deduce from Table 5.4 and Figure 5.8 that:

$$P(0 < Z < 1.54) = (0.5 - 0.5) + (0.5 - 0.062) = 0.438$$

Also, because of symmetry

$$P(-1.54 < Z < 0) = 0.438 \text{ and } P(Z < -1.54) = 0.062$$

The following probabilities can also be deduced by noting that the area to the right of 1.54 is 0.062:

$$P(-1.54 < Z < 1.54) = (0.5 - 0.062) + (0.5 - 0.062) = 0.876$$
$$P(Z < 1.54) = 0.938$$
$$P(Z > -1.54) = 0.938$$

TABLE 5.4
Standard Normal Cumulative Probability in Right-Hand Tail (Area Under Curve for Specified Values of z_o, Area = $P(Z \geq z_o)$)

The Standard Normal Distribution

z	0.00	0.01	0.02	0.03	0.04	0.05	0.06	0.07	0.08	0.09
0.0	0.500	0.496	0.492	0.488	0.484	0.480	0.476	0.472	0.468	0.464
0.1	0.460	0.456	0.452	0.448	0.444	0.440	0.436	0.433	0.429	0.425
0.2	0.421	0.417	0.413	0.409	0.405	0.401	0.397	0.394	0.390	0.386
0.3	0.382	0.378	0.374	0.371	0.367	0.363	0.359	0.356	0.352	0.348
0.4	0.345	0.341	0.337	0.334	0.330	0.326	0.323	0.319	0.316	0.312
0.5	0.309	0.305	0.302	0.298	0.295	0.291	0.288	0.284	0.281	0.278
0.6	0.274	0.271	0.268	0.264	0.261	0.258	0.255	0.251	0.248	0.245
0.7	0.242	0.239	0.236	0.233	0.230	0.227	0.224	0.221	0.218	0.215
0.8	0.212	0.209	0.206	0.203	0.200	0.198	0.195	0.192	0.189	0.187
0.9	0.184	0.181	0.179	0.176	0.174	0.171	0.189	0.166	0.164	0.161
1.0	0.159	0.156	0.154	0.152	0.149	0.147	0.145	0.142	0.140	0.138
1.1	0.136	0.133	0.131	0.129	0.127	0.125	0.123	0.121	0.119	0.117
1.2	0.115	0.113	0.111	0.109	0.107	0.106	0.104	0.102	0.100	0.099
1.3	0.097	0.095	0.093	0.092	0.090	0.089	0.087	0.085	0.084	0.082
1.4	0.081	0.079	0.078	0.076	0.075	0.074	0.072	0.071	0.069	0.068
1.5	0.067	0.066	0.064	0.063	0.062	0.061	0.059	0.058	0.057	0.056
1.6	0.055	0.054	0.053	0.052	0.051	0.049	0.048	0.047	0.046	0.046
1.7	0.045	0.044	0.043	0.042	0.041	0.040	0.039	0.038	0.038	0.037
1.8	0.036	0.035	0.034	0.034	0.033	0.032	0.031	0.031	0.030	0.029
1.9	0.029	0.028	0.027	0.027	0.026	0.026	0.025	0.024	0.024	0.023
2.0	0.023	0.022	0.022	0.021	0.021	0.020	0.020	0.019	0.019	0.018
2.1	0.018	0.017	0.017	0.017	0.016	0.016	0.015	0.015	0.015	0.014
2.2	0.014	0.014	0.013	0.013	0.013	0.012	0.012	0.012	0.011	0.011
2.3	0.011	0.010	0.010	0.010	0.010	0.009	0.009	0.009	0.009	0.008
2.4	0.008	0.008	0.008	0.008	0.007	0.007	0.007	0.007	0.007	0.006
2.5	0.006	0.006	0.006	0.006	0.006	0.005	0.005	0.005	0.005	0.005
2.6	0.005	0.005	0.004	0.004	0.004	0.004	0.004	0.004	0.004	0.004
2.7	0.003	0.003	0.003	0.003	0.003	0.003	0.003	0.003	0.003	0.003
2.8	0.003	0.002	0.002	0.002	0.002	0.002	0.002	0.002	0.002	0.002
2.9	0.002	0.002	0.002	0.002	0.002	0.002	0.002	0.001	0.001	0.001

Source: Adapted from Dell, Compare distribution tables, Tulsa, OK. http://www.statsoft.com/textbook/sstable.html.

Table 5.4 also can be used to determine probabilities concerning normal random variables that are not standard normal variables. The required probability is first converted to an equivalent probability about a standard normal variable. For example if T, the time to failure, is normally distributed with mean $\mu = 200$ and standard deviation $\sigma = 4$ then $(T - 200)/4$ is a standard normal variable. One may write:

$$P(T_1 < T < T_2) = P\left(\frac{T_1 - \mu}{\sigma} < \frac{T - \mu}{\sigma} < \frac{T_2 - \mu}{\sigma}\right) = P\left(\frac{T_1 - \mu}{\sigma} < Z < \frac{T_2 - \mu}{\sigma}\right) \quad (5.35)$$

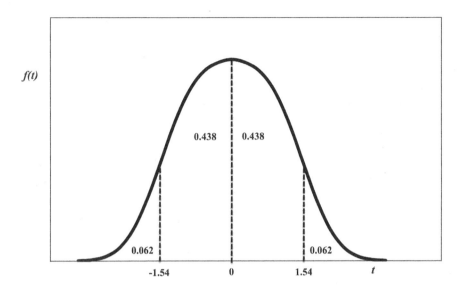

FIGURE 5.8 Areas under a standard normal curve.

where

$$\frac{T-\mu}{\sigma} = \frac{T-200}{4} = Z = \text{the standard normal variable for this example.}$$

Therefore if the probability of failure between 196 and 208 hr is to be determined, the following calculation can be carried out:

$$P(196 < T < 208) = P\left(\frac{196-200}{4} < \frac{T-200}{4} < \frac{208-200}{4}\right) = P(-1 < Z < 2)$$
$$= 0.341 + 0.477 = 0.818 = 81.8\%$$

For any random variable X that is normally distributed with mean μ and standard deviation σ, the following can be now written:

$$P(\mu - \sigma < X < \mu + \sigma) = P\left(-1 < \frac{X-\mu}{\sigma} < 1\right) = P(-1 < Z < 1) = 0.68$$

$$P(\mu - 2\sigma < X < \mu + 2\sigma) = P\left(-2 < \frac{X-\mu}{\sigma} < 2\right) = P(-2 < Z < 2) = 0.95$$

$$P(\mu - 3\sigma < X < \mu + 3\sigma) = P\left(-3 < \frac{X-\mu}{\sigma} < 3\right) = P(-3 < Z < 3) = 0.997$$

This is often referred to as the 68-95-99.7 rule. The rule states that in a *normally* distributed population, 68% the of population falls within one standard deviation of the mean, 95% falls within two standard deviations of the mean, and 99.7% falls within three standard deviations of the mean.

Note again that because the standard normal curve is symmetric about 0, the area to the right of 0 is equal to the area to the left of 0, namely, 0.5. Because the area over (0, 1) is 0.3413, the area over (1, ∞) is 0.5 − 0.3413 = 0.1587. By symmetry, the area over (−1, 0) is 0.3413, and the area over (−∞, −1) is 0.1587. These areas are bounded by ordinates erected at −1, 0, and 1.

Additional probabilities are represented by areas under the standard normal curve over certain intervals. Thus, one can obtain the following additional probabilities:

1. $P(-1 < Z < 1)$ = Area over (−1, 1) = 0.6826
2. $P(Z > -1)$ = Area over (−1, ∞) = 0.3413 + 0.5 = 0.8413
3. $P(Z < 1)$ = Area over (−∞, 1) = 0.5 + 0.3413 = 0.8413

The normal distribution is used to obtain probabilities concerning the mean X of a sample of n observations on a random variable X, if X is normally distributed with mean μ and standard deviation σ. For example, suppose X is normally distributed with mean 100 and standard deviation 2. Then, $\bar{X}$, the mean of a sample of 16 observations on X, is normally distributed with mean 100 and standard deviation 0.5. To calculate the probability that X is greater than 101, one would write:

$$P(\bar{X} > 101) = P\left(\frac{\bar{X} - 100}{0.5} > \frac{101 - 100}{0.5}\right); Z = \frac{\bar{X} - 100}{0.5}$$

$$P(\bar{X} > 101) = P(Z > 2) = 0.023 = 2.3\%$$

If X is not normally distributed, then $\bar{X}$ of a sample of n observations on X is approximately normally distributed with mean μ and standard deviation $\sigma/\sqrt{n}$, provided the sample size n is large (>30). This result is based on an important theorem in probability called the *central limit theorem.*

Actual (experimental) data have shown many physical variables to be normally distributed. Examples include physical measurements on living organisms, molecular velocities in an ideal gas, scores on intelligence tests, average temperatures in a locality, height of men belonging to a certain race, experimental measurement subject to random errors, and time for a delivery truck to travel along a particular route. Other variables, though not normally distributed per se, sometimes approximate a normal distribution after an appropriate transformation. Examples include taking the logarithm or square root of the original variable. The normal distribution also has the advantage that it is tractable mathematically. Consequently, many of the techniques of statistical inference have been derived under the assumption of underlying normal variants.

Because of the prominence (and perhaps the name) of the normal distribution, it is sometimes assumed that a random variable is normally distributed, unless proven otherwise. This notion could lead to incorrect results. For example, the normal distribution is generally inappropriate in a model of time to failure. Frequently a normal distribution provides a reasonable approximation of the main part of the distribution but is inadequate at one or both of the tails. Finally, certain phenomena are not symmetrically distributed as is required for normality. The error in incorrectly assuming

Probability Distributions

normality depends on the use to which the assumption is applied. Many statistical models and methods derived under this assumption remain valid under moderate deviations from it. On the other hand, if the normality assumption is used to determine the proportion of *items* above or below some extreme limit (e.g., at the tail of the distribution), serious errors might result.

Summarizing, one reason the normal distribution is so important is that a number of natural phenomena are normally distributed or closely approximate it. In fact, many experiments, when repeated many times, will approach the normal distribution curve. In its pure form, the normal curve is a continuous, symmetrical, smooth curve as shown earlier. Naturally, a finite distribution of discrete data can only approximate this curve.

5.3.3 Exponential Distribution

Exponents, exponential functions, exponential derivatives, and exponential integrals find application in engineering and science. Their use in environmental risk calculations finds even wider applications (this last statement particularly applies to the Weibull distribution that was discussed above).

The exponential distribution is an important distribution in that it represents the distribution of the time required for a single event from a Poisson process to occur. In particular, in sampling from a Poisson distribution with parameter µ, the probability that no event occurs during (0, t) is $e^{-\lambda t}$. Consequently, the probability that an event will occur during (0, t) is:

$$F(t) = 1 - e^{-\lambda t} \qquad (5.36)$$

This represents the cdf of t. One can therefore show that the pdf is:

$$f(t) = e^{-\lambda t} \qquad (5.37)$$

Note that the parameter $1/\lambda$ (sometimes denoted as µ) is the expected value. Normally, the reciprocal of this value is specified and represents the expected value of $f(t)$.

Because the exponential function appears in the expression for both the pdf and cdf, the distribution is justifiably called the *exponential distribution*. A typical pdf of x plot is provided in Figure 5.9. Alternatively, the cumulative exponential distribution can be obtained from the pdf (with x replacing t) as follows:

$$F(x) = \int_0^x \lambda e^{-\lambda x} dx = 1 - e^{-\lambda x} \qquad (5.38)$$

All that remains is a simple evaluation of the negative exponent in Equation 5.38.

In statistical and reliability applications one often encounters a random variable's conditional failure density or hazard function, $g(x)$. In particular $g(x)\,dx$ is the

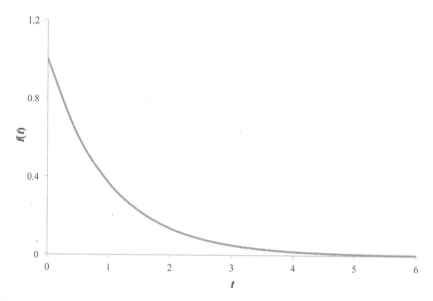

FIGURE 5.9 Exponential distribution.

probability that a "product" will fail during $(x, x + dx)$ under the condition that it had not failed before time x. Consequently,

$$g(x) = \frac{f(x)}{1 - F(x)} \tag{5.39}$$

If the probability density function $f(x)$ is exponential, with parameter λ, it follows from Equations 5.37 and 5.38 that:

$$g(x) = \frac{\lambda e^{-\lambda x}}{1 - \left(1 - e^{-\lambda x}\right)} = \frac{\lambda e^{-\lambda x}}{e^{-\lambda x}} = \lambda \tag{5.40}$$

Equation 5.40 indicates that the failure probability is constant, irrespective of time. It implies that the probability that a component whose time-to-failure distribution is exponential fails in an instant during the first hour of its life is the same as its failure probability during an instant in the thousandth hour, presuming it has survived up to that instant. It is for this reason that the parameter λ is usually referred to in life-test applications as the *failure rate*. This definition generally has meaning only with an exponential distribution.

This natural association with life-testing and the fact that it is very tractable mathematically makes the exponential distribution attractive as representing the life distribution of a complex system or several complex systems. In fact, the exponential distribution is as prominent in reliability analysis as the normal distribution is in other branches of statistics.

It has been shown theoretically that this distribution provides a reasonable model for systems designed with a limited degree of redundancy and made up of many components, none of which has a high probability of failure. This is especially true when low component failure rates are maintained by periodic inspection and replacement or in

Probability Distributions

situations in which failure is a function of outside phenomena rather than a function of previous conditions. On the other hand, the exponential distribution often cannot represent individual component life and it is sometimes questionable even as a system model.

It should be noted that many systems, including those described by exponential and Weibull distributions, consisting of several components, can be classified as *series*, *parallel*, or a combination of both. However, the majority of industrial and process plants (units and systems) have a series of parallel configurations. A series system is one in which the entire system fails to operate if any one of its components fails to operate. If such a system consists of n components that function independently, then the reliability of the system is the product of the reliabilities of the individual components. If R_s denotes the reliability of a series system and R_i denotes the reliability of the ith component where $i = 1,..., n$, then:

$$R_s = R_1 R_2 \ldots R_n = \prod_{i=1}^{n} R_i \qquad (5.41)$$

A *parallel system* is one that fails to operate only if all its components fail to operate. If R_i is the reliability of the ith component, then $(1 - R_i)$ is the probability that the ith component fails; $i = 1,..., n$. Assuming that all n components function independently, the probability that all n components fail is $(1 - R_1)(1 - R_2)...(1 - R_n)$. Subtracting this product from unity yields the following formula for R_P, the reliability of a parallel system:

$$R_P = 1 - (1 - R_1)(1 - R_2)\ldots(1 - R_n) = 1 - \prod_{i=1}^{n}(1 - R_i) \qquad (5.42)$$

The reliability formulas for series and parallel systems can be used to obtain the reliability of a system that combines features of a series and a parallel system. Consider, for example, the system diagrammed in Figure 5.10. Components A, B, C, and D have for their respective reliabilities 0.90, 0.80, 0.80, and 0.90. The system fails to operate if A fails, if B and C both fail, or if D fails. Components B and C constitute a parallel subsystem connected in series to components A and D. The reliability of the parallel subsystem is obtained by applying Equations 5.42, which yields:

$$R_P = 1 - (1 - 0.80)(1 - 0.80) = 0.96$$

The reliability of the system is then obtained by applying Equation 5.41, which yields:

$$R_S = (0.90)(0.96)(0.90) = 0.78$$

5.3.4 Log-normal Distribution

Before proceeding to the presentation of the log-normal distribution, the interests of the reader would be better served with a short introduction to logarithms. The logarithm of a number is the exponent of that power to which another number, the base, must be raised to give the number first named. Any positive number greater than

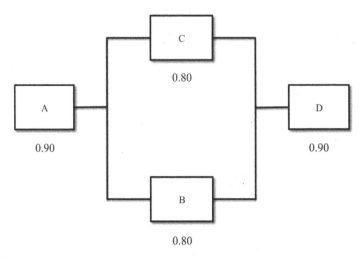

FIGURE 5.10 System with parallel and series components.

1 might serve as a base. Two have been selected by the technical community. One base, 2.718 denoted by the letter e, gives rise to a system of logarithms that are conveniently applied in engineering and science. These are referred to as Napierian or hyperbolic or natural logarithms. The other base used is 10, giving logarithms particularly adapted for use in computation, referred to by some as common or Briggsian logarithms. Tables of logarithms given without designation invariable refer to base 10. Since most numbers are irrational powers of 10, a common logarithm, in general, consists of an integer, which is called the characteristic and an endless decimal, the mantissa (Theodore and Taylor 1996; Shaefer and Theodore 2007).

The use of logarithmic scales on graphs also finds applications. Log-log coordinates are useful in plotting equations of the frequently occurring form of $y = bx^a$. If logarithms (natural logarithms also apply) are taken on both sides of this equation, one obtains $\log y = a \log x + \log b$. If $\log y$ is plotted as a function of $\log x$, or equivalently, if y is plotted as a function of x on log-log coordinates, a straight line of slope a and intercept $\log b$ will result for log-normal distributions.

A nonnegative random variable X has a log-normal distribution whenever $\ln X$, the natural logarithm of X, has a normal distribution. The pdf of a random variable X having a log-normal distribution is specified by:

$$f(t) = \frac{1}{\sqrt{2\pi}\beta} x^{-1} \exp\left[-\frac{(\ln x - \alpha)^2}{2\beta^2}\right]; x > 0; = 0 \text{ elsewhere} \tag{5.43}$$

The mean and variance of a random variable X having a log-normal distribution are given, respectively, by:

$$\mu = e^{\alpha + \beta^2/2}$$

$$\sigma^2 = e^{2\alpha + \beta^2}\left(e^{\beta^2} - 1\right)$$

Probability Distributions

Probabilities concerning random variables having a log-normal distribution can be calculated using tables of the normal distribution. If X has a log-normal distribution with parameters α and β, then $\ln X$ has a normal distribution with $\mu = \alpha$ and $\sigma = \beta$. Probabilities concerning X can therefore be converted into equivalent probabilities concerning $\ln X$. Suppose, for example, that X has a log-normal distribution with $\alpha = 2$ and $\beta = 0.1$. Then

$$P(6 < x < 8) = P(\ln 6 < \ln X < \ln 8) = P\left[\frac{\ln 6 - 2}{0.1} < \frac{\ln X - 2}{0.1} < \frac{\ln 8 - 2}{0.1}\right]$$
$$= P(-2.08 < Z < 0.79) = 0.78 = 78\%$$

Estimates of the parameters α and β in the pdf of a random variable X having a log-normal distribution can be obtained from a sample of observations on X by making use of the fact that $\ln X$ is normally distributed with mean α and standard deviation β. Therefore, the mean and standard deviation of the natural logarithms of the sample observations on X furnish estimates of α and β. To illustrate the procedure, suppose the time to failure T, in thousands of hours, was observed for a sample of five electric motors. The observed values of T were 8, 11, 16, 22, and 34. The natural log of these observations are 2.08, 2.40, 2.77, 3.09, and 3.53, respectively. Assuming that T has a log-normal distribution, the estimates of the parameters α and β in the pdf are obtained from the mean and standard deviation of the natural logs of the observations on T (Shaefer and Theodore 2007):

$$\mu = \sum_{i=1}^{5} \frac{\ln T_i}{5} = 2.77 = \text{Estimate of } \alpha$$

$$\sigma = \sqrt{\sum_{i=1}^{5} \frac{(\ln T_i - 2.77)}{4}} = 0.57 = \text{Estimate of } \beta$$

Using these α and β values, as estimate can be made of the likelihood that the motors will last less than 5,000 hr. The probability of T is expressed as the probability of $\ln T$ and the following can be written:

$$P(T < 5) = P(\ln T < \ln 5) = P(\ln T < 1.61)$$

Treating $\ln T$ as a random variable that is normally distributed with a mean of α and standard deviation of β, this probability can be determined using Table 5.4 as:

$$P(\ln T < 1.61) = P\left(\frac{\ln T - \alpha}{\beta} < \frac{1.61 - \alpha}{\beta}\right) = P\left(Z < \frac{1.61 - 2.77}{0.57}\right)$$
$$= P(Z < -2.04) = 0.021 = 2.1\%$$

One can also calculate the probability that a pump lasts more than 5,000 hr as follows:

$$P(T > 5) = P(\ln T > \ln 5) = P(\ln T < 1.61) = 1 - P(T < 5) = 1 - 0.021 = 0.979 = 97.9\%$$

Finally once can calculate the probability that the pumps last more than 10,000 hr as follows:

$$P(T > 10) = P(\ln T > \ln 10) = P(\ln T > 2.303)$$

Then

$$P(\ln T > 2.303) = P\left(\frac{\ln T - \alpha}{\beta} > \frac{2.303 - \alpha}{\beta}\right) = P\left(Z > \frac{2.303 - 2.77}{0.57}\right)$$
$$= P(Z > -0.82) = 1 - 0.209 = 0.791 = 79.1\%$$

A modified form of a log-normal distribution is the so-called "Probit" function. In its basic form, the probit model is expressed as:

$$P = a_0 + a_1 \log(C^{a_2} t) \tag{5.44}$$

where P = percent, fraction or number of individuals affected; a_0, a_1, a_2 = constants; C = chemical concentration, consistent units; and t = time, consistent units.

Although this model is employed at times to represent toxicological data, it finds its main application in emergency/accidental chemical exposures.

The log-normal distribution has been employed to characterize occupational risk with time. It describes a distribution of data where there are many measurements with lower values and fewer measurements with high values. This distribution can be used to describe a relatively constant measurement, which is occasionally punctuated by higher values due to cyclic variations that arise in epidemiology studies, particularly for dose-response analysis. Since the log-normal distribution is also characterized by a geometric mean and a geometric standard deviation, the 95th percentile has been used as an indicator of peak values. This 95th percentile value is usually an important statistic for those chemicals that produce primarily acute toxicological effects.

The log-normal distribution, as noted, has been employed as an appropriate model in a wide variety of situations from environmental management to biology to economics. Additional applications include the distribution of personal incomes, inheritances, bank deposits, and the distribution of organism growth subject to many small impurities. Rosenkrantz (2009) has discussed the log-normal distribution in terms of a model for the distribution of stock prices. The unusual feature of this model is that it is the logarithm of the stock price, not the stock price itself, that is normally distributed. This author provides an extensive treatment of this topic plus illustrative examples. Perhaps the primary environmental application of the lognormal distribution has been to represent the distribution of particle sizes in gaseous emissions from many industrial processes (Theodore 2008).

5.3.5 OTHER CONTINUOUS PROBABILITY DISTRIBUTIONS

In addition to the probability distribution discussed above, there are several other well-known distributions that can be used in risk analysis calculations. These

Probability Distributions

TABLE 5.5
Additional Probability Distributions

Statistical Class	Distribution
Discrete	Geometric
Continuous	Extreme Value
	Gamma
	Rayleigh
	Rectangular

distributions may be either discrete or continuous functions and are listed in Table 5.5. Descriptions of each of these distributions not covered above can be found in the literature (Theodore and Taylor 1996; Shaefer and Theodore 2007; Rosenkrantz 2009).

5.4 ILLUSTRATIVE EXAMPLES

Seven illustrative examples complement the material presented above.

5.4.1 ILLUSTRATIVE EXAMPLE 1

The probability that exposure to a toxic chemical will be fatal is 0.80. Find the probability of the following events for a group of 15 workers:

1. At least nine will die.
2. From four to eight will die.

Solution: This can be solved considering fatalities to be described by a binomial distribution. For Event 1 using Equation 5.6

$$P(\text{at least 9 will die}) = P(X \geq 9); p = 0.8, q = 0.2$$

$$P(X \geq 9) = \sum_{x=9}^{15} \frac{15!}{x!(15-x)!}(0.8)^x(0.2)^{15-x}$$

$$= 0.430 + 0.1032 + 0.1876 + 0.2501 + 0.2309 + 0.1319 + 0.0352 = 0.982$$

This calculation can be performed by longhand or obtained directly form binomial table (Woonacott and Woonacott 1985).

For Event 2,

$$P(4 \leq X \leq 8) = 1.0 - P(X \geq 9) - P(0 \leq X \leq 4)$$

One notes almost immediately that:

$$P(X \leq 4) = \sum_{x=0}^{4} \frac{15!}{x!(15-x)!}(0.8)^x (0.2)^{15-x}$$

$$= \left(\frac{15!}{0!(15)!}(0.8)^0 (0.2)^{15}\right) + \left(\frac{15!}{1!(14)!}(0.8)^1 (0.2)^{14}\right) + \left(\frac{15!}{2!(13)!}(0.8)^2 (0.2)^{13}\right)$$

$$+ \left(\frac{15!}{3!(12)!}(0.8)^3 (0.2)^{12}\right) + \left(\frac{15!}{4!(11)!}(0.8)^4 (0.2)^{11}\right)$$

$$= (1)(1)(3.3 \times 10^{-11}) + (15)(0.8)(1.6 \times 10^{-10}) + (105)(0.64)(8.2 \times 10^{-10})$$

$$+ (455)(0.512)(4.1 \times 10^{-9}) + (1365)(0.41)(2.0 \times 10^{-8}) = 3.3 \times 10^{-11}$$

$$+ 1.9 \times 10^{-9} + 5.5 \times 10^{-8} + 9.6 \times 10^{-7} + 1.1 \times 10^{-5} \approx 0$$

$$P(0 \leq X \leq 4) \approx 0$$

Therefore,

$$P(4 \leq X \leq 8) = 1.0 - 0.982 - 0.0 = 0.018$$

5.4.2 Illustrative Example 2

A nanoparticle production unit contains (for cooling purposes) 20 independent sprays, each of which fails with a probability of 0.10. The system fails only if four or more of the sprays fail. What is the probability that the unit will fail?

Solution: Let X denote the number of sprays that fail. The term X has a binomial distribution with $n = 20$ and $p = 0.10$. The probability that the system fails is given by:

$$P(X \geq 4) = \sum_{x=4}^{20} \frac{20!}{x!(20-x)!}(0.1)^x (0.9)^{20-x} = 1 - \sum_{x=0}^{3} \frac{20!}{x!(20-x)!}(0.1)^x (0.9)^{20-x}$$

$$= 1 - \left(\frac{20!}{0!(20)!}(0.1)^0 (0.9)^{20}\right) + \left(\frac{20!}{1!(19)!}(0.1)^1 (0.9)^{19}\right)$$

$$+ \left(\frac{20!}{2!(18)!}(0.1)^2 (0.9)^{18}\right) + \left(\frac{20!}{3!(17)!}(0.1)^3 (0.9)^{17}\right)$$

$$= 1 - (1)(1)(0.122) + (20)(0.1)(0.135) + (190)(0.01)(0.15) + (1140)(0.001)(0.167)$$

$$= 0.122 + 0.27 + 0.285 + 0.19 = 1 - 0.867 = 0.133 = 13.3\%$$

5.4.3 Illustrative Example 3

Over the last 10 years, a local hospital reported that the average number of deaths per year due to temperature inversions (air pollution) was 0.5. What is the probability of exactly three deaths in a given year assuming these deaths can be described by a

Probability Distributions

Poisson distribution? What is the probability of three or more deaths being attributed to temperature inversions?

Solution: Calculate the annual probabilities using Equation 5.24. First, the probability of exactly three deaths is:

$$P(X=3) = P(3) = \frac{e^{-0.5}(0.5)^3}{3!} = \frac{0.607(0.125)}{(3)(2)(1)} = \frac{0.076}{6} = 0.013 = 1.3\%$$

The probability of three or more deaths is written as follows:

$$P(X \geq 3) = \sum_{x=3}^{\infty} \frac{e^{-0.5}(0.5)^x}{x!} = 1 - \sum_{x=0}^{2} \frac{e^{-0.5}(0.5)^x}{x!} = 1 - \frac{e^{-0.5}(0.5)^0}{0!} - \frac{e^{-0.5}(0.5)^1}{1!} - \frac{e^{-0.5}(0.5)^2}{2!}$$
$$= 1 - 0.6065 - 0.3033 - 0.0758 = 0.0144 = 1.4\%$$

5.4.4 Illustrative Example 4

Consider the system shown in Figure 5.11. Determine the reliability, R, if the operating time for each unit is 5,000 hr. Components A and B have exponential failure rates, λ, of 3×10^{-6} and 4×10^{-6} failures per hour, respectively, where $R_i = e^{-\lambda_i t}$; and t = time, hr. The term λ may be viewed as the reciprocal of the average time to failure.

Solution: Because this is a series system

$$R_S = R_A R_B$$

As indicated earlier, for an exponential failure rate:

$$R = e^{-\lambda t}; t = \text{time, hr}$$

so that

$$R_A = e^{(3 \times 10^{-6})(5,000)} = e^{0.015} = 0.9851$$

and

$$R_B = e^{(4 \times 10^{-6})(5,000)} = e^{0.02} = 0.9802$$

FIGURE 5.11 Exponential failure rate: series system.

Therefore,

$$R_S = (0.9851)(0.9802) = 0.9656 = 96.56\%$$

5.4.5 ILLUSTRATIVE EXAMPLE 5

Estimate the probability that a pump will survive at least three times, five times, and ten times its expected life. Assume the exponential distribution applies. Also

$$P(T) = e^{-\lambda t}$$

with $\lambda = 1/a$ and $t = na$, where n is the number of a = expected lives of the pump.

Solution: The calculation for three times the expected pump life, $n = 3$, is:

$$P(T > 3a) = e^{-\frac{1}{a}(3a)} = e^{-3} = 0.0498 = 4.98\%$$

For five times the expected pump life, $n = 5$:

$$P(T > 5a) = e^{-\frac{1}{a}(5a)} = e^{-5} = 0.0067 = 0.67\%$$

For ten times the expected pump life, $n = 10$:

$$P(T > 10a) = e^{-\frac{1}{a}(10a)} = e^{-10} = 0.000045 = 0.0045\%$$

As expected, the probability of survival decreases with increasing survival time.

5.4.6 ILLUSTRATIVE EXAMPLE 6

The concentration of a particular toxic substance in a wastewater stream is known to be normally distributed with mean $\mu = 100$ parts per million and a standard deviation $\mu = 2.0$ parts per million. Calculate the probability that the toxic concentration, C, is between 98 and 104.

Solution: Because C is normally distributed with $\mu = 100$ and a standard deviation $\sigma = 2.0$, then $(C - 100)/2$ is a standard normal variable and using Equation 5.35, the following can be written:

$$P(98 < C < 104) = P\left(\frac{98-100}{2} < \frac{C-100}{2} < \frac{104-100}{2}\right) = P(-1 < Z < 2)$$

Referring to Figure 5.8 and from the values in the standard normal table, Table 5.4, the following probability can be calculated:

$$P(98 < C < 104) = (0.5 - 0.159) + (0.5 - 0.023) = (0.341) + (0.477) = 0.818 = 81.8\%$$

5.4.7 ILLUSTRATIVE EXAMPLE 7

The failure rate per year, Y, of a coolant recycle pump in a wastewater treatment plant has a log-normal distribution. If $\ln Y$ has a mean of 2.0 and variance of 1.5, find $P(0.175 < Y < 1)$.

Solution: If Y has a log-normal distribution, $\ln Y$ has a normal distribution with mean 2 and standard deviation $\sigma = 1.5^{1/2} = 1.22$. Therefore, referring to Figure 5.8 and from the values in the standard normal table, Table 5.4, the following probability can be calculated:

$$P(0.175 < Y < 1) = P(\ln 0.175 < \ln Y < \ln 1) = P\left[\frac{\ln 0.175 - 2}{1.22} < \frac{\ln Y - 2}{1.22} < \frac{\ln 1 - 2}{1.22}\right]$$

$$= P(-3.07 < Z < -1.64) = (0.5 - 0.001) - (0.5 - 0.051) = 0.499 - 0.449 = 0.05 = 5\%$$

PROBLEMS

5.1 The temperature of a polluted estuary during the summer months is normally distributed with mean 56°F and standard deviation 3.0°F. Calculate the probability that the temperature is between 55°F and 62°F.

5.2 The life (time to failure) of a machine component has a Weibull distribution. Determine the probability that the component lasts at least 25,000 hr if t is measured in thousands of hours. Outline how to determine the probability that the component lasts a given period of time if the failure rate is $t^{-1/2}$.

5.3 Assuming a failure rate relationship can be defined by a Weibull distribution, write the Weibull pdf (Equation 5.29) if the failure rate, $Z(t)$ (Equation 5.28) for the system is given by:
   a. $t^{-0.25}$
   b. $t^{-\sqrt{1/2}}$

5.4 The life (time to failure) of a machine component has a Weibull distribution. Determine the probability that the component lasts at least 25,000 hr if t is measured in thousands of hours. Outline how to determine the probability that the component lasts a given period of time if the failure rate is $t^{-0.5}$.

5.5 The life of an electronic component is a random variable having a Weibull distribution with $\alpha = 0.025$ and $\beta = 0.50$. What is the probability that the component will last more than 4,000 hr?

5.6 The time in days to failure of each sample of ten electronic components is observed as follows: 71, 40, 90, 149, 127, 53, 106, 36, 18, 165. Assuming a Weibull distribution applies, estimate the parameters α and β.

5.7 The average number of breakdowns of personal computers during 1,000 hr of operation of a computer control center is 3. What is the probability of only one breakdown during a 10-hr work period? Assume a Poisson distribution describes the failure rate relationship for this system.

5.8 The concentration of a toxic specialty chemical produced in a pilot plant is normally distributed with a mean of 48 µg/L and a standard deviation of 3 µg/L. What is the probability that an unstable concentration of 55 µg/L will be produced?

REFERENCES

Bury, K. 1975. *Statistical Models in Applied Science*. Hoboken, NJ: John Wiley & Sons.

Dupont, R.R., and Theodore, L. 2011. *Calculating hazard probabilities using the Weibull distribution*, Paper #125. Proceedings of the 104th Annual Air and Waste Management Association Conference, June 21–24, Orlando, FL.

Dupont, R.R., McKenna, J., and Theodore, L. 2013. *Baghouse failures: applying the Weibull distribution to estimate bag failure emissions as a function of time*, Paper #112. Proceedings of the 106th AWMA Conference, Chicago, IL.

Dupont, R.R., Ricci, F., and Theodore, L. 2014. *An improved failure rate model applied to baghouse failures*, Paper #32932. Proceedings of the 107th AWMA Conference, Long Beach, CA, 2014.

Dupont, R.R., Ricci, F., and Theodore, L. 2015. *Replacing the Weibull distribution failure model*, Paper #118. Proceedings of the 108th AWMA Conference, Raleigh, NC, 2015.

Rosenkrantz, W. 2009. *Probability and Statistics Applications in Environmental Science*. Boca Raton, FL: CRC Press/Taylor & Francis Group.

Shaefer, S., and Theodore, L. 2007. *Probability and Statistics Applications in Environmental Science*. Boca Raton, FL: CRC Press/Taylor & Francis Group.

Theodore, L. 2008. *Air Pollution Control Equipment Calculations*. Hoboken, NJ: John Wiley & Sons.

Theodore, L., and Ricci, F. 2011. *Mass Transfer Operations for the Practicing Engineer*. Hoboken, NJ: John Wiley & Sons.

Theodore, L., and Taylor, F. 1996. *Probability and Statistics*. Theodore Tutorials, originally published by US EPA, RTP, NC, East Williston, NY.

Woonacott, R.J., and Woonacott, T.H. 1985. *Introductory Statistics*, 4th Edition. Hoboken, NJ: John Wiley & Sons.

Part II

Chemical Process Industries

The rapid growth and expansion of the chemical process industry has been accompanied by not only a spontaneous rise in chemical emissions to the environment but also human, material, and property losses because of fires, explosions, hazardous and toxic spills, etc. Concern over the potential health consequences of these massive emissions, particularly from chemical, petrochemical, and utility plants, has sparked interest at both the industrial and regulatory levels in obtaining a better understanding of the main subject of this book. The writing of this chemical process industries section was undertaken in part to address this growing concern. Following Chapter 6 concerned with definitions and a glossary of terms related to chemical processes, the remaining five chapters in Part II are split among chemical process history (Chapter 7), chemical process equipment and process fundamentals (Chapters 8 and 9), industry-specific processes (Chapters 10), and emergency planning and response (Chapter 11).

6 Definitions/Glossary of Chemical Process Terms

Emma Parente

6.1 INTRODUCTION

As one might suppose, the reader is primarily introduced to chemical/chemistry processes and industry-related terms. As noted in the preface, this book is concerned with health and hazard risk and as such, this chapter primarily addresses health and hazard risk-related terms. It has been written not only for academic use in colleges and universities but also for both engineers and scientists who work in the chemical process industry's (CPI's) related field. This glossary may be used whenever and wherever information is needed about words and/or terms in the CPI.

Some additional points deserve mentioning.

1. Each definition avoids technical jargon.
2. No mathematical equations – in any form – are employed in the definition. In some instances, where necessary, common scientific and engineering units have been included.
3. Only key words or terms used in practice are provided. As is the case in preparing a text, particularly a dictionary, the problems of what to include and what to omit have been particularly difficult.
4. Only one spelling is used for words with multiple accepted spellings, e.g., modeling vs. modelling.
5. Some important acronyms are also included with a one-sentence definition.
6. As with nearly every glossary, the terms have been alphabetized.

This chapter defines many – but not all – of the terms that the reader will encounter in this book. The following list is therefore not a complete glossary of all terms that appear in the field. It should also be noted that many of the terms have come to mean slightly different things to different people; this will become evident as one delves deeper into the literature (Theodore, Reynolds and Morris 1997; Shaefer and Theodore 2007; Theodore and Dupont 2012; Theodore and Theodore 2021).

In addition to providing terms and definitions, this chapter provides eight illustrative examples to elaborate on materials presented throughout the chapter.

6.2 TERMS AND DEFINITIONS

As these terms and definitions are reviewed, the reader should carefully note the difference between the definitions associated with *health* and *hazard risk*.

Absolute humidity: The amount of water vapor present in a unit mass of air, which is usually expressed in kg water vapor/kg dry air or lb water vapor/lb dry air.

Absolute pressure: The actual pressure exerted on a surface that is measured relative to zero pressure; it equals the gauge pressure plus the atmospheric pressure.

Absolute temperature: The temperature expressed in degrees of Kelvin (K) or Rankine (°R).

Absorbate: A substance that is taken up and retained by an absorbent.

Absorbent: Any substance that takes in or absorbs other substances.

Absorber: A device in which a gas is absorbed by contact with a liquid.

Absorption: The process in which one material (the absorbent) takes up and retains another (the absorbate) to form a homogeneous solution; it often involves the use of a liquid to remove certain gas components from a gaseous mixture.

Actual cubic feet per minute (acfm): A unit of flow rate measured under actual pressure and temperature conditions.

Acute (risk): Risks associated with short periods of time. For health risk, it usually represents short exposures to high concentrations of an agent.

Adiabatic: A term used to describe a system in which no gain or loss of heat is allowed to occur.

Adiabatic flame temperature: The maximum temperature that a combustion system can reach.

Adiabatic lapse rate: The rate at which temperature of a moving air parcel decreases in the atmosphere as height above the surface increases when no heat is added or subtracted from the moving air parcel; the adiabatic lapse rate is 10°C/km.

Adsorbent: A substance (e.g., activated carbon, activated alumina, silica gel) that has the ability to condense or hold molecules of other substances on its surface.

Adsorber: An apparatus in which molecules of gas or liquid are retained on the surface of an adsorbent.

Adsorption: The physical or chemical bonding of molecules of gas, liquid, or dissolved solid to the external or internal (if porous) surface of a solid (adsorbent); it is a method of treating waste that is employed to remove odor, color, or organic matter from a system.

Afterburner: A secondary burner located so that combustion gases from the primary incinerator are further burned to remove smoke, odors, and other pollutants.

Agent: A biological, physical, or chemical entity capable of causing disease or adverse health effects.

Alpha (α) particle: A positively charged helium nucleus (i.e., two protons and two neutrons) that is emitted spontaneously from the decay of radioactive elements.

Asphyxiant: A vapor or gas that has little or no positive toxic effect but which can bring about unconsciousness and death by replacing air and thus depriving an organism of oxygen.

Aspirator: A hydraulic device that creates a negative pressure by forcing liquid through a restriction, thus increasing the velocity head.

Atmospheric dispersion: The mixing of a gas or vapor (usually from a discharge point) with air in the lower atmosphere. The mixing is the result of convective motion and turbulent eddies.

Atmospheric stability: A measure of the degree of atmospheric turbulence, often defined in terms of the vertical temperature gradient in the lower atmosphere.

Atomic fission: The breaking down of a large atom into smaller atoms or elements, involving the liberation of heat, gamma rays, alpha particles, and beta particles.

Audit: The examination of something with intent to check, verify, or inspect a particular subject matter.

Autoignition: The starting of a fire without the addition of an external source such as a flame, spark, or heat.

Autoignition temperature: The lowest temperature at which a flammable gas in air will ignite without an ignition source.

Average rate of death (ROD): The average number of fatalities that can be expected per unit time (usually on an annual basis) from all possible risks and/or incidents.

Baffle: A flow-regulating device usually consisting of a plate placed horizontally across a pipe or channel to restrict or divert the passage of a fluid, usually used for the purpose of providing a uniformly dispersed flow.

Ball joint: A flexible pipe joint formed in the shape of a ball or a sphere.

Ball valve: A nonreturn valve consisting of a ball resting on a cylindrical seat within a fluid passageway or pipe.

Barometric pressure: The pressure of the ambient air in the atmosphere at a particular point on or above the surface of the earth.

Basic event: A fault tree event (FTE) that is sufficiently basic that no further explanation or development of additional events is necessary.

Batch process: A process that is not continuous; its operations are carried out with discrete quantities of material added and removed from it at appropriate time intervals.

Beta (β) particle: A charged particle emitted from a radioactive atomic nucleus; it has moderate penetrative power and is able to damage living tissue.

Bias: The systematic distortion of data; it is the tendency of a sample to be unrepresentative of all the cases involved in a study.

Bleeding: The gradual release of material and/or reduction of pressure from a system or process (e.g., by a valve or leak).

Blowdown: The cyclic or constant removal of a portion of any process flow to maintain the constituents of the flow at a desired level.

Blower: A fan employed to force or move air or gas under pressure.

Brownian movement: The constant, random movement of small, suspended particles due to collisions with other molecules.

Buffer: A solution containing both a weak acid and its conjugate weak base which is employed to stabilize the pH value in a solution, and anything that acts to diminish and/or regulate changes in a system or process.

Bulk density: The mass per unit volume of a solid in a mixture such as a packed bed or soil mass; unlike the real solid particle density, the pore space is included in the volume for this calculation.

Bulk sample: A small portion of material that is collected and sent to a laboratory for analysis.

Butterfly damper: A plate or blade installed in a duct, flue, breeching, or stack that rotates on an axis to regulate the flow of gases.

Butterfly valve: A flow control valve containing a disk supported by a shaft on which it rotates.

By-pass: The avoiding of a particular portion of a process or pollution control system.

By-pass valve: A valve arranged to cause the fluid which it controls to flow past some part of its normal path (e.g., to allow a liquid to avoid a filter through which it usually passes).

By-product: A material that is not one of the primary products of a production process and is not solely or separately produced by the production process.

C (ceiling): The term used to describe the maximum allowable exposure concentration of a hazardous agent related to industrial exposures to hazardous vapors.

Calibration: The determination, checking, or adjustment of the accuracy of any instrument that gives quantitative measurements.

Cancer: A tumor formed by mutated cells.

Carcinogen: A cancer-causing chemical.

Carrier gas: A gas that acts as the mobile phase in gas chromatography.

Carryover: The entrainment of liquid or solid particles in the vapor evolved by a boiling liquid or from a process unit.

Catalyst: A substance whose presence changes (normally increasing) the rate of a chemical reaction without itself undergoing permanent change in its composition.

Catalytic cracking: The breaking of a carbon-carbon bond with the aid of a catalyst; it is an essential process in the refining of petroleum.

Catastrophe: A major loss in terms of death, injuries, and/or property damage.

Cause-consequence: A method for determining the possible consequences or outcomes arising from a logical combination of input events or conditions that determine a cause.

Cavitation: The formation of vapor bubbles in a liquid when subjected to localized low-pressure regions causing severe mechanical damage to the surface of metals exposed to it.

Centrifugal pump: A device that increases the pressure of a liquid by using centrifugal force.

Check valve: A one-way valve that prevents flow through a pipe in an undesired direction; it opens in the direction of the normal flow and closes with a reversal of flow.

Chemical Abstract Service (CAS) numbers: CAS numbers are used to identify chemicals and mixtures of chemicals.

Chemical agent: An element, compound, or mixture that coagulates, dissolves, neutralizes, solubilizes, oxidizes, concentrates, makes a pollutant mass more rigid, or otherwise facilitates the lessening of harmful effects or the removal of a pollutant from a fluid.

Definitions/Glossary of Chemical Process Terms

Chemical Process Quantitative Risk Analysis (CPQRA): The process of hazard identification, followed by numerical evaluation of incident consequences and frequencies, and their combination into an overall measure of risk when applied to the CPI. Ordinarily applied to episodic events. Related to Probabilistic Risk Assessment (PRA) used in the nuclear industry.

Chemical equation: A representation of a chemical reaction using symbols to show the molar relationship between the reacting substances and the products.

Chronic (risk): Risks associated with long-term chemical exposure duration, usually at low concentrations.

Closed-loop: A term used to describe an enclosed, recirculating process.

Closed-loop recycling: The reclaiming or reusing of wastewater for non-potable purposes in an enclosed process; it may also be applied to other streams including gaseous ones.

Cocurrent: A term used to describe a flow in which materials travel in the same direction.

Combustion: A reaction at a high temperature with oxygen that produces carbon dioxide, water, and energy in the form of light and heat; it is a basic cause of air pollution.

Concurrent: A term used to describe a situation in which two or more controls or systems exist in an operated condition at the same time.

Condensate: Any liquid resulting from the cooling of a gas or vapor.

Conditional probability: The probability of occurrence of an event given that a precursor event has occurred.

Confidence interval: A range of values of a variable with a specific probability that the true value of the variable lies within this range. The conventional confidence interval probability is the 95% confidence interval, defining the range of a variable in which its true value falls with 95% confidence.

Confidence limits: The upper and lower range of values of a variable defining its specific confidence interval.

Consequences: A measure of the expected effects of an incident outcome or cause.

Continuous release: Emissions that are of a continuous duration.

Convection: The transfer of heat through a fluid by the movement of the fluid; and the vertical movement of air leading to cooling.

Cooling tower: A hollow, vertical structure with internal baffles to disperse water so it is cooled by flowing air and by evaporation at ambient temperature.

Corona: An electrical discharge effect that causes ionization of oxygen and the formation of ozone; it is found in electrostatic precipitators.

Corrosion: The deterioration or destruction of a material by chemical action.

Countercurrent: A term used to describe the flow pattern within equipment in which two streams travel in opposite directions.

CPQRA: The acronym for chemical process quantitative risk analysis; it is analogous to a hazard risk assessment (HZRA).

Cracking: A refining process involving decomposition and molecular recombination of organic compounds to form molecules of smaller sizes that are suitable for fuels.

Crawl space: An area beneath the floor of a house or a building that allows access to utilities and other services.

Critical temperature: The temperature above which a gas or vapor cannot be liquefied by an increase in pressure alone.

Cryogenics: The production and utilization of extremely low temperatures.

Crystallization: The change of state of a substance from a liquid to a solid by the phenomenon of crystal formation by nucleation and accretion (e.g., the freezing of water onto ice).

Cutback: A coating substance or varnish that has been diluted or thinned.

Damper: A manually or automatically controlled valve or plate in a breeching, duct, or stack that is employed to regulate a draft or the rate of flow of air or other gases.

Dead time: The time interval, after a response to one signal or event, during which a system is unable to respond to another signal or event.

Dehumidifier: A device incorporated into many air conditioning systems to dry incoming air by passing it across a bed of a hygroscopic substance or through a spray of very cold water.

Delphi method: A polling of experts that involves the following: (1) Select a group of experts (usually three or more). (2) Solicit, in isolation, their independent estimates of the value of a particular parameter and their reason for choice. (3) Provide initial analysis results to all experts and allow them to then revise their initial values. (4) Use the average of the final estimates as the best estimate of the parameter and use the standard deviation of the estimates as a measure of uncertainty. The procedure is iterative, with feedback between iterations.

Demister: A device composed of plastic threads, wire mesh, or glass fibers employed to remove liquid droplets entrained in a gas stream.

Dermal: Applied to the skin.

Desorption: A process of removing an absorbed material from a solid on which it is adsorbed by increasing the temperature or reducing the pressure, or both.

Detention time (detention period): The average time that a unit volume of a fluid is retained in a unit during a flow process.

Detoxification: The destruction of the toxic quality of a substance.

Dew point: The temperature at which the first droplet of water forms on the progressive cooling of a mixture of air and water vapor; at the dew point, the air becomes saturated with water.

Dike: An embankment that restricts the movement and provides the containment of a liquid.

Dilution ventilation: The mixing of contaminated air with uncontaminated air for the purpose of controlling potential airborne health hazards, fire and explosion conditions, odors, and nuisance type contaminants.

Dispersion coefficient: The standard deviation, σ, in a specified direction used in a Gaussian plume atmospheric dispersion model.

Distillation: A process of separating the constitutes of a liquid mixture by means of partial vaporization of the mixture and the separate recovery of vapor and residue.

Definitions/Glossary of Chemical Process Terms

Domino effect: The triggering of secondary events; usually considered when a significant escalation of the original incident could result.

Dose: A weight (or volume) of a chemical agent, usually normalized to a unit of body weight.

Downcomer: A pipe or flue that conveys gases, vapors, or condensate downward in blast furnaces, distillation towers, and refineries.

Downdraft: A current of air in a stack with a bulk downward motion.

Downstream: The direction in which a fluid stream is flowing.

Downtime: The lost production time during which a piece of equipment is not operating correctly due to a breakdown, maintenance, or power failure.

Dry bulb: A thermometer bulb maintained dry; it is used with a wet bulb to measure humidity.

Dry-bulb temperature: The temperature of air, usually used in conjunction with the wet-bulb temperature to measure humidity.

Duct: A round or rectangular conduit, usually metal or fiberglass, employed to transport fluids.

Economizer: A device that uses the heat of combustion gas to raise the temperature of boiler feedwater prior to entry into the boiler.

Ejector: A device for moving a fluid or solid by entraining it in a high-velocity stream of air or water jet.

Elbow: A pipe fitting that connects two pipes at a 90° angle.

Electrostatic field: A region in which a stationary electrically charged particle is subjected to a force of attraction or repulsion as a result of another stationary electric charge.

Electrostatic precipitator (ESP): A device that separates particles from a gas stream by passing the carrier gas between two electrodes across which a high voltage is applied.

Elutriator: An air-sampling device, widely used in cotton-dust area air sampling, that uses gravitational forces to remove dust that is unfit for breathing from the air sample, and then collects the dust on a filter for further analysis.

Endothermic: A term used to describe a process or change that occurs with the input of heat.

Entrainment: The carryover of drops of liquid during a process such as distillation, absorption, scrubbing, or evaporation.

Environmental audit: An independent assessment of the current status of a company's compliance with applicable environmental requirements.

Episode: An air pollution incident in a given area caused by an increase in atmospheric pollutant concentration in response to meteorological conditions (inversion) that may result in a significant increase in illness or in death.

Episodic release: A massive release of limited or short duration, usually associated with an accident.

Event: An occurrence associated with an incident either as the cause or a contributing cause of the incident, or as a response to an initiating event.

Event sequence: A specific sequence of events composed of initiating events and intermediate events that may lead to a hazard or an incident.

Event tree analysis (ETA): A graphical logic model that identifies and attempts to quantify possible outcomes following an initiating and subsequent event.

Excursion: An unintentional occurrence, such as a discharge of pollutants above the permitted amount, due to reasons beyond human control.

Exhaust: A duct for the escape of gases, fumes, and odors from an enclosure.

Exothermic: A term used to describe a process or change that occurs with the production of heat.

Expansion joint: A joint installed in a structure to provide for changes (often in length) due to expansion or contraction resulting from changes in temperature, without distortion of the structure.

Exposure period: The duration of an exposure.

External event: A natural or man-made event; often an accident that may serve as an initiating event in a sequence of subsequent events in a process leading to a release of hazardous materials.

Extraction: The process of dissolving and separating out particular constituents of a solid or liquid using an immiscible solvent.

Extrapolation: An estimation of unknown values by extending or projecting from known values.

Failure frequency: The frequency (relative to time) of failure.

Failure mode: A symptom, condition, or manner in which a failure occurs.

Failure probability: The probability that failure will occur, usually in a given time interval.

Failure rate: The number of failures divided by the total elapsed time during which these failures occur.

Fallout: The radioactive debris or material that settles to the earth after a nuclear explosion.

Fatal accident rate (FAR): The estimated number of fatalities per 10^8 exposure hours (roughly 1,000 employee working lifetimes).

Fatigue: The incremental weakening of a material as a result of repeated cycles of stresses that are far lower than its breaking load; the end result is failure.

Fault: A fracture in the earth along which there has been displacement parallel to the fault plane; an error or failure in a process that can lead to a hazardous event.

Fault tree: A method for representing the logical combinations of adverse hazard events that lead to a particular outcome (top event).

Fault tree analysis (FTA): A logic model that identifies and attempts to quantify possible causes of a hazard event.

Federal Register: A daily publication of laws and regulations promulgated by the US Federal Government.

Feedforward control system: A system in which changes are detected at the process input and a corrective signal is applied before process output is affected.

Fire point: The lowest temperature at which a liquid evolves vapor fast enough to support continuous combustion; it is usually close to the flash point.

Fission: The splitting of atomic nuclei into smaller nuclei, accompanied by the release of significant quantities of energy; it is induced by bombardment with neutrons from an external source and continued by the neutrons that are released.

Definitions/Glossary of Chemical Process Terms

Fixed-bed operation: An operation in which the additive material (e.g., catalyst, absorbent, filter media) remains stationary in the chemical reactor.

Fixed carbon: The ash-free, carbonaceous material that remains after volatile matter is driven off a dry solid sample.

Flammability: The ease with which a material (gas, liquid, or solid) will ignite spontaneously (autoignition) from exposure to a high-temperature environment, or from a spark or open flame.

Flange: A projecting rim, edge, lip, or rib used to connect piping or conduit.

Flap valve: A valve that is hinged at one edge and that opens and shuts by rotating about the hinge.

Flare: A tall stack that is employed to burn small, discrete quantities of undesirable gases.

Flash distillation: A separation process in which an appreciable portion of a liquid is quickly converted to vapor by changing the temperature and/or pressure; this method is widely employed for the desalination of seawater.

Flash point: The minimum temperature at which a liquid or solid gives off enough vapor to form a flammable mixture with the air near the surface of the liquid or solid.

Flotation: A treatment process using air, injected into a waste stream, that adheres to the solids and causes them to rise, allowing for their removal at a tank surface; it takes advantage of the differences in specific gravities of the air-associated solids and waste liquid.

Flow diagram: A chart or line drawing employed by engineers to indicate successive steps in the production of a chemical or treatment of a waste stream; it includes materials input and output, by-products, wastes, and other relevant data.

Fluidization: A technique in which a finely divided solid is caused to behave like a fluid by suspending it in a moving gas or liquid.

Forced draft: The positive pressure created by the action of a fan or blower.

Free convection: The motion and mixing of a fluid caused solely by density differences within the fluid.

Frequency: Number of occurrences of an event per unit time.

Friction head: The pressure differential required to overcome the frictional resistance to flow.

Fugitive source: Any source of emissions that is not confined to an identifiable point of discharge.

Fusion: The process in which the nuclei of two light elements (chiefly the hydrogen isotopes) combine to form the nucleus of a heavier element with the release of substantial amounts of energy.

Gamma decay: A nuclear reaction in which the nucleus of an atom emits a pulse of energy in the form of gamma radiation.

Gate valve: A valve in which the closing element consists of a disk that slides vertically over the opening or cross-sectional area through which liquid passes; it is suitable for on-off control but not for throttling.

Gauge: An instrument for measuring and indicating such process variables as pressure, liquid level, thickness, volumetric flow, etc.

Gaussian model: A plume dispersion model of mixing and turbulence in the lower atmosphere based on the assumption that a pollutant concentration within the plume is normally distributed vertically when sampled perpendicular to the direction of flow.

Grab sample: A single sample that is collected at such a time and place so that it is ideally representative of a total discharge.

Gradient: The rate of change of a quantity or variable.

Guide vane: A fixed or adjustable device intended to direct fluid flow in a conduit channel.

Half-life: The time required for a chemical concentration or quantity to decrease by half its current value.

Hazard (problem): An event associated with an accident which has the potential for causing damage to people, property, or the environment.

Hazard and operability study (HAZOP): A technique to identify process hazards and potential operating problems using a series of guide words that key on process deviations.

Hazard risk assessment (HZRA): A technique associated with quantifying the risk of a hazard employing probability and consequence information.

Head: The height of the free surface of fluid above any point in a hydraulic system; it is a measure of the pressure or force exerted by a fluid.

Health (problem): A problem normally associated with and arising from the continuous emission of a chemical into the environment causing adverse health effects in humans.

Health risk assessment (HRA): A technique associated with quantifying the risk of a health problem employing toxicology (dose-response) and exposure information.

Hearth: The bottom of a furnace on which waste materials or fuels are exposed to a flame.

Heat of combustion: The heat liberated in a combustion reaction.

Heat of vaporization: The heat required to convert a liquid or solid to a vapor.

Heterogeneous: A term used to describe a mixture of different phases (e.g., liquid-vapor, liquid-vapor-solid, etc.).

Holdup: A volume of material held or contained in a process vessel or line.

Hood: A canopy or collecting structure over a process area or piece of equipment; it is employed to collect fumes, gases, vapors, or fine particulates which are then exhausted or treated in some manner.

Human error: Actions by engineers, operators, managers, etc., that may contribute to or result in accidents.

Human error probability: The ratio between the number of human errors and the number of opportunities for human error.

Human factors: Factors attempting to match human capacities and limitations.

Human reliability: A measure of human errors.

Humidification: A process for increasing the water content of air; it is usually incorporated into an air conditioning system.

Humidity: The measure of the amount of water vapor in the air at any given time.

Hybrid: An offspring of two organisms that differ in at least one gene, and any equipment that differs from another in at least one respect or accomplishes two or more objectives in a process.

Hydrogen cracking: The decomposition of petroleum or other hydrocarbons by heat to give lower-boiling materials such as gasoline, kerosene, other motor fuels, domestic fuel oil, and other products.

Hydroelectric power: The electricity produced by turbines capturing the energy of moving water.

In situ: A term used to describe any reaction or analysis occurring in place; and a term used to describe a fossil, mineral, or rock found in its original place of deposition, growth, or formation.

Incident: An event.

Incubation: Controlled environmental conditions (e.g., temperature and moisture) for the growth and development of microorganisms.

Individual risk: The risk to an individual from a hazardous chemical or event.

Induced draft: The negative pressure created by the action of a fan, blower, or other gas-moving device.

Inert: A term used to describe a material that is chemically inactive; they can be ingredients added to mixtures chiefly for bulk and weight purposes.

Ingestion: The intake of a chemical through the mouth.

Initiating event: The first event in an event sequence.

Instantaneous release: Emissions that occur over a very short duration.

Intermediate event: An event that propagates or mitigates the initiating event during an event sequence.

Ion: A charged molecule or atom that has gained or lost one or more electrons; the migration of an ion affects the transport of electricity through an electrolyte or, to a certain extent, through a gas.

Isobaric: A term used to describe a process that occurs at constant pressure.

Isokinetic sampling: A technique for collecting air pollutants, constructed so that the gas entering the sampling nozzle is at the same speed as the surrounding air.

Isopleth: A plot of uniform concentrations, usually downwind from a release source.

Isotherm: A line connecting point of equal temperature that is employed on climactic maps or in groups of thermodynamic relations.

Isotropic: A term used to describe anything that exhibits uniform physical properties throughout and in all directions.

Laminar flow: One of the two types of flow that occur in fluids (the other being turbulent) in which the fluid follows a smooth, well-defined path; this type of flow occurs at low Reynolds numbers.

Lapse rate: The rate at which temperature decreases in the atmosphere as height above the surface increases; the adiabatic lapse rate is 10°C/km.

Latent heat: The heat released or absorbed by a change of state.

Leaching: The process by which soluble material dissolves and is removed from a solid in a solution.

Leakage: An undesired and gradual escape or entry of a quantity.

Lethal concentration (LC): The concentration of a chemical that will kill test animals, usually based on 1- to 4-hr-exposure duration.

Lethal concentration 50 (LC_{50}): The concentration of a chemical that will kill 50% of test animals, usually based on 1- to 4-hr-exposure duration.

Lethal dose (LD): The quantity of a chemical that will kill a test animal, usually normalized to a unit of body weight.

Lethal dose 50 (LD_{50}): The quantity of a chemical that will kill 50% of test animals, usually normalized to a unit of body weight.

Level of concern (LOC): The concentration of a chemical above which there may be adverse human health effects.

Life cycle: The phases, changes, or stages that an organism passes through during its lifetime; the duration used to evaluate costs of a process based on its operational life prior to replacement.

Likelihood: A measure of the expected probability or frequency of the occurrence of an event.

Liner: A relatively impermeable barrier designed to prevent leachate from leaking from a landfill; and an insert or sleeve for pipes or conduits to prevent leakage or infiltration.

Liquid trap: The sumps, well cellars, and other traps employed in association with oil and gas production, gathering, and extraction operations for the purpose of collecting oil, water, and other liquids.

Liquid-level control: A control device used to monitor liquid level in a reactor or container designed to actuate to limit fluid entering the reactor or container to maintain fluid level at a set value.

Mach number: The ratio of the speed of an object to the speed of sound in the undisturbed medium through which the object is traveling; anything traveling supersonically has a Mach number greater than one.

Make-up solvent: The solvent introduced into a process that compensates for solvent lost from the process during operation.

Malfunction: A failure in the normal operation of a system or process.

Malignant: A cancerous tumor.

Manifest: A document employed for identifying the quantity, composition, origin, routing, and destination of material during transportation, treatment, storage or disposal.

Manifold: A pipe fitting with numerous branches to convey fluids among a large pipe and several smaller pipes or to permit the choice of diverting flow from one of the several sources or to one of the many discharge points.

Manometer: An instrument for measuring pressure that usually consists of a U-shaped tube containing a liquid.

Mathematical model: A mathematical representation of a real process, expressed as a set (one or more) of equations or algorithms.

Maximum individual risk: The highest individual risk in an exposed population subjected to a health or hazard risk.

Metastable: A term used to describe an intermediate state in between stability and instability.

Miscibility: The ability of a liquid or gas to dissolve uniformly in another liquid or gas.

Definitions/Glossary of Chemical Process Terms

Mist eliminator: A control device employed to remove water/liquid droplets entrained in a gas stream.

Mole fraction: A ratio employed in expressing concentrations of solutions and mixtures; the mole fraction of any component of a mixture is defined as the number of moles of that component divided by the total number of moles of the mixture.

Molecular sieve: A microporous structure composed of crystalline aluminosilicates that are chemically similar to clays and feldspars and belong to a class of materials known as zeolites.

Monomer: A molecule that is capable of conversion to polymers, synthetic resins, or elastomers by combination with itself or other similar molecules; it usually contains carbon and is of relatively low molecular weight.

Monte Carlo method: A method that constructs an artificial stochastic model of a process on which sampling experiments are performed; it usually involves the use of random numbers in its calculations.

Mutagen: A chemical capable of changing a living cell.

Natural draft: The negative pressure created by the height of a stack or chimney and/or the difference in temperature between flue gases and the atmosphere.

Neutral: A term used to describe a particle without an electric charge; and a term used to describe a solution that is neither acidic nor basic.

Noble: A term used to describe an element that is completely unreactive or reacts only to a limited extent with other elements.

Nonconservative: A term used to describe a substance that can react.

Normal boiling point: The boiling point when the ambient pressure is 1 atmosphere.

Nuclear reaction: A reaction that involves a change in the nucleus of an atom, as opposed to a chemical reaction in which only the electrons take part.

Nucleus: The small, positively charged central mass of an atom that contains essentially the entire mass of the atom in the form of protons and neutrons; the central portion of a living cell that controls its functions; and any small particle that can serve as the basis for crystal growth.

Off-gas: A gaseous product from the chemical, biological, or thermal decomposition of a material.

Optimization: The adjustment of a system or process to make it as effective or functional as possible.

Organic: A term used to describe anything that contains carbon and is thus derived from living organisms.

Orifice: An opening with a closed perimeter in a plate, wall, or partition, through which fluid may flow; it is generally employed for the purpose of measuring or controlling a fluid.

Osmosis: The passage of a pure liquid (usually water) through a semipermeable membrane from a solution of low concentration into a solution of a higher concentration (e.g., the flow of pure water into a solution of salt and water).

Oxidizing agent: Any material that removes hydrogen or electrons from or adds oxygen to an element or compound.

Packed tower: A device that forces gas through a tower packed with metal, plastic, ceramic, or crushed rock, while liquid is sprayed over the packing material.

Packing: A collar or gasket employed to seal mechanical devices to prevent leakage of liquid, and the inert material employed in absorption towers and distillation columns.

Parallel: A term used to describe two straight lines that are everywhere equidistant from each other, and the arrangement of two or more units that perform the same function and share a common input and a common output, e.g., two pumps operating in parallel.

Parameter: A quantitative or characteristic property that describes the physical, chemical, or biological conditions of a system.

Partial pressure: The pressure exerted by one of the several components of a gaseous mixture.

Partial vacuum: The description of a space condition in which the pressure is less than atmospheric.

Partial volume (pure component volume): The volume that would be occupied by one component of a gaseous mixture if it were alone and under the total pressure and temperature of the mixture.

Parts per billion (ppb): The fraction (ppb_m for mass fraction and ppb_v for volume fraction) multiplied by 10^9; it is a unit used to measure extremely small quantities of a substance.

Parts per million (ppm): The fraction (ppm_m for mass fraction and ppm_v for volume fraction) multiplied by 10^6; it is a unit used to measure extremely small quantities of a substance.

PEL: The acronym for Permissible Exposure Limit; the permissible exposure limit of a chemical in air, established by the Occupational Safety and Health Administration (OSHA).

Periodic: Phenomena that repeat after regular intervals of time.

Personal protection equipment (PPE): Material/equipment worn to protect a worker from exposure to hazardous agents.

Photon: A quantum of electromagnetic radiation with zero mass; it is the unit particle of light.

Physical model: The simulation of real process by a physical experiment which models the important features of the original process that are the object of study.

Pilot plant: A trial assembly of small-scale reactions and processing equipment.

Pitot tube: A device that is inserted into a pipe to measure the velocity of a flowing fluid.

Pneumatic: A term used to describe anything of or containing wind, air, or gases.

Polymerization: A chemical reaction, usually carried out with a catalyst, heat, and often under high pressure, in which a large number of relatively simple molecules combine to form chain-like macromolecules.

Pore: Any interstitial or void space in a solid material.

Porosity: The ratio of the volume of open space in a bed to the total bed volume.

Potable water: Any water that is safe for drinking.

Power plant: A unit constructed for the conversion of stored energy (usually in fossil fuels) into another desired form of energy.

Precision: The degree of "exactness" of repeated measurements.

Preheater: A unit employed to heat the air needed for combustion by absorbing heat from the hot flue gases.

Definitions/Glossary of Chemical Process Terms

Pressure drop: A measure of the difference in pressures measured immediately upstream and downstream of a unit or process.

Pressure relief valve: A valve that opens automatically to a sufficient area when the pressure reaches an assigned limit in order to relive the stress on a pipeline or vessel.

Primary pollutant: A pollutant emitted directly from a process stack.

Probability: An expression for the likelihood of occurrence of an event or an event sequence, usually over an interval of time.

Probe: A tube employed to measure pressures at a distance from the actual measuring equipment, and a general term used to describe a sampling port.

Process: A manufacturing procedure or the equipment employed in the chemical, pharmaceutical, food, metalworking, or other industry.

Process control: The manipulation of the conditions of a process to effect a desired change in the output characteristics of the process.

Process safety audits: An inspection of a plant or process unit, drawings, procedures, emergency plans, and/or management systems, etc., usually by an independent, impartial team.

Process Safety Management (PSM): A management system that is focused on prevention of, preparedness for, mitigation of, response to, and restoration from catastrophic releases of chemicals or energy from a process associated with a facility.

Process vent: Any open-ended pipe in a process or stack that is vented to the atmosphere.

Protective system: Systems, such as pressure vessel relief valves, that function to prevent or mitigate the occurrence of an accident or incident.

Protocol: The plan and procedures that are to be followed in conducting a test or project.

Proton: The positively charged fundamental unit of matter with about the same mass as a neutron; it is present in all atomic nuclei.

ppm: The parts per million of a chemical in air, water, or solid – almost always on a volume basis in air designated as ppm_v as opposed to on a mass basis in water and solid designated as ppm_m.

ppb: The parts per billion of a chemical in air, water, or solid – almost always on a volume basis in air, designated as ppb_v as opposed to on a mass basis in water and solid designated as ppm_m.

Psychrometric chart: A chart employed to determine the properties of moist air.

Purging: A cleansing or removal of impurities, foreign matter, or undesirable elements from a vessel or reactor.

Pyrolysis: The chemical decomposition of organic matter through the application of heat in the absence of oxygen.

Quality assurance/quality control (QA/QC): A system of procedures, checks, audits, and corrective actions to ensure that all research, design and performance, environmental monitoring and sampling, and other technical and reporting activities are of the highest achievable quality.

Quenching: A rapid cooling of metals or alloys by immersion in cold water or oil, or air by exposure to cold water.

Quiescent: A term used to describe a body at rest.
Radiation: The emission and propagation of energy through space or through a material medium, usually in the form of electromagnetic waves.
Radioactive: A term used to describe any substance that emits radiation either naturally or as a result of chemical manipulation.
Reboiler: An auxiliary heating unit for a distillation column, designed to supply additional heat to the lower portion of the column.
Recirculation: The repeated flow of a fluid around a closed system.
Rectification: The enrichment or purification of a vapor during a distillation process by contact or interaction with a countercurrent stream of liquid condensed from the vapor.
Reducing agent: Any material that adds hydrogen or electrons to an element or compound.
Refining: A separation process in which undesirable components are removed from various types of mixtures to yield a more purified product.
Reflux: The liquid that has condensed from a rising vapor and is allowed to flow back down a column in a distillation process.
Refractory: An inert, ceramic material with the ability to retain its physical shape and chemical identity when subjected to extremely high temperatures.
Refrigerant: A substance that is suitable as the working medium of a cycle of operations producing refrigeration (e.g., liquid ammonia, Freon).
Regression: A data analysis method used to measure the extent to which two variables increase or decrease together or the extent to which one increases as the other decreases.
Relief valve: A pressure relief device such as pressure valves, rupture disks, and other pressure relief systems employed to protect process components from possible explosions or other damage.
Retrofit: The addition or removal of a piece of equipment, or a required adjustment, connection, or disconnection of an existing piece of equipment, often for the purpose of reducing emissions or optimizing a process.
Reverse flow: A flow in a direction opposite to normal flow.
Reverse osmosis: A water treatment process employed to separate water from pollutants by the application of pressure to force the water through a semipermeable membrane against the osmotic pressure of the solution.
Reynolds number: A dimensionless number used in fluid flow calculations to describe the level of turbulence within a system.
Risk: A measure of economic loss or human injury in terms of both the incident likelihood and the magnitude of the loss or injury.
Risk analysis: The structured evaluation of incident consequences, frequencies, and risk assessment results.
Risk assessment: The process by which risk estimates are made.
Risk contour: Lines on a risk graph that connect points of equal risk.
Risk estimation: Combining the estimated consequences and likelihood of a risk.
Risk management: The application of management policies, procedures, and practices in analyzing, assessing, and controlling risk.

Definitions/Glossary of Chemical Process Terms 117

Risk perception: The perception of risk that is a function of age, race, sex, personal history and background, familiarity with the potential risk, dread factors, perceived benefits of the risk-causing action, marital status, residence, etc.

Rotameter: A device composed of a float inside a tapered glass tube employed for measuring fluid flow.

Rotary dryer: A long, steel cylinder, slowly rotating, with a slightly inclined axis, through which a solid or semi-solid material passes to be dried by hot air.

Rotary valve: A valve consisting of a casing, more or less spherical in shape, and a gate that turns through 90° when opening or closing; it has a cylindrical opening of the same diameter as that of the pipe it serves.

Rupture disk: A thin piece of metal between flanges that breaks at a certain pressure to prevent dangerous pressure buildup within a pressure vessel.

Safety data sheet (SDS): A compilation of information required under OSHA communication standard on the identity of hazardous chemicals, health and physical hazards, exposure limits, and precautions.

Safety valve: A valve that automatically opens when prescribed conditions, usually of pressure, are exceeded in a pipeline or other closed receptacle containing liquid and/or gases.

Salinity: The amount of salt in water.

Salting out: A reduction in the water solubility of a solid, liquid, or gas by adding a salt to an aqueous solution of the substance.

Saturated steam: Steam that is in equilibrium with liquid water at a given temperature or pressure.

Scale-up: The calculations and planning involved in increasing operations from the pilot plant stage to the large-scale production stage.

Scheduled maintenance: Any periodic procedure that is necessary to maintain the integrity or reliability of a system, which can be anticipated and scheduled in advance.

Scope of work: A document detailing program requirements.

Scrubber: A device that uses a liquid spray to remove particulates and gaseous components from an air stream; the gases are removed by absorption or a chemical reaction and the particulates are removed through contact and capture by the liquid droplets.

Sensible heat: The heat that, when added or removed, results in a change of temperature.

Serendipity: An unexpected scientific discovery that turns out to be more important than the project being researched.

Shutdown: The cessation of operation of an affected facility for any purpose.

Side reaction: A secondary reaction accompanying or following a primary reaction.

Sink: A receptacle for the materials moved through a system.

Site inspection: An on-site investigation to determine whether there is a release or potential release and the extent and severity of hazards posed by the release.

Slaking: The process of mixing with water so that a chemical combination occurs, as in slaking of lime.

Sludge: The thick, solid, or semisolid waste that accumulates as a result of the settling which occurs during water and wastewater treatment processes, especially sedimentation.

Slurry: A mixture of solid matter and liquid.

Smelter: A facility that melts or fuses ore.

Societal risk: A measure of risk to a group of individuals.

Solute: The substance that is dissolved in a solvent to form a solution.

Solution: A homogeneous mixture of two or more substances constituting a single phase.

Source term: The estimation of the mass or volumetric release of a (hazardous) agent from a specific source.

Space velocity: The volume of gas or liquid, measured at a specified temperature and pressure, passing through a unit volume in a certain unit time.

Sparger: A perforated pipe through which steam, air, water, or other fluid is injected into a liquid during a reaction, e.g., during fermentation.

Spent: A term used to describe any material that has been used and, as a result of contamination, can no longer serve the purpose for which it was produced without further processing.

Stand-alone system: An independent system or process.

State: One of the three forms that matter can take: solid, liquid, and gas.

Static head: The pressure in a fluid due to the height of fluid above a given point.

Steady flow: A flow that does not vary with time; the mass flow rate is constant and all other quantities (e.g., temperature, pressure, velocity) are independent of time.

Steam drum: A vessel in a boiler in which the saturated steam is separated from the steam-water mixture and into which the feedwater is introduced.

Stoichiometric: A term used to describe the elements of a compound in exactly the proportion represented by the compound's chemical formula; and a term used to describe the minimum amount of a chemical necessary to completely react with another reactant.

Stoker: A mechanical device employed to feed solid fuel or solid waste to a furnace.

Stop valve: A valve installed in a pipeline to shut off flow for the purpose of inspection or repair.

Stripper: A column in which one or more components of a liquid stream are removed by being transferred into a gas stream.

Sump: A pit or tank that catches liquid runoff for drainage or disposal.

Supercooled liquid: A liquid cooled below its normal freezing point without solidification.

Superficial mass velocity: The quantity obtained when the mass rate of flow is divided by the total cross-sectional area, regardless of the presence of any obstruction to flow.

Superheated steam: Steam at a temperature above its boiling point.

Synergism: The cooperative interaction of two or more chemicals or other phenomena producing a greater total effect than the sum of their individual effects.

Synthesis: The combination of parts to form a whole; and the creation of a substance that either duplicates a natural product or is a unique material not found in nature.

Definitions/Glossary of Chemical Process Terms

Systemic: A term used to describe something that affects the entire system or body.

Tailings: The residues of raw materials or waste separated out during the processing of mineral ores.

Tee: A pipe fitting that has the outlets at right angles to the inlet.

Theoretical air: The quantity of air, calculated from the chemical composition of a waste, that is required to burn/combust a waste/feedstock completely so that no oxygen remains.

Throat velocity: The gas or liquid velocity through the throat of a contraction.

Throttle valve: A valve designed to control the rate of fluid flow.

Time of failure: The time when a duty or intended function associated with a component or entire system is no longer able to be performed.

TLV: The acronym for the Threshold Limit Value (established by the American Council of Government Industrial Hygienists, ACGIH). The concentration of a chemical in air that may be breathed in without harmful effects for five consecutive 8-hr working days.

TLV-C: The ceiling exposure limit representing the maximum concentration of a chemical in air that should never be exceeded in any part of the working exposure.

TLV-STEL: The acronym for Short-Term Exposure Limit, a 15-minute, time-weighted average concentration to which workers may be exposed up to four times per day with at least 60 minutes between successive exposures with no ill effect if the TLV-TWA is not exceeded.

TLV-TWA: The allowable time weighted average concentration of a chemical in air for an 8-hr workday/40-hr work week that produces no adverse health effects in exposed individuals.

Top event: The accident, event, or incident at the "top" of a fault tree that is traced downward to more basic failures using logic gates to determine their causes.

Toxic dose: The combination of concentration and exposure duration for a toxic agent to produce a specific harmful effect.

Transient: A term used to describe anything that changes with time.

Triple point: The temperature and pressure at which all three states (i.e., solid, liquid, gas) of a substance exist together.

Tsunami: A sea wave caused by an underwater seismic disturbance such as sudden faulting, land sliding, or volcanic activity.

Turbine: A machine that converts the energy in a stream of fluid into mechanical energy by passing the stream through a system of fan-like blades, causing them to rotate.

Turbulence: The fluid property that is characterized by irregular variations in the speed and direction of the movement of individual particles or elements of flow.

UEL/UFL: The upper explosive/flammability limit, the highest concentration of a chemical in air that will produce an explosion or flame if ignited.

Uncertainty: A measure, often quantitative, of the degree of doubt or lack of certainty associated with an estimate.

Unit operation: One of many operations employed in chemical engineering in the industrial production of various chemicals in which a physical change and not a chemical change takes place.

Unit process: One of many operations employed in chemical engineering in the industrial production of various chemicals in which chemical change and not physical change take place.

Unstable: A term used to describe a chemical that tends to move toward decomposition or other unwanted chemical change during normal handling or storage.

Unsteady flow: A flow that is characterized by a mass flow rate and/or other quantities that vary with time.

Upset: A disturbance in the functioning, fulfillment, or completion of a process or material.

Upstream: The direction from which a stream is flowing.

Uptake: The act of taking up, drawing up, or absorbing.

Useful life: The period during which a piece of equipment or a process operates as per their intended function.

Valve: A device for controlling the flow of a fluid through a pipe or tube.

Vane: A device that pivots on a rooftop or other elevated location for the purpose of determining the wind direction.

Vent: An opening through which a fluid or gas is ejected from a system.

Venturi scrubber: A unit using a liquid, usually water, to remove particulate and gaseous pollutants from the air.

Viscosity: The internal resistance to flow exhibited by a fluid; it is the ratio of shearing stress to rate of shear.

Watershed: The area surrounding a stream that supplies it with water runoff.

Weir box: A box installed in a narrow open channel upstream from a weir to provide an enlargement of the flow area; the velocity of approach of the weir is consequently reduced.

Wet bulb: A thermometer bulb maintained wet with distilled water; it is employed with the dry bulb to measure of humidity.

Zeolite: Any of the hydrated aluminum complex silicates, either natural or synthetic, that possesses cation exchange properties.

6.3 ILLUSTRATIVE EXAMPLES

Eight illustrative examples complement the material presented above regarding terms and definitions related to health and hazard risk in the CPI.

6.3.1 ILLUSTRATIVE EXAMPLE 1

Define failure and reliability in laymen terms and note the difference between the two.

Solution: Failure represents an inability to perform some required function while *reliability* is the probability that a system or one of its components will perform its intended function under certain conditions for a specified period. The reliability of a system and its probability of failure are complementary in the sense that the sum of

Definitions/Glossary of Chemical Process Terms 121

these two probabilities is unity. The basic concepts and theorems of probability find application in the estimation of both failure and reliability.

6.3.2 Illustrative Example 2

From an environmental perspective, list and define several guidelines that should be followed when selecting a site for a plant.

Solution: The following guidelines should be followed when selecting a site for a plant:

1. *Topography*. A fairly level site is needed to contain spills and prevent spills from migrating and creating more of a hazard. Firm soil above water level is recommended.
2. *Utilities and water supply*. The water supply must be adequate for fire protection and cooling. The sources for electricity should be reliable to prevent unplanned shutdowns.
3. *Roadways*. Roadways should allow access to the site by emergency vehicles such as ambulances and fire engines in the event of an emergency.
4. *Neighboring communities and plants*. Population density and proximity to the plant should be considered for the initial site and in anticipation of a possible future expansion.
5. *Waste disposal*. Waste disposal systems containing flammable, corrosive, or toxic materials should be a minimum distance of 250 ft from plant equipment.
6. *Climate and natural hazards*. Lighting arrestors should be installed to reduce/eliminate ignition sources in flammable areas. Storm drainage systems should be maintained.
7. *Emergency services*. Emergency services should be readily available, well trained, and appropriately equipped.
8. Air and water quality standards. The location and operation of the facility should be consistent with efforts to maintain or protect air and water resources.

6.3.3 Illustrative Example 3

Do chemical companies and other industries keep track of employees' illnesses and deaths?

Solution: Many large companies do. Some companies keep extremely detailed medical records on employees. Larger companies may have an epidemiologist who studies the incidence of disease in their workers. Smaller companies may not have as detailed medical records on employees as larger companies, and they may not keep them over an extended period of time.

6.3.4 Illustrative Example 4

Convert gaseous concentration units to parts per million by volume (ppm_v) from mg/m^3 at a standard temperature and pressure of 0°C and 1.0 atm.

Solution: Set C as the concentration in mg/m³ at 0°C and 1.0 atm. Apply the appropriate conversion factors:

$$\text{ppm}_v = C\left(\text{mg/m}^3\right)\left(1\,\text{m}^3/10^3\,\text{L}\right)\left(22.4\,\text{L/gmol}\right)\left(\text{g}/10^3\,\text{mg}\right)\left(\text{gmol}/MW\,(\text{g/gmol})\right) \quad (6.1)$$

where MW = molecular weight of the compound of interest. Therefore,

$$\text{ppm}_v = C\left(22.4 \times 10^{-6}\right)(MW) \quad (6.2)$$

Thus, one can carry out the required conversion only if the molecular weight (MW) of the gas is known.

6.3.5 ILLUSTRATIVE EXAMPLE 5

Provide a comment – in laymen terms – of the difference between engineers and scientists.

Solution: Simply put, scientists generally discover things while engineers make them work.

6.3.6 ILLUSTRATIVE EXAMPLE 6

Describe the difference between bench-scale tests and pilot plant studies.

Solution: The former are tests conducted in a laboratory while pilot plant studies are conducted at a scale larger than bench-scale in order to establish full-scale design requirements.

6.3.7 ILLUSTRATIVE EXAMPLE 7

Define the difference between a homogeneous reaction and a heterogeneous reaction.

Solution: The former are reactions that occur uniformly throughout a fluid so that the rate of reaction at any point within the fluid is the same. The latter reactions occur between one or more constituents that exist as different phases.

6.3.8 ILLUSTRATIVE EXAMPLE 8

Define the difference between reaction rate and reaction order.

Solution: Reaction rate describes the rate of a chemical reaction. Reaction order is used to describe the relationship between reaction rate and concentrations of reactants in a given reaction. For example, a zero-order reaction is described by reactions unaffected by reactant concentrations. A first-order reaction is described by overall mass reaction rates that are proportional to reactant concentrations, i.e., the rate of reaction increases with an increase in reactant concentration and decreases with a decrease in reactant concentration.

PROBLEMS
6.1 Describe the difference between a degraded failure, a revealed failure, and a catastrophic failure.
6.2 Is exposure to a virus or toxin a health problem or a hazard problem?
6.3 Explain the term: Failure Modes, Effects, and Criticality Analysis (FMECA).
6.4 Discuss the significance of Process Safety Audits (PSAs) and Process Safety Management (PSM) in chemical process start-up activities.

REFERENCES
Shaefer, S., and Theodore, L. 2007. *Probability and Statistics for Environmental Science.* Boca Raton, FL: CRC Press/Taylor & Francis Group.

Theodore, L., and Dupont, R.R. 2012. *Environmental Health Risk and Hazard Risk Assessment: Principles and Calculations.* Boca Raton, FL: CRC Press/Taylor & Francis Group.

Theodore, L., Reynolds, J., and Morris, K. 1997. *Dictionary of Concise Environmental Terms.* Amsterdam, The Netherlands: Gordon and Breach Science Publishers.

Theodore, M.K., and Theodore, L. 2021. *Introduction to Environmental Management*, 2nd Edition. Boca Raton, FL: CRC Press/Taylor & Francis Group.

7 History

7.1 INTRODUCTION

This second chapter of Part II is primarily concerned with the history of the chemical process industry (CPI). Before proceeding this industry needs to be clearly described. However, any definition or description of the CPI is bound to be incomplete. Most processes in the chemical industry involve a chemical change where the term "chemical change" should be interpreted to include not only chemical reactions but also physical changes such as the separation and purification of the components of a mixture. Purely mechanical changes are usually not considered part of the chemical process unless they are essential to later chemical changes. For example, the manufacture of polyethylene, using ethylene produced from petroleum or natural gas, involves a chemical process. On the other hand, the molding and fabrication of the resulting plastic resin into final shapes for consumer products would not be considered part of the chemical process. Thus, a satisfactory definition of a true CPI is an industry whose principal products are manufactured based upon chemical and physical principles. The entire CPI is large and accounts for approximately 15% of the US annual gross national product. It is also dynamic and constantly expanding.

This chapter initially presents the early history of the CPI, discusses the role of science in the chemical industry of today, and provides a brief history of engineering and the sources used by engineers for information critical to the design, operation, and maintenance of systems in the CPI. Four illustrative examples are provided to demonstrate concepts presented throughout the chapter.

7.2 EARLY HISTORY

The early history of the chemical industry dates to prehistoric times when humans first attempted to control and modify their environment. The work was largely empirical, with little understanding of basic chemistry. The industry developed as any other trade or craft. Since those who practiced the craft were generally not of the literate or learned class, few of the remaining reports of chemical processes were written by people who were actually engaged in the craft. More adequate records are available from the beginning of the 16th century onward. With no knowledge of chemical science and no means of chemical analysis, early chemical craftsman had to rely on both previous art and superstition. Progress was slow and confusion reigned. For example, in the 16th century "oil of vitriol" was made from blue or green vitriol (now known as cupric or ferrous sulfate). "Oil of sulfur" was made by burning sulfur. The latter was initially much more expensive than the former, even though they were the same compound, i.e., sulfuric acid.

Probably the oldest CPI is fermentation, although in its earliest times it was more a folkcraft than an industry. Fermentation was known to the most primitive man, possibly because of the ease of accidental discovery. Ale is the oldest fermented liquor, dating to the Egyptians before 3,000 BCE. Distillation began in the 1st-century AD, and by the 13th century, the distillation of fermented liquors to concentrate alcohol was commonplace in Europe.

Recovery and use of metals began before 4,000 BCE. The first metals found were gold and silver, and since they commonly occur in metallic form were easily recognized. The first reported metal to be extracted from its ore was copper in Egypt and Mesopotamia in approximately 3,500 BCE. Other early metals include tin, lead, zinc, and iron. Methods of smelting were developed by the 16th century. Early iron production involved the heating of the oxide ore with charcoal to reduce it. Evidence of forced drafts has been found in ruins of Roman furnaces.

Glass, cements, and ceramics were also known in early times. The earliest cement was wet clay, Egyptians used gypsum mortar, and the Greeks and Romans used lime mortar. Little basic improvement was made until the discovery of Portland cement in the 19th century. Ceramics began with baked-clay bricks and pottery as early as 5,000 BCE. Many early civilizations used glazed and colored ceramics. It is believed that porcelain was developed in the 8th or 9th century in China.

Vinegar (dilute acetic acid) was the earliest known acid since it was formed from the oxidation of fermented liquors. No historical references are made to mineral acids until the 14th century. At this time nitric acid was made from saltpeter (KNO_3) and ferrous sulfate by heating the mixture and condensing the distilled nitric acid. The first industrial application of nitric acid was in the separation of gold from silver in the 16th century. Industrial use of sulfuric acid developed later and became important only in the late 18th century when it was used to produce chlorine for bleaching cloth. Hydrochloric acid, another important industrial acid, was discovered in the 17th century. In the early 19th century a process for making soda ash (Na_2CO_3) also produced large quantities of hydrochloric acid, and it became the cheapest and most widely used mineral acid of the period.

Early cleansing agents included many natural materials, such as the alkali found in wood ashes. True soap was first mentioned in the 1st century. It was probably made by boiling ashes with a natural oil or fat. The chemistry of soapmaking was not understood until the early 19th century.

7.3 THE ROLE OF SCIENCE

Progress and growth of the chemical industry during its earliest history was slow because there was little understanding of the scientific principles underlying the various processes. The alchemists who might have contributed to an understanding of the processes were preoccupied with attempting to convert base metals into gold and in attempting to find the elixir of life. Alchemists were also superstitious and mysterious, and often fraudulent and deceitful.

In late 18th-century Europe, interest in the scientific principles of chemistry rapidly increased and great progress was made. Oxygen was discovered around 1770. At the same time the role of oxygen in combustion was clarified, thereby refuting the popular phlogiston theory, the scientific theory that postulated the existence of a fire-like element called phlogiston contained within combustible bodies and released during combustion. This work in oxidation and other chemical fields did much to place chemistry on a sound quantitative footing. It recognized that organic compounds contained carbon and hydrogen and developed methods for their determination. In the early 1800s the Englishman John Dalton developed a workable atomic theory that helped explain how elements combined into molecules, thereby making it possible to consider chemical reactions on a quantitative basis.

This increased understanding of chemical science led to improvements and new developments in chemical processing. The principal chemical industries of the early 19th century manufactured alkalies, acids, and metals.

The organic chemical industry began to develop after 1850. Although many natural products had been purified for centuries, the lack of understanding of organic chemistry prevented the synthesis of organic compounds. Cellulose was treated with nitric acid to produce nitrocellulose explosives. Rubber was introduced in the early 19th century and was first vulcanized in the 1840s. Ether and chloroform were made on a small scale for use as anesthetics. The birth of synthetic organic chemistry brought the synthesis of urea from inorganic ammonium cyanate. This also led to the development of the aniline dye industry, the first large-scale synthetic organic chemical industry in history. The synthetic drug industry developed in Germany after 1880. Attempts were made to synthesize a drug which would have action similar to the natural product quinine in reducing fever. Although early attempts to synthesize quinine were unsuccessful, a similar acting compound kairine was ultimately synthesized at the time.

Germany almost completely monopolized the synthetic drug and dye industries until the First World War when the US chemical industry became a major force. The American chemical industry developed slowly in colonial days, and somewhat more rapidly after the Revolution. Industry in the United States grew with development in European industry, but America remained dependent on many European chemical manufactures until the First World War.

A few chemical products, such as wood tar, pot-ashes, bricks, and glass, were exported from the Jamestown Colony in the early 1600s. American industry was to some extent restricted by British regulations. After the Revolution, US industry expanded in an attempt to become independent of British manufactures. Sulfuric acid was first produced in America in 1793. In 1802 a young refugee of the French revolution Eleuthère Irénée duPont de Nemours, who had learned to make gunpowder, established a powder works in Delaware. As was the case in Europe, the American chemical industry concentrated on processing natural products, such as sugar, textile fibers, and coal, on metals, and on a few inorganic chemicals, such as alkalies, acids, and bleaching agents.

Most basic chemistry was of European origin until the 20th century. The first thermosetting plastic, a phenol-formaldehyde polymer called Bakelite, was developed around 1900. This was the beginning of the world plastic industry. The impetus for a self-sufficient American chemical industry came in the First World War, when the United States was cut off from German chemicals. At the time, Germany was the undisputed world leader in the chemical industry. After the First World War, the US chemical industry expanded rapidly until it came to dominate in the global CPI arena.

7.4 THE MODERN CHEMICAL PROCESS INDUSTRY

The chemical industry in the United States today is a sprawling complex of raw material sources, manufacturing plants and distribution facilities which supplies the country and the world with thousands of chemical products, most of which were unknown 150 years ago. As discussed above, no clear limit can be used to define the CPI, and within the industry it is not possible to classify completely the various industries. This section considers the major companies in the CPI over the past 150 years. Some of these leading chemical companies in the United States are listed in Table 7.1.

The scientific explosion in more recent years has had an enormous influence on society. Processes were developed for synthesizing completely new substances that were either better than the natural ones or could replace them more cheaply. As the complexity of synthesized compounds increased, entirely new products appeared. Plastics and new textiles were developed, energy use increased, and new drugs conquered whole classes of disease.

The progress brought forth by both engineers and scientists in recent years has been spectacular, although the benefits of this progress have included corresponding liabilities. The most obvious risks have come from nuclear weapons and radioactive materials, with their potential for producing cancer(s) in exposed individuals and

TABLE 7.1
Major Chemical Companies in the United States (1875–2020)

Company Name
Air Reduction Company
Allied Chemical Corporation
American Cyanamid Company
American Potash & Chemical Corporation
Atlas Powder Company
Columbian Carbon Company
Commercial Solvents Corporation
Diamond Alkali Company
Dow Chemical Company
E. I. du Pont de Nemours & Company
W.R. Grace and Company
Hercules Powder Company
Hooker Chemical Corporation
Monsanto Chemical Corporation
Olin Mathieson Chemical Corporation
Reinchold Chemicals, Inc.
Rohm and Haas Company
Stauffer Chemical Company
Thiokol Chemical Corporation
Union Carbide Corporation
Witco Chemical Company
Wyandotte Chemical Corporation

TABLE 7.2
Common Chemical Conversion Reactions

Acylation	Dehydrogenation	Neutralization
Alcoholysis	Electrolysis	Nitration
Alkylation	Esterification	Oxidation
Aromatization or cyclization	Fermentation	Polymerization
Calcination	Halogenation	Pyrolysis
Carboxylation	Hydrogenation	Reduction
Combustion	Hydrolysis	Sulfonation
Condensation	Ion exchange	
Dehydration	Isomerization	

mutations in their children. In addition, certain pesticides have potentially damaging effects. This led to the emergence of a new industry that is now defined as *environmental engineering*. Mitigating these negative effects is one of the challenges that the science community continues to strive to meet.

The chemical industry brings numerous new products to the marketplace each year. There are several dozen chemical conversion "reactions" that are an integral part of all of these processes. A partial list of these chemical conversions is provided in Table 7.2.

7.5 THE HISTORY OF ENGINEERING

In the terms of the history of engineering, the engineering profession as defined today is usually considered to have originated shortly after 1800. However, many of the "processes" in the CPI were developed in antiquity. As noted above, filtration operations were carried out approximately 5,000 years ago by the Egyptians. Operations such as crystallization, precipitation, and distillation soon followed. Others evolved from a mixture of craft, mysticism, incorrect theories, and empirical guesses during this period.

In the latter half of the 19th century an increased demand arose for individuals trained in the fundamentals of the CPI. This demand was ultimately met by engineers. The technical advances of the 19th century greatly broadened the field of engineering and introduced a large number of engineering specialties. The rapidly changing demands of the socioeconomic environment in the 20th and 21st centuries have widened the scope even further.

A related field of engineering, health, safety, and accident management received wide attention in the late 1970s and 1980s when the safety of nuclear reactors was questioned following accidents caused by operator errors, design failures, and malfunctioning equipment. "Human factors" engineering seeks to establish criteria for the efficient, human-centered design of, among other things, the large, complicated control panels that monitor and govern nuclear reactor operations (Theodore and McGuinn 1992).

The first attempt to organize the principles of chemical processing and to clarify the professional area of chemical engineering was made in England by George E. Davis, who organized a Society of Chemical Engineers in 1880 and presented a series of lectures in 1887 which were later expanded and published in 1901 as *A Handbook of Chemical Engineering*. In 1888, the first course in chemical engineering in the United States was organized at the Massachusetts Institute of Technology (MIT) by Lewis M. Norton, a professor of industrial chemistry. The course applied aspects of chemistry and mechanical engineering to chemical processes.

Chemical engineering began to gain professional acceptance in the early years of the 20th century. The American Chemical Society (ACS) was founded in 1876 and, in 1908, organized a Division of Industrial Chemists and Chemical Engineers while authorizing the publication of the *Journal of Industrial and Engineering Chemistry*. A group of prominent chemical engineers also met in 1908 in Philadelphia and founded the American Institute of Chemical Engineers (AIChE). The mold for what is now called *chemical engineering* was fashioned at the 1922 meeting of the AIChE. A timeline of the history of chemical engineering between the profession's founding and the present day is provided in Theodore, Dupont and Ganesan (2017).

7.6 SOURCES OF INFORMATION FOR THE CPI

The fields engineering and science encompass many diverse fields of activity. It is therefore important that means be available to place widely scattered information in the hands of those practicing in the field. One need only examine the wide variety of problems used in these practices in order to appreciate the need for the ready availability of a vast variety of information. Such problems are almost always approached by checking on and reviewing all sources of information. Most of this information can be classified in the following source categories:

1. Traditional
2. Engineering and science
3. Internet
4. Personal experience

Each topic is briefly discussed below.

7.6.1 TRADITIONAL SOURCES

The library is the major repository of traditional information. To efficiently use the library, one must become familiar with its classification system. In general, environmental engineers and scientists using the technical section of a library seek information on a certain subject and not material by a particular author in a specific book. For this reason, classification systems are based on subject matter. The question confronting the user then is how to find and/or obtain information on a particular subject. Two basic subject matter related classification systems are employed in US libraries: the Dewey Decimal system and the Library of Congress system. These systems have changed somewhat with the advent of computers and the Internet. A library (particularly a technical one) usually contains all of the following:

1. General books (fiction, etc.)
2. Reference books

History 131

 3. Handbooks
 4. Journals
 5. Transactions
 6. Encyclopedias
 7. Periodicals
 8. Dictionaries
 9. Trade literature
10. Catalogs

7.6.2 Engineering and Science Sources

Engineering and science technical literature generally includes textbooks, handbooks, periodicals, magazines, journals, dictionaries, encyclopedias, and industrial catalogs. There are a host of textbooks, including some of the classic works. The three major engineering handbooks include those by Theodore (2014), Green and Perry (2019), Kirk and Othmer (2001), and Albright (2008). Journals and magazines are abundant in the literature.

7.6.3 Internet Sources

The advent of the Internet has significantly changed the means to access technical information. The Internet serves as the electronic backbone that connects almost all computers, storage devices, servers, websites, cellular devices, etc., throughout the world. In addition to information, the Internet is regularly used for online shopping, map searches, accessing libraries and databases, forums, communication, employment searching, etc.

Virtually every university, business, government (state, local, national), and organization has its own website, which can provide plentiful sources of information. For instance, federal agencies such as the US Environmental Protection Agency and the Department of Energy have comprehensive information available. Mapping details can be provided by a number of sources, from Google Earth to Geographic Information Systems where available.

Comprehensive search engines make navigating the Internet for information significantly simpler. Common search engines include Google, Yahoo!, Bing, etc. Such search engines also have more in-depth search engines, such as Google Scholar, which provides a web search for technical and research articles. In addition, websites designed to provide information may also have a search engine embedded within them, in order to create targeted searches within their databases (although some may require payment or membership access).

The advent of cloud storage systems and online servers (such as Drop Box and Google Drive) allows for users to share files anywhere in the world and even update documents in real time. The idea of sharing information and constant updating have opened numerous possibilities for large-scale concerted efforts and projects, one that receives special mention is Wikipedia. Derived from the phrase "What I Know Is …", "Wikipedia is an online site that provides free and open information. It is a collaborative encyclopedia where one can obtain or apply information to a host of topics.

Note that sites such as Wikipedia themselves are not generally advised to be used as scholarly citations in technical papers due to their nature as crowdsourced information hubs subject to rapid changes or bias. However, this rarely detracts from their accuracy in engineering applications, and they are extremely helpful as encyclopedic information for personal use and review of information. In addition, the sources and sites that they list are normally of sufficient quality for reference.

7.6.4 Personal Experience

Another information source, and a frequently underrated one, is personal experience, personal files, and company experience files. Although the traditional engineering and science and Internet sources described above are important, other fields or the particular field with which the reader is directly involved can also provide useful information. And, though engineering is generally known for being technically up-to-date on new processes, in truth most equipment designs, process changes, and new-plant designs in the CPIs are based on either prior experience, company files, or both.

7.7 ILLUSTRATIVE EXAMPLES

Four illustrative examples complement the material presented above regarding the history of the CPI.

7.7.1 Illustrative Example 1

Provide a layman's definition of engineering.

Solution: One solution for this comes from the *Simple English Wikipedia* (Wikipedia contributors 2021) and is as follows.

Engineering is the use of science and math to design or make things and solve technical problems. People who do engineering are called engineers. They learn engineering at a college or university. There are different types of engineers that design everything from computers and buildings to watches and websites. People have been engineering things for thousands of years.

7.7.2 Illustrative Example 2

Provide a technical definition of environmental engineering.

Solution: The following definition and description of environmental engineering is adapted from Nathanson (2020).

Environmental engineering is charged with the development of processes and infrastructure for the supply of water, the disposal of waste, and the control of pollution of all kinds. These endeavors protect public health by preventing disease transmission, and they preserve the quality of the environment by averting contamination and degradation of air, water, and land resources.

Environmental engineering draws on such disciplines as chemistry, ecology, geology, hydraulics, hydrology, microbiology, economics, and mathematics. It was traditionally a specialized field within civil engineering and was called sanitary engineering until the mid-1960s, when the more accurate name environmental engineering was adopted.

Projects in environmental engineering involve the treatment and distribution of drinking water; the collection, treatment, and disposal of wastewater; the control of air and noise pollution; municipal solid waste and hazardous waste management; the cleanup of hazardous waste sites; and the preparation of environmental assessments, audits, and impact studies. Mathematical modeling and computer analysis are widely used to evaluate and design the systems required for such tasks. Chemical and mechanical engineers may also be involved in environmental engineering projects. Environmental engineering functions include applied research and teaching; project planning and management; the design, construction, and operation of facilities; the sale and marketing of environmental control equipment; and the enforcement of environmental standards and regulations.

The education of environmental engineers usually involves graduate level course work, though some colleges and universities allow undergraduates to specialize or take elective courses in the environmental field. Programs offering associate (two-year) degrees are available for training environmental technicians. In the public sector, environmental engineers are employed by national and regional environmental agencies, local health departments, and municipal engineering and public works departments. In the private sector, they are employed by consulting engineering firms, construction contractors, water and sewerage utility companies, and manufacturing industries.

7.7.3 Illustrative Example 3

Describe what is meant by natural science.

Solution: Science in general can be defined as the systematized knowledge derived from and tested by recognition and formulation of a problem, and collection of data through observation and experimentation. Social science deals with the study of people and how they live together as families, tribes, communities, and nations. Natural science is concerned with the study of nature and its impact on the physical world. Natural science includes such diverse disciplines as biology, chemistry, geology, and physics.

7.7.4 Illustrative Example 4

Describe what is meant by environmental science.

Solution: Although the disciplines as biology, chemistry, mathematics, and physics focus on a particular aspect of natural science, environmental science encompasses all of the fields of natural science. The historical focus of environmental scientists has been, of course, the natural environment which includes the atmosphere, the land, the water and their inhabitants, and the impact of the "built" environment on these natural systems.

PROBLEMS

7.1 Describe the difference between environmental engineering and chemical engineering.
7.2 Interview a practicing engineer and request their description of the profession and what their field of engineering dies.
7.3 Comment on the various traditional sources of information available for the practicing engineer.
7.4 Search the internet and report on the careers of at least three major players in the CPI field.

REFERENCES

Albright, L. 2008. *Albright's Chemical Engineering Handbook*. Boca Raton, FL: CRC Press/Taylor & Francis Group.

Green, D., and Perry, R. (editors). 2019. *Perry's Chemical Engineers' Handbook*, 9th Edition. New York, NY; McGraw-Hill.

Kirk, R.E., and Othmer, D.F. 2001. *Encyclopedia of Chemical Technology*, 4th Edition. Hoboken, NJ: John Wiley & Sons.

Nathanson, J.A. 2020. Environmental engineering. *Encyclopedia Britannica*. https://www.britannica.com/technology/environmental-engineering (accessed September 3, 2021).

Theodore, L. 2014. *Chemical Engineering: The Essential Reference*. New York, NY: McGraw-Hill.

Theodore, L., Dupont, R.R., and Ganesan, K. 2017. *Unit Operations in Environmental Engineering*. Beverly, MA: Scrivener-Wiley.

Theodore, L., and McGuinn, Y. 1992. *Health, Safety, and Accident Management; Industrial Applications*. Theodore Tutorials, originally published by U.S. EPA/APTI, RTP, NC. East Williston, NY.

Wikipedia contributors. *Engineering, Wikipedia, The Free Encyclopedia*, https://simple.wikipedia.org/w/index.php?title=Engineering&oldid=7741487 (accessed September 3, 2021).

8 Chemical Process Equipment

8.1 INTRODUCTION

This chapter provides details on a number of commonly used process units: reactors, heat exchangers, columns of various types (distillation, absorption, adsorption, evaporation, and extraction), dryers, and grinders. As one would expect, the treatment of topics in this chapter is somewhat superficial. Details of other process unit operations are available in the literature cited later in the chapter.

It is fair to say that the chemical process industry (CPI) is involved with the development and application of chemical or certain physical changes. In the course of the development of the CPI, many of the physical processes involved in a chemical process were, since about 1900, called *unit operations*. This classification was instituted in the early years by a group of chemical engineering professors at Massachusetts Institute of Technology headed by H. Walker. The name "unit operations," however, was suggested by A. D. Little, who had been an industrial partner of Walker's. For 30 or more years, the concept of unit operations was a tremendous stimulus, because it classified and unified these physical changes. This resulted in filtration, heat transfer, and many other changes to be studied individually, and the findings in application to one industry were then applied to many other such operations in other industries. Shortly after 1930 a similar classification was introduced called *unit processes*. These generally pertain to chemical changes, just as unit operations pertain to physical changes.

This chapter presents information on chemical reactors, heat exchangers, mass transfer equipment, and fluid flow equipment, all essential in the CPI. Additional information and discussion is provided on ancillary equipment, material transportation and storage equipment, instrumentation and controls, and flowsheets necessary for the proper design, operation, and maintenance of CPI systems. Five illustrative examples are provided to demonstrate concepts presented throughout the chapter.

8.2 CHEMICAL REACTORS

8.2.1 Reactor Definition

The reactor is often the heart of a chemical process (Theodore 2012). It is the place in the process where raw materials are usually converted into products, and reactor design is therefore a vital step in the overall design of a process.

The treatment of reactors in this section is restricted to a discussion of the appropriate reactor types for a specific process. The design of an industrial chemical reactor must satisfy requirements in the following four main areas:

1. *Chemical factors.* These involve mainly the kinetics of the reaction. The design must provide sufficient residence time for the desired reaction to proceed to the required degree of conversion.
2. *Mass transfer.* The reaction rate of heterogeneous reactions may be controlled by the rates of diffusion of the reacting species, rather than the chemical kinetics.
3. *Heat transfer factors.* These involve the removal, or addition, of the heat of reaction.
4. *Safety factors.* These involve the confinement of any hazardous reactants and products, as well as the control of the reaction and the process conditions.

The need to satisfy these interrelated, and often contradictory, factors makes reactor design a complex and difficult task. However, in many instances, one of the factors predominates, hence determining the choice of reactor type and the design method for the overall process or system being evaluated. More details of reactor selection are provided in Theodore (2012).

8.2.2 Reactor Type

The characteristics normally used to classify reactor designs are the following:

1. Mode of operation, that is, batch or continuous
2. Phases present, that is, homogeneous or heterogeneous
3. Reactor geometry, that is, flow pattern and manner of contacting the phases

The five major classes of reactors are as follows:

1. Batch reactor
2. Stirred tank reactor
3. Tubular reactor
4. Packed bed (fixed) reactor
5. Fluidized bed reactor

In a batch process, all the reagents are added at the beginning of the reaction, the reaction proceeds, and the compositions change within the reactor over time. The reaction is stopped and the product is withdrawn when the required conversion has been reached. Batch processes are suitable for small-scale production and for processes that use the same equipment to make a range of different products or grades. Examples include pigments, dyestuffs, pharmaceuticals, and polymers.

In continuous processes, the reactants are fed to the reactor and the products are withdrawn continuously, and the reactor is usually operated under *steady-state* conditions. Continuous production normally entails lower production costs than batch production does but lacks the flexibility of batch production. Continuous stirred tank reactors (CSTRs) are usually selected for large-scale production. Processes that do not fit the definition of batch or continuous are often referred to as semicontinuous and semibatch processes. In a semibatch reactor, some of the reactants may be added to, or some of the products withdrawn from, the batch as the reaction proceeds. A semicontinuous process is basically a continuous process that is interrupted periodically, for instance, for the regeneration of catalyst (Coulson, Richardson and Skinnott 1983; Theodore 2012).

Homogeneous reactions are those in which the reactants, products, and any catalysts used form one continuous phase, gaseous or liquid. Homogeneous gas phase reactors are almost always operated continuously, whereas liquid phase reactors may be batch or continuous. Tubular (pipeline) reactors are normally used for homogeneous gas phase reactions. Both tubular and stirred tank reactors are used for homogeneous liquid phase reactions.

In a heterogeneous reaction, two or more phases exist, and the overriding problem in reactor design is to promote mass transfer between/among the phases. The possible combinations of phases are as follows:

1. Liquid–liquid, with immiscible liquid phases.
2. Liquid–solid, with one or more liquid phases in contact with a solid; the solid may be a reactant or catalyst.
3. Liquid–solid-gas, where the solid is normally a catalyst.
4. Gas–solid, where the solid may take part in the reaction or act as a catalyst.

The reactors used for established processes are usually unique and complex designs that have been developed over a period of years to suit the requirements of a particular process. However, it is convenient to classify flow reactors into broad categories, discussed below (Coulson, Richardson and Skinnott 1983; Theodore 2012).

A stirred tank (agitated) reactor consists of a tank fitted with a mechanical agitator and (usually) a cooling jacket or coils. These are operated as batch or continuous reactors. Several reactors may be used in series. The stirred tank reactor can be considered the basic chemical reactor, modeling on a large scale a conventional laboratory flask. Tank sizes range from a few liters to several thousand liters. This equipment is used for homogeneous and heterogeneous liquid–liquid and liquid–gas reactions and for reactions that involve finely suspended solids that are held in suspension by agitation. Since the degree of agitation is under the designer's control, stirred tank reactors are particularly suitable for reactions that require high mass transfer or heat transfer efficiencies. When operated as a continuous process, the composition in the reactor is constant and the same as the product stream. Except for very rapid reactions, this limits the conversion that can be obtained in a one-stage stirred tank reactor.

Tubular reactors are not only generally used for gaseous reactions but are also suitable for some liquid phase reactions. If high heat transfer rates are required, small diameter tubes are used to increase the ratio of surface area to volume within the reactor. Several tubes may be arranged in parallel, connected to a manifold, or fitted into a tube sheet in an arrangement similar to a shell and tube heat exchanger (see the following section, "Heat Exchangers"). For high-temperature reactions, the tubes may be placed in a furnace.

There are two basic types of packed bed reactors: those in which the solid is a reactant and those in which the solid is a catalyst. Many examples of the first type can be found in the extractive metallurgical industries. In the CPI, the designer normally employs the second type, that is, catalytic reactors. Industrial packed bed catalytic reactors range in size from units with small tubes, a few centimeters in diameter, to large diameter packed beds. Packed bed reactors are used for gas and gas–liquid reactions. Heat transfer rates in large diameter packed beds are poor, and where high transfer rates are required, fluidized beds should be considered.

The essential feature of a fluidized bed reactor is that the solids are held in suspension by the upward flow of the reacting fluid. This promotes high mass and heat transfer rates and good mixing. Heat transfer coefficients in the order of 200 W/m-°C for jackets and internal coils are typically obtained. The solids may be a catalyst, a reactant (in some fluidized combustion processes), or an inert powder added to promote heat transfer. Although the principal advantage of a fluidized bed over a fixed bed is the higher heat transfer rate, fluidized beds are also useful when large quantities of solids must be transported as part of the reaction processes.

Operational factors that contribute to waste and potentially hazardous and toxic emissions in chemical reactors include incomplete conversion resulting from inadequate temperature control, by-product formation resulting from inadequate mixing, and catalyst deactivation resulting from poor feed control or poor reactant purity control. Wastes are often generated in a chemical reactor if the reactor is improperly designed or the catalyst selection is not correct.

8.3 HEAT EXCHANGERS

Heat transfer is often only one of several unit operations involved in a process, and the interrelations among the different unit operations involved must be understood. Heat transfer requires that there be a driving force present for heat to flow. This driving force is the temperature difference between a warmer and colder body (Theodore 2011).

Practically all of the operations that are carried out in industry involve the production or absorption of energy in the form of heat. The laws governing the transfer of heat and the types of equipment that consider heat flow as one of their main objectives are therefore of great importance. Heat may flow by one or more of the following three basic mechanisms:

1. *Conduction.* When heat flows through a body by the transference of the momentum of individual atoms or molecules *without* mixing, it is said to flow by conduction. For example, the flow of heat through the wall of a furnace or the metal shell of a boiler takes place by conduction.
2. *Convection.* When heat flows by actual mixing of warmer portions with cooler portions of the same material, the heat transfer mechanism is known as convection. Convection is restricted to the flow of heat in fluids. Rarely does heat flow through fluids by pure conduction without some convection resulting from the eddies caused by the changes of density of the fluid with

Chemical Process Equipment

changes in temperature. For that reason, the terms *conduction* and *convection* are often used together, although in many cases, heat transfer is primarily by convection. For example, the heating of water flowing by a hot surface is an example of heat transfer due primarily to convection. Heat transfer by convection due to density differences is defined as *natural convection*.

3. *Radiation.* A body emits radiant energy in all directions. If this energy strikes a receiver, a portion of it may be transmitted, a portion may be reflected, and a portion may be absorbed. It is this absorbed portion that represents heat transfer in the form of radiation, that is, the transfer of energy through space by means of electromagnetic waves. Radiation passing through empty space is not transformed to heat or any other form of energy. If, however, matter appears in its path, the radiation will be transmitted, reflected, or absorbed by that matter. For example, fused quartz transmits practically all the radiation that strikes it. A polished opaque surface or mirror will reflect most of the radiation impinging on it, while a black, dull, or matted surface will absorb most of the radiation it receives. The absorbed energy will be transformed quantitatively into heat. Note that it is only the absorbed energy that appears as heat.

The transfer of heat to and from process fluids is an essential part of most chemical processes. The CPI uses four principal types of heat exchangers.

1. *Double-pipe exchanger* – the simplest type, used for cooling and heating
2. *Shell and tube exchangers* – used for all applications
3. *Plate and frame exchangers (plate heat exchangers)* – used for heating and cooling
4. *Direct or contact exchangers* – used for cooling and quenching

The word *exchanger* applies to all types of equipment in which heat is exchanged but is often used specifically to denote equipment in which heat is transferred between two process streams. An exchanger in which a process fluid is heated or cooled by a plant service stream is referred to as a heater or cooler.

One of the simplest and cheapest types of heat exchangers is the concentric pipe arrangement known as the *double-pipe heat exchanger.* Such equipment can be made from standard fittings and is useful where only a small heat transfer area is required. Several units can be connected in series to extend their capacity.

The *shell and tube exchanger* is by far the most commonly used type of heat transfer equipment in the chemical and allied products industries. The advantages of this type of heat transfer device include the following:

1. Large surface area in a small volume
2. Good mechanical layout, that is, good shape for pressure operations
3. Reliance on well-established fabrication techniques
4. Wide range of construction materials available
5. Easily cleaned equipment
6. Well-established design procedures

Essentially, shell and tube exchangers consist of a bundle of tubes enclosed in a cylindrical shell. The ends of the tubes are fitted into tube sheets, which separate the shell-side and tube-side fluids. Baffles are provided in the shell to direct the fluid flow and increase heat transfer.

In *direct contact heat exchange*, there is no wall to separate hot and cold streams, and high rates of heat transfer are achieved. Applications include reactor off-gas quenching, vacuum condensers, de-superheating, and humidification. Water cooling towers are particularly common examples of direct contact heat exchangers. In direct contact cooler condensers, the condensed liquid is frequently used as the coolant.

Use of direct contact heat exchangers should be considered whenever the process stream and coolant are compatible. The equipment is simple and cheap and is suitable for use with heavily fouling fluids and with liquids containing solids. Spray chambers, spray columns, and plate and packed columns are used for direct contact heat exchanger reactors.

Heat exchangers contribute to waste generation by the presence of cling (process side) or scale (cleaning side). This can be corrected by designing for lower film temperature and high turbulence.

8.4 MASS TRANSFER EQUIPMENT

There are numerous types of mass transfer equipment employed in the manufacturing industry (Theodore and Ricci 2011). Some of the more common pieces of equipment are presented here. The reader is referred to the literature (Theodore and Ricci 2011; McCabe, Smith and Harriot 2014; Green and Southard 2019) for additional details on not only the mass transfer equipment listed here but also equipment not covered in this section.

8.4.1 DISTILLATION

Distillation is probably the most widely used separation process in the chemical industry. Its applications range from the rectification of alcohol, which has been practiced since antiquity, to the fractionation of crude oil. The separation of liquid mixtures by distillation is based on differences in volatility among components. The greater the relative volatilities, the easier the separation. Vapor flows up a column and liquid flows counter currently down a column. The vapor and liquid are brought into contact on plates or inert packing material. A portion of the condensate from the condenser is returned to the top of the column to provide liquid flow above the feed point (reflux), and a portion of the liquid from the base of the column is vaporized in the reboiler and returned to provide the vapor flow.

In the stripping section, which lies below the feed, the more volatile components are stripped from the liquid. Above the feed, in the enrichment or rectifying section, the concentration of the more volatile components is increased. Figure 8.1a shows a *distillation column* producing two product streams, referred to as tops and bottoms, from a single feed. These columns are occasionally used with more than one feed, and with side streams withdrawn at points throughout the column (Figure 8.1b). This does not alter the basic operation, but it does complicate the analysis of the process

Chemical Process Equipment

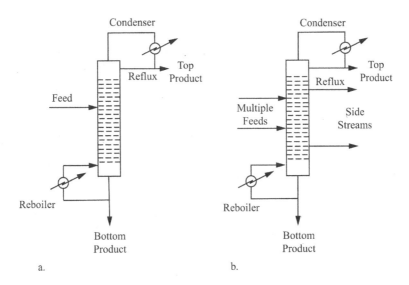

FIGURE 8.1 Common distillation column configurations. (a) Single feed column and (b) multi-feed column.

to some extent. If the process requirement is to strip a volatile component from a relatively nonvolatile solvent, the rectifying section may be omitted, and the column is then called a *stripping column*.

In some operations where the top product required is a vapor, the liquid condensed is sufficient only to provide the reflux flow to the column, and the condenser is referred to as a *partial condenser*. When the liquid is totally condensed, the liquid returned to the column will have the same composition as the top product. In a partial condenser, the reflux will be in equilibrium with the vapor leaving the condenser. Virtually pure top and bottom products can be achieved by using multiple distillation stages or, sometimes, additional columns.

8.4.2 Adsorption

In the *adsorption* process, one or more components in a mixture are preferentially removed from the mixture by a solid (referred to as the adsorbent) (Theodore 2008). Adsorption is influenced by the surface area of the adsorbent, the nature of the compound being adsorbed, the pressure of the operating system, and the temperature of operation. These are important parameters to be aware of when designing or evaluating an adsorption process since the possibility for an explosion or fire exists in the adsorption column.

The adsorption process is normally performed in a column. The column is run as either a packed or fluidized bed reactor. The adsorbent, after it has reached the end of its useful life, can be either discarded or regenerated. This operation can be applied to either a gas mixture or a liquid mixture. The reader is directed to the literature for further information on the adsorption process (Theodore and Ricci 2011; McCabe, Smith and Harriot 2014; Green and Southard 2019).

8.4.3 Absorption

The process of *absorption* conventionally refers to the intimate contacting of a mixture of gases with a liquid so that part of one or more of the constituents of the gas will dissolve in the liquid (Theodore 2008). The contact usually takes place in some type of packed column.

Packed columns are used for the continuous contact between liquid and gas. The countercurrent packed column is the most common type of unit encountered in gaseous pollutant control for the removal of the undesirable gases, vapors, or odors. This type of column has found widespread application in the chemical industry. The gas stream moves upward through the packed bed against an absorbing or reacting liquid that is injected at the top of the packing. This results in the highest possible efficiency. Since the concentration in the gas stream decreases as it rises through the column, there is constantly fresher liquid available for contact. This provides a maximum average driving force for the diffusion process throughout the bed. The reader is directed to the literature for further information on the adsorption process (Theodore and Ricci 2011; McCabe, Smith and Harriot 2014; Green and Southard 2019).

8.4.4 Evaporation

The processing industry has given the operations involving heat transfer to a boiling liquid the general name *evaporation*. The most common application of evaporation is the removal of water from a processing stream. Evaporation is used in the food, chemical, and petrochemical industries, and it usually results in an increase in the concentration of a solution until it forms a thickened slurry or syrup. Applications of evaporation for material processing include sugar slurry thickening, the concentration of dispersed kaolin clay, desalination, and bentonite clay dewatering. The factors that affect the evaporation process are concentration in the liquid, compound solubility, system pressure and temperature, scaling, and materials of construction.

An *evaporator* is a type of heat transfer device designed to induce boiling and evaporation of a liquid. The major types of evaporators include the following:

1. Open kettle or pan evaporators
2. Horizontal tube natural convection evaporators
3. Vertical-tube natural convection evaporators
4. Forced-convection evaporators

The efficiency of an evaporator can be increased by operating the equipment in a single effect or multi-effect mode. Triple-effect evaporators are implemented in kaolin clay processing. In this process, the steam utilized for evaporating the solution liquid is passed by the slurry three times before exiting the system. This approach decreases the volume or pressure of steam required to meet the target evaporation rates, thus reducing energy and materials used in the process.

8.4.5 EXTRACTION

Extraction (sometimes called *leaching*) encompasses liquid–liquid as well as liquid–solid systems. Liquid–liquid extraction involves the transfer of solutes from one liquid phase into another liquid solvent. It is normally conducted in mixer-settlers, plate and agitated tower contracting equipment, or packed or spray towers. Liquid–solid extraction, in which a liquid solvent is passed over a solid phase to remove some solute, is carried out in fixed-bed, moving-bed, or agitated-solid columns.

8.4.6 DRYING

Drying involves the removal of relatively small amounts of water or organic liquids from a solid phase. This can be contrasted to evaporation processes that remove large amounts of water from starting liquid solutions. In many applications, such as in corn processing, drying equipment follows an evaporation step to provide an ultrahigh solid content final product stream.

Drying, in either a batch or continuous process, removes liquid as a vapor by passing warm gas (usually air) over or indirectly heating the solid phase. The drying process is carried out in one of four basic dryer types. The first type is a *continuous tunnel dryer*. In a continuous dryer, trays with wet solids are moved through an enclosed system and warm air is blown over the trays. Similar in concept to the continuous tunnel dryer, *rotary dryers* consist of an inclined rotating hollow cylinder. The wet solids are fed in one side and hot air is usually passed counter currently over the wet solids. The dried solids then pass out the opposite side of the dryer unit. *Indirect drying* equipment heats either a drum or paddle surface that is in contact with the wet solids. The solids are fed across the outside of the hot transfer surface, dried, and discharged continuously. The final type of dryer is a *spray dryer*. In spray dryers a liquid or slurry is sprayed through a nozzle, and fine droplets are dried by a hot gas passed either concurrently, counter currently, or concurrently/counter currently past the falling droplets. Despite the differing methods of heat transfer, the continuous tunnel, indirect, rotary, and spray dryers can all reduce the moisture content of solids to less than 0.01% when designed and operated properly.

8.5 FLUID FLOW EQUIPMENT

The discussion in this section begins with a consideration of devices for conveying gases and liquids to, from, or between units of process equipment. This equipment can be a significant source of materials loss in an industrial facility, and analysis of such equipment often leads to the discovery of significant pollution prevention and risk reduction opportunities. Some of the devices discussed are simply conduits for the movement of material (e.g., pipes, ducts, fittings, and stacks); others control the flow of material (e.g., dampers and valves); still others provide the mechanical driving force for flow (e.g., fans, pumps, and compressors).

Pipes, ducts, fittings, pump selection, valves, fans, compressors, and so on generally are all part of a chemical process. Selection of proper equipment in the construction and design phase of a conveyance system is therefore important. Some of the more important conveyance-related equipment is discussed here.

8.5.1 Pipes and Tubing

The most common conduits for fluids are *pipes* and *tubing*. Both usually have circular cross sections, but pipes tend to have larger diameters and thicker walls than does tubing. Because of their heavier walls, pipes can be threaded, whereas tubing cannot. Process systems usually handle large flow rates that require the larger diameters associated with pipes.

Tubing and pipes are manufactured from a large variety of materials. The initial selection of the material for piping primarily depends on the compatibility of the fluid with the piping material in terms of corrosion and the normal system operating pressure. If special piping is required to accommodate corrosive liquids or high standards of purity, stainless steel, nickel alloys, or materials of high resistance to heat and mechanical damage are used. Steel pipe can be lined with tin, plastic, rubber, lead, cement, or other coatings for special purposes. If corrosion problems or contamination are controlling factors, the use of a nonmetallic pipe such as glass, porcelain, thermosetting plastic, or hard rubber is often acceptable.

There are several techniques used to join pipe sections. For small pipes, threaded connectors are the most common; for larger pipes (typically above 2½ in nominal), flanged fittings, or welded connections, are normally employed.

8.5.2 Ducts

While pipes and tubing are used as conduits for the conveyance of liquids or gases, *ducts* are used only for gases. Pipes, with their thicker walls, can be used for flows at higher pressures whereas ducts are relatively thin walled (1/16 in max.) and are generally employed for gas flow pressures below 15 psig. Pipes are usually circular in cross section. Ducts come in many shapes (circular, oval, rectangular, etc.). In general, ducts are much larger in cross section than pipes are because the gases typically transported have low densities and require high volumetric flow rates. Ducts are often constructed of field fabricated galvanized sheet steel, although other materials such as fibrous glass board, factory fabricated round fiberglass, spiral sheet metal, and flexible duct materials are becoming increasingly popular. Other duct construction materials include black steel, aluminum, stainless steel, plastic, plastic coated steel, cement, asbestos, and copper. All duct work can be fitted with jacketed insulation or sprayed with a fibrous insulation to minimize any heat transfer occurring between the process gas flow and the outside environment. The duct can also be fabricated with acoustical insulation to reduce the noise that may be generated from high gas velocities in the duct.

Duct fittings are similar to pipe fittings in that they allow connections to be made between duct sections and enable flow to be diverted where needed. Typical duct fittings include long and short radius elbows of all degrees, transitions, reducers, "Ts," and "Ys." Joining duct work of a relatively small cross-sectional area, typically below 500 in^2 (e.g., 32 in × 16 in nominal), requires only an inert sealing compound and external restraining guides and clamps (commonly referred to as "slips" and "slides").

Chemical Process Equipment 145

The preferred sealing compound for low temperature (below 120°F) and/or noncorrosive gas transport applications is (room-temperature-vulcanizing) RTV silicon. Those gas transport systems that require higher temperatures and/or corrosion resistance should use an elastomer for joint sealer. Joining ducts of larger sizes requires the use of thicker sheet metal to fabricate a "flanged" surface as part of the duct. The flanged surfaces of ducts that are joined are typically sealed with an inert gasket material appropriate for the intended service (e.g., red rubber below 120°F or "Viton" below 350°F) and securely fastened to each other.

In order to properly support a system of ducts and prevent seam leaks, all four sides of the duct work should be creased and at least one support should be located at or near every connection (typically every 6–8 ft).

Dampers act as flow control valves for gas flow. They are typically actuated by air piston positioners controlled by another system variable or a system "mode" selector switch. Economizers act as regenerative heat exchangers for a duct system. Typically, a hot process flow line or a waste gas (e.g., furnace flue discharge) is placed on the "shell" side, and a cooler gas that requires preheating (e.g., furnace combustion air intake) is passed through the "tube" side of the economizer to take advantage of available low grade heat.

8.5.3 FITTINGS

A fitting is a piece of equipment that has one or more of the following functions:

1. The joining of two pieces of straight pipe (e.g., couplings and unions)
2. The changing of pipeline diameter (e.g., reducers and bushings)
3. The changing of pipeline direction (e.g., elbows)
4. The changing of pipeline direction and diameter (e.g., reducer elbows and street elbows)
5. The splitting of a stream into multiple streams (e.g., Ts, Ys, and manifolds)
6. The joining of multiple streams (e.g., Ts, Ys, and manifolds)
7. The mixing of multiple streams (e.g., blender)

A *coupling* is a short piece of pipe whose ends are threaded on the inside (some plastics are not) used to connect straight sections of pipe. A *union* is also used to connect two straight sections but differs from a coupling in that it can be opened conveniently without disturbing the rest of the pipeline, whereas when a coupling is opened, a considerable amount of piping must usually be dismantled. A *reducer* is a coupling for two pipe sections of different diameters. A *bushing* also connects pipes, and/or fittings, of different diameters, but unlike the reducer, it is threaded on the outside of one end and on the inside of the other. This allows a larger pipe to screw onto the outside and the smaller pipe to screw into the inside of the bushing. An *elbow* is an angled coupling or bushing used to change flow direction, usually by 30°, 45°, or 90°. An elbow that is an angled bushing is referred to as a "street" elbow.

A "T" is also used to change flow direction and/or allow future system add-ons but is more often used to split one stream (supply) into two and later recombine two streams into one (return). A "Y" is similar to the T in that it is used to split a stream and/or later combine two streams. A *blender* is a fitting that introduces one stream of a liquid into another to achieve a homogenous mixture of the two constituents. Multiple T and Y sections can be combined to form manifolds of various configurations to split and/or combine multiple process streams as necessary for a given process.

8.5.4 Valves

Valves have one main function in a pipeline, that is, to control and/or redirect the amount of flow passing through various sections of system piping. There are many different types of valves, but the two most commonly used are the *gate valve* and the *globe valve*. The gate valve contains a disk that slides at right angles to the flow direction. This type of valve is used primarily for on–off control of liquid flows. Typically, 70% of system flow will occur in the first 30% of valve stem travel because small adjustments in disk travel cause extreme changes in the flow cross-sectional area. Therefore, this type of valve is not suitable for adjusting flow rates.

Unlike the gate valve, the globe valve is designed for flow control. Liquid passes through a globe valve via a somewhat torturous route. In one form, the globe valve seal is a horizontal ring into which a plug with a beveled edge is inserted when the valve is shut. Better control of flow is achieved with this type of valve because stem movement results in relatively small changes in cross-sectional flow area. However, the pressure drop across a fully open globe valve is greater than that in a comparable gate valve.

Other available valve types include the following: *check valves*, which permit flow in one direction only; *butterfly valves*, which operate in a damper-like fashion by rotating a flat plate to either a parallel or a perpendicular position relative to the flow; *plug valves*, in which a rotating tapered plug provides on–off service; *needle valves*, a variation of the globe valve, which give improved flow control; *diaphragm valves*, specially designed to handle very viscous liquids, slurries, or corrosive liquids that might clog the moving parts of other valves; and *ball valves*, valves that provide easy on–off service in addition to good flow control with only small pressure drops when the valves are fully open.

All these valves may be obtained with a packing gland, bellows seal, or packing-less sealing system. Valves that pass a relatively nontoxic fluid, such as water/steam or nitrogen, should have a packing gland. Valves that are used to control the flow of highly toxic materials should be purchased with a bellows seal so that any packing leak that may occur will be contained and diverted away from plant personnel into a collection area. Systems using fluids that can easily leak through conventional packing systems and cause a safety concern, such as fine oils or hydrogen gas, should be provided with packing free valves.

8.5.5 Fans and Blowers

The terms *fan* and *blower* are often used interchangeably, and only one major distinction is made between them in the discussion that follows. Fans are used for low-pressure drop operation, generally below 2 psig. Fans are usually classified as either the centrifugal or the axial-flow type. In centrifugal fans, a gas is introduced into the

Chemical Process Equipment

center of the revolving wheel (the eye) and is discharged at angles from the rotating blades. In axial-flow fans, the gas moves directly through the axis of rotation of the fan blades. Both types of fans are used widely throughout industry.

Blowers are generally employed when pressure heads in the range of 2–15 psig are required. Blowers are only offered in axial-flow configurations and may require more than one stage to boost system pressures. However, if operation at higher pressures is required, large-volume centrifugal or reciprocating compressors may be required.

8.5.6 Pumps

Pumps may be classified as *reciprocating, rotary,* or *centrifugal.* The first two are referred to as *positive-displacement pumps* because, unlike the centrifugal type, the liquid or semiliquid flow is divided into small portions as it passes through these pumps.

Reciprocating pumps operate by the direct action of a piston on the liquid contained in a cylinder within the pump. As the liquid is compressed by the piston, the higher pressure forces it through discharge valves to the pump outlet. As the piston retracts, the next batch of low-pressure liquid is drawn into the cylinder and the cycle is repeated.

The rotary pump combines rotation of the liquid with positive displacement. The rotating elements mesh with elements of the stationary casing in much the same way that two gears mesh. As the rotation elements come together, a pocket is created that first enlarges, drawing in liquid from the inlet or suction line. As rotation continues, the pocket of liquid is trapped, reduced in volume, and then forced into the discharge line at a higher pressure. The flow rate of liquid from a rotary pump is a function of the pump size and speed of rotation and is slightly dependent on the discharge pressure. Unlike reciprocating pumps, rotary pumps deliver nearly constant flow rates. Rotary pumps are used on liquids of almost any viscosity if the liquids do not contain abrasive solids.

Centrifugal pumps are widely used in the process industry because of their simplicity of design, low initial cost, low maintenance, and flexibility of application. Centrifugal pumps have been built to move as little flow as a few gallons per minute against a pressure of several hundred pounds per square inch. In its simplest form, this type of pump consists of an impeller rotating within a casing. Fluid enters the pump near the center of the rotating impeller and is thrown outward by centrifugal force. The kinetic energy of the fluid increases from the center of the impeller to the tips of the impeller vanes. This high velocity is converted to a high pressure as the fast-moving fluid leaves the impeller and is driven into slower moving fluid on the discharge side of the pump.

8.5.7 Compressors

Compressors operate in a manner like that of pumps and have the same classification, that is, rotary, reciprocating, and centrifugal. Compressors are used for gas conveyance and processing, and an obvious difference between compressors and pumps is the large decrease in volume resulting from the compression of a gaseous stream as compared with the negligible change in volume caused by the pumping of a liquid stream.

Centrifugal compressors are employed when large volumes of gases are to be handled at low- to-moderate pressure increases (0.5–50 psig). *Rotary compressors* have smaller capacities than centrifugal compressors do but can achieve discharge pressures up to 100 psig. *Reciprocating compressors* are the most common type of compressors used in industry and are capable of compressing small gas flows to as much as 3,500 psig. With specially designed compressors, discharge pressures as high as 25,000 psig can be reached, but these devices are capable of handling only very small gas volumes and do not work well for all gases.

8.5.8 STACKS

Gases are discharged into the ambient atmosphere by means of *stacks* (referred to as chimneys in some industries) of several types. *Stub or short stacks* are usually fabricated of steel and extend a minimum distance up from the discharge of an induced draft fan. These are constructed of steel plate, either unlined or refractory lined, or entirely of refractory and structural brick. *Tall stacks*, which are constructed of the same materials as short stacks, provide a driving force (draft) of greater pressure difference than that resulting from the shorter stacks. In addition, tall stacks ensure more effective dispersion of gaseous and particulate effluent into the atmosphere. Some chemical and utility applications use metal stacks that are made of a double wall with an air space between the metal sheets. The insulating air pocket created by the double wall prevents condensation on the inside of the stack, thus avoiding corrosion of the metal.

8.6 ANCILLARY EQUIPMENT

Today, the word *utilities* generally designates the ancillary services needed in the operation of any production process. These services are normally supplied from a central location on-site or from off-site utility providers and usually include the following:

1. Electricity
2. Steam for process heating
3. Water for cooling, potable use, and steam production
4. Refrigeration
5. Compressed air
6. Inert gas supplies

The production and use of these utility services can have a significant impact on the energy demand of a given process or facility, and the audit of energy demands is an important part of a complete efficiency assessment of a chemical process.

8.6.1 ELECTRICITY

The power required for processes (motor drives, lighting, and general use) may be generated on-site, but more often, it is purchased from a local utility. Evaluation of the efficiency of motors, lighting, and so on is an important part of a process assessment, as less power demand can often be equated to less overall pollutant generation by a facility. In addition, reduced power costs by reducing both average and peak power demand provide direct monetary benefits to the facility in reduced utility bills, providing quantitative incentives for implementing energy reduction options.

8.6.2 STEAM

The steam for process heating is generated in either fire- or water-tube boilers, using the most economical fuel available. The process temperatures required can usually be obtained with low-pressure steam (typically 25 psig), with higher steam pressures needed only for high process temperature requirements. Energy and cost reduction opportunities may exist in a facility's steam generation system, as many facilities operate at higher than necessary steam pressures, increasing their energy demands for steam production and the cost of producing this steam. In addition, repair and replacement of leaking steam and condenser lines will prevent the wasting of steam and the associated energy needed to produce this wasted steam.

8.6.3 WATER

A range of water needs typically exist within a facility, all of which may have different water quality demands for specific uses. Significant opportunities exist for the reuse of water in a facility as many liquid streams can be reused for multiple purposes, that is, using cooling water blowdown for part rinsing or boiler blowdown for cooling water, and so on. Water quality needs for specific unit operations within a facility should be clearly defined so that these reuse opportunities can be effectively evaluated.

8.6.3.1 Cooling Water

Natural and forced-draft cooling towers are generally used to provide the cooling water required at a site unless water can be drawn from a convenient river or lake in sufficient quantity. Seawater to brackish water can be used at coastal sites but, if used directly, necessitates more expensive materials of construction for heat exchangers because of potential corrosion problems resulting from the high dissolved solid content of this water. Often, cooling water does not have to be of high purity and can be taken from blowdown streams from boilers or process lines that require much more stringent water quality conditions.

8.6.3.2 Potable and General Use Water

The water required for general purposes on a site is usually taken from a local municipal or private supplier, unless a cheaper source of suitable quality water (e.g., a river, lake, or well) is available. If the cost of this supply is low, incentives for water use reduction will also be low. It should be remembered, however, that the cost of the supply may be a small portion of the overall cost of managing this water, particularly if it is used in process rinsing, general facility cleaning and wash-down, and so on. More water use requires more energy to convey it from the source and within a plant. When the water becomes contaminated, costs for its handling increase exponentially. In addition, waste treatment becomes progressively less efficient and more costly as a waste stream is diluted, and load penalties may be incurred based on the volume of waste discharged to publicly owned treatment works (POTWs). It should be evident, then, that numerous benefits can be associated with increased water use efficiency within a plant.

8.6.3.3 Demineralized Water

Water from which all the minerals have been removed by ion exchange must be used where ultrapure water is needed to meet process demands and strict boiler feedwater requirements. Mixed and multiple-bed ion exchange units are used for this purpose, with resins exchanging multivalent cations for hydrogen. Boiler water as condensed steam and process water must be removed on a routine basis (blow down) to prevent the build-up of unwanted constituents within these systems, and this ultrapure water may be effectively reused within a plant for other unit operations that demand much less stringent water quality characteristics, that is, cooling water, process rinse water, and so on.

8.6.4 REFRIGERATION

Refrigeration is needed for processes that require temperatures below those that can be economically obtained with cooling water. Chilled water can be used to lower process temperatures down to approximately 10°C. For lower temperatures, to –30°C, salt brines (NaCl and $CaCl_2$) are used to distribute the "refrigeration" around the site from a central refrigeration unit. Vapor compression units are normally used for this purpose. As with boilers, evaluation of the operating conditions of a chiller can lead to significant pollution prevention opportunities. Consider, for example, that a 1% improvement in chiller efficiency can be expected for each 1°F increase in the chiller site setpoint (Sprague 1999).

8.6.5 COMPRESSED AIR

Compressed air is needed for general use and for pneumatic controllers that usually serve as chemical process plant controllers. Air is often distributed at a pressure of 100 psig. Rotary and reciprocating single-stage or two-stage compressors are normally used to generate compressed air within a facility. Instrument air must be dry (–20°F dew point) and clean (free from oil). Compressed air also represents potentially significant cost, energy saving, and pollution prevention opportunities as the production of compressed air is highly inefficient (≈90% of the energy used is converted into heat, while only 10% is used to compress the air) (Sprague 1999). Significant improvements in the production of compressed air can result from modifications of distribution pressure, repair of airline leaks, use of outside air at the intake, and so on.

8.6.6 INERT GAS SUPPLIES

Large quantities of inert gas are often required in a chemical production facility for the inert blanketing of tanks and for reactor, tank, and line purging. This gas is usually supplied from a central facility. Nitrogen is normally used and can be manufactured on-site in an air liquefaction plant or purchased as liquid in tankers. There should be general considerations for the management of gaseous materials to minimize the amount of inert gas supplies that must be used per unit of product generated. These considerations should include: pressure reduction to the lowest practical level to minimize fugitive emissions via leaks; insulation of tanks, lines, and so on to lower vapor pressure and potential fugitive emissions; inspection for and repair of line leaks; etc.

8.7 MATERIAL TRANSPORTATION AND STORAGE EQUIPMENT

This section deals with equipment needed for the handling, storage, and transportation of gases, liquids, and solids. Each type of material is considered individually.

8.7.1 GASES

The type of equipment best suited for the transportation of gases depends on the differential and operating pressures and flow rates required. In general, fans are used where the pressure drop in the transportation system is small and operating pressures are low. Axial-flow compressors are employed for high flow rates and moderate differential pressures, while centrifugal compressors are chosen for applications requiring high flow rates and, by staging, high differential pressures. Reciprocating compressors can be used over a wide range of pressures and capacities but are normally specified in preference to centrifugal compressors only where high pressures are required at relatively low flow rates.

Gases are stored at high pressures to meet a process requirement and to reduce the required storage volumes. The volume of some gases can be further reduced by liquefying them by pressure or refrigeration. Cylindrical and spherical vessels (Horton spheres) are normally used for pressurized gas storage.

8.7.2 LIQUIDS

The transportation of liquids is usually accomplished with pumps. The type of pump used, that is, centrifugal, reciprocating, diaphragm, or rotary (gear or sliding vane), depends on the operating pressures and capacity range needed in each application. Liquids are usually stored in bulk in vertical cylindrical steel tanks, although horizontal cylindrical tanks and rectangular tanks can also be used for storing relatively small quantities of liquids.

Fixed and floating-roof tanks can be used for liquid storage. In fixed-roof tanks, toxic or volatile liquids will tend to emit hazardous vapors through a vent line when they are filled. Various configurations can be used to prevent, or at least limit, these releases to the environment. Such designs include routing the vapors to a pollution control system, using a flare to thermally destroy the emissions, or using a compressor to recover vapors as a condensed liquid. In a floating-roof tank, a movable cover floats on the surface of the liquid and is sealed to the tank walls using a variety of configurations.

Floating-roof tanks are used to eliminate evaporation losses and, for flammable liquids, to obviate the need for inert gas blanketing to prevent the formation of an explosive mixture above the liquid, as can occur with a fixed-roof tank.

8.7.3 Solids

Solids are usually more expensive to move and store than liquids or gases. The best equipment to use depends on several factors including material throughput, length of travel, change in elevation, and nature of the solids (size, bulk density, angle of repose, abrasiveness, corrosiveness, wet or dry, etc.).

Belt conveyors are the most used type of equipment for the continuous transport of solids. They can carry a wide range of materials economically over both short and long distances, either horizontally or at an appreciable angle, depending on the angle of repose of the solids. *Screw conveyors*, also called *worm conveyors*, are used for free-flowing materials. The modern conveyor consists of a helical screw rotating in a U-shaped trough. This type of device can be used horizontally or, with some loss of capacity, at an incline to lift materials.

Where a vertical lift is required, the most widely used equipment is the *bucket elevator*, consisting of buckets fitted to a chain or belt, which passes over a driven roller or sprocket at the top extent of travel. Bucket elevators can handle a wide range of solids, from heavy "lumps" to fine powders, and are also suitable for use with wet solids and slurries.

The simplest way to store solids is to pile them on the ground in the open air. This is satisfactory for the long-term storage of materials that do not deteriorate on exposure to the elements, for example, coal, which is seasonally stockpiled at utilities, and for which fugitive emissions from the pile are not an air quality concern. For large stockpiles, permanent facilities are usually used for distributing and reclaiming the material. At permanent facilities, traveling cranes, grabs, and drag scrapers are used to feed belt conveyors. Where the cost of recovery from the stockpile is large compared with the value of the stock held, storage in silos or bunkers should be considered.

Overhead bunkers, also called bins or hoppers, are normally used for the short-term storage of materials that must be readily available within the process. The units are arranged so that the material can be withdrawn at a steady rate from the base of the bunker onto a suitable conveyor for addition where needed within the process.

8.8 INSTRUMENTATION AND CONTROLS

In chemical processes it is essential that the operator and/or production engineer verify that any given process is functioning properly (Liptak 2006; Theodore 2014). To perform this job function properly, the individual must have access to numerous types of process data. In typical chemical processes, many instruments measure, indicate, and record process conditions using process data, such as flow rates, compositions, pressures, and temperatures.

To best aid an operator in controlling a given chemical process, it is often desirable to employ automatically controlled systems. This allows the instruments to not only measure, indicate, and record a variable but also to maintain a system variable at a predetermined value. Automatic controllers establish a given system parameter at a set value using either "open" or "closed" feedback loops.

8.8.1 FEEDBACK LOOP INSTRUMENTATION AND CONTROL SYSTEMS

Open feedback loop instrumentation and control (I&C) systems are typically less expensive to purchase and much easier to install and maintain when compared to *closed-loop I&C systems*. However, open-loop systems cannot be relied upon to correct a given out-of-range process condition; it can only maintain the status quo of a given variable. This is because an open-loop system measures only one variable, not its effect on any other system parameters, and controls the variable based on a provided input. Consider the following example: it has been calculated that a heat exchanger requires 100 gpm of 75°F river water to cool an important process liquid to 120°F. An open-loop system design could be used to modulate the heat exchanger cooling water outlet based on the output of a cooling water flow transmitter. Therefore, if the flow transmitter measured 90 gpm of cooling water, the cooling water valve would automatically stroke more open until the flow transmitter increased to 100 gpm and vice versa if the flow transmitter measured 110 gpm.

However, problems with this open-loop system design arise whenever the river water temperature changes. This results in process liquid temperatures being higher than 120°F when the river warms up and cooler than 120°F when the river gets colder. Since the automatic valve positioner is concerned only with the amount of flow that passes through the valve regardless of the important process liquid final temperature, the feedback system is deemed "open."

Closed feedback loop I&C systems are typically more expensive to purchase and more difficult to install and maintain. However, closed feedback I&C systems are simply much more reliable for ensuring that a process is operating at optimum conditions. These systems control a variable by directly measuring its effects on another system variable. Consider the previous example: it is critical that an important process liquid be cooled to 120°F to allow it to safely react with another constituent. A closed-loop system design could automatically modulate the heat exchanger cooling water outlet valve based on the outlet temperature of the process liquid, not the flow rate of cooling water. Therefore, as process liquid temperature increases above 120°F, the cooling water valve would stroke more open and vice versa if the process liquid temperature decreased below 120°F. Since the automatic valve positioner is only concerned with the final temperature of the important process liquid regardless of what amount of cooling water flow it requires to remain at 120°F, the feedback system is deemed "closed."

Automatic systems have been coupled using computers and make it possible to run entire processes without any direct human control. A process engineer determines the best operating parameters for a given set of conditions and the computer controls all the open- and closed-loop systems to produce a product at optimum process operating efficiency. If the systems are not operating at optimum, as compared to the standards inputted by the process engineer, the computer initiates corrective actions, thereby effectively operating the process as one closed-loop system. However, each process should still be analyzed for all normal and emergency conditions of operations and the suitable I&C provided accordingly.

Once an instrument operating range, duty, and environment have been calculated and determined, the instrument should be selected with the following design specifications in mind. First, all I&C must be of the "fail safe" type. Second, the I&C must be constructed of a corrosion-resistant material and/or encased in an "environmentally" qualified enclosure, when appropriate. Third, I&C must be relatively easy to access for calibration and maintenance. Fourth, separate indicators should be used for each critical point of operation. Fifth, I&C needs to provide both audible and visual alarms to permit operators to identify any type of process deviation occurring. Lastly, all I&C must be completely tested prior to initial service and periodically during plant life to ensure that all the automatic functions (e.g., process equipment sequencing, alarms, and automatic system trips) operate as designed.

Following these specifications will allow I&C schemes to maintain process variables within the known operating limits and detect and mitigate any hazards if they develop. Instruments must be installed to maintain the key process variables within their design and operating ranges. Critical process variables, such as temperature, pressure, and flow, should be coupled with automatic alarms. Training programs should stress the following concerning alarms: alarms should not be ignored; appropriate operator action(s) should be known and should be immediately taken to correct the problem; and management should be immediately notified of the problem whether it was corrected or not. All critical processes should also be equipped with a system "trip" signal if the operator cannot respond quickly enough to correct the critical system alarm condition.

8.8.2 AUTOMATIC TRIP SYSTEMS AND INTERLOCKS

There are three basic components to an automatic trip system. First, a sensor monitors the process control variable of interest and sends an output signal when a preset value for that variable is exceeded (e.g., pump discharge pressure goes above design pressure). Second, pneumatic and/or electric relays transfer the signal to an actuator and/or contactor (e.g., the pump discharge pressure switch actuates a relay in the pump motor breaker). Lastly, the receiving device carries out the appropriate response (e.g., the motor breaker goes open, deenergizing a pump motor).

The automatic safety trip may be designed as a separate system, or it may be part of a process control loop. However, since adding a trip system to a control loop may present a greater potential for failure (due to having more components), it is safer to design for a separate system. In either case, these safety systems should be inspected and/or tested on a routine, periodic basis.

System designers should also integrate the use of *interlocks* into a given safety scheme. Interlocks ensure that an operator follows the required sequence of actions in a simple, complex, or rarely performed process, from simple batching up to and including plant start-up/shutdown. Examples of interlocks include: ensuring that a centrifugal pump is started with its discharge valve shut to prevent a dangerous pump runout condition; to not allow a pump to start without the suction valve being full open; or to not allow a positive displacement pump to start until all the valves in its flow path are opened.

Using system process alarms, automatic trips, and interlocks, both design and process engineers can be reasonably assured that a well-employed I&C system will ensure system integrity, personnel safety, optimum product yield, and reduced environmental and health risk inside and outside the facility.

8.9 PROCESS DIAGRAMS

The complete design specification for a medium to large-sized chemical process includes process diagrams; energy and material balance tables; chemical, mechanical, electrical, metallurgical, and civil engineering design considerations and plans and specifications; and process flowsheets. The process flowsheet is the key instrument for defining, refining, and documenting a chemical process. The process flow diagram is the authorized process blueprint, serves as the framework for specifications used in equipment designation and design, and is the single, authoritative document employed to define, construct, and operate a chemical process.

In a sense, process diagrams are the international language of the engineer, particularly the practicing engineer. Chemical engineers conceptually view a chemical plant as consisting of a series of interrelated building blocks that are defined as *units* or *unit operations*. The process diagram ties together the various pieces of equipment that make up the process. Flow schematics follow the successive steps of a process by indicating where the pieces of equipment are located, and the material streams entering and leaving each unit.

Beyond equipment symbols and process stream flow lines, there are several essential components contributing to a *process flowsheet*. These include equipment identification numbers and names, temperature and pressure designations, utility designations, volumetric or molar flow rates for each process stream, and a material balance table pertaining to process flow lines. The process flowsheet may show additional information such as energy requirements, major instrumentation, and physical properties of the process streams. When properly assembled and employed, this type of process schematic provides a coherent picture of the overall process. The flowsheet symbolically and pictorially represents the interrelations among the various flow streams and equipment and permits easy calculations of material and energy balances. A number of symbols are universally employed to represent equipment, equipment parts, valves, piping, and so on. These symbols obviously reduce, and in some instances replace, detailed written descriptions of the process.

A flowsheet usually changes over time with respect to the degree of sophistication and the details it contains. A crude flowsheet may initially consist of a simple, freehand block diagram offering information about the equipment only. A later version (described in the next section) may include line drawings with pertinent process data, such as overall and componential flow rates, utility and energy requirements, and instrumentation. During the later stages of a design project, the flow sheet usually consists of a highly detailed piping and instrumentation diagrams (P&IDs), which are covered in a later section of this chapter. For information on aspects of the design procedure, which are beyond the scope of this text, the reader is referred to the literature (Treybal 1980; Coulson, Richardson and Skinnott 1983).

One can conceptually view a chemical plant as consisting of a series of interrelated building blocks that are defined as units or unit operations. The process flowsheet ties together the various pieces of equipment that make up the process. Flow schematics follow the successive steps of a process by indicating where the pieces of equipment are located and where the material streams enter and leave each unit.

There are five basic types of schematic diagrams in general use in process engineering. These are:

1. Block diagrams
2. Graphic flow diagrams
3. Process flow diagrams
4. Process P&IDs
5. Tree diagrams

8.9.1 BLOCK DIAGRAMS

The *block diagram* is the simplest but least descriptive of the schematic diagrams in common use. As the name implies, it consists of neat rectangular blocks which usually represent a single unit operation in a plant or an entire section of the plant. These blocks are connected by arrows indicating the flow sequence. The block diagram is extremely useful in the early stages of a process design and is particularly valuable in presenting the results of economic or operating studies since the significant data can be placed within the blocks. Four different types of operations can be described by block diagrams. As indicated in Figure 8.2, these include: *pure inventory operations, separation operations, assembly operations,* and *chemical operations.*

Increases in inventory represent accumulation; decreases represent depletion. An operation such as this is defined as *a pure inventory operation*. The flow diagram for

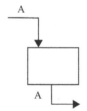

a. *Pure inventory operations*
No chemical change
No phase change
No separation or combination
of materials

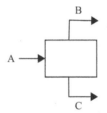

b. *Separation*
No chemical change
Physical separation of materials
occurs with or without phase
change

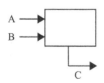

c. *Assembly*
No chemical change
Physical combination of materials
occurs with or without phase
change

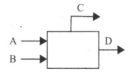

d. *Chemical change*
Reactants (A and B) undergo
molecular rearrangement to form
products (C and D), with or
without phase change

FIGURE 8.2 Fundamental operations described by block diagrams. A, B, C, and D represent individual material flows.

this operation is shown in its most generalized form in Figure 8.2a. If no chemical change takes place, a starting material can be physically separated into more than one product. This is by definition a *separation operation*. The generalized flow diagram for this operation is shown in Figure 8.2b. It would apply equally well to the recovery of the two components in an apparatus or the partial separation of an entering feed water into exiting steam and exiting hot water. Two or more material streams that are combined or assembled without chemical reaction are, by definition, an *assembly operation*. A generalized flow diagram for this operation is presented in Figure 8.2c. With more material streams involved, it would apply equally well to an automobile assembly line or the creation of an emulsion by the intimate mixing of oil, water, and an emulsifying agent. These three types of operation have one important common characteristic, the absence of chemical change. This sets them apart from the fourth type, the *chemical operation*. In Figure 8.2d two reactants, A and B, are shown entering a chemical reaction producing products C and D. These chemical operation flow diagrams will vary with the number of reactants and products involved.

8.9.2 Graphic Flow Diagrams

Graphic flow diagrams are used most frequently in advertising, company financial reports, and technical reports in which certain features of the flow diagram require extra emphasis. It should present the desired information clearly and in an eye-catching fashion that is both novel and informative.

8.9.3 Process Flow Diagrams

The *process flow diagram*, or PFD, is a pictorial description of the process. It provides the basic processing scheme, the basic control concept, and process information from which equipment can be specified and designed. As described earlier, it provides the basis for the development of the P&I diagram (P&ID), equipment design, and specifications and serves as a guide for the design, construction, and operation of a chemical process.

The process flow diagram usually includes:

1. Material balance data (may be on separate sheets)
2. Flow scheme, equipment, and interconnecting streams
3. Basic control instrumentation
4. Temperature and pressure at various points
5. Any other important parameters unique to each process

Data on spare and parallel equipment are often omitted as is valving. A valve is shown only where its specification can aid in understanding intermittent or alternate flows. Instrumentation is indicated to show the location of variables being controlled and the location of an actuating device, usually a control valve. To help the reader better understand the process flow sheet, a list of commonly used symbols for valves are presented in Figure 8.3 (Theodore, Reynolds and Taylor 1989).

Various symbols are universally employed to represent equipment, parts, valves, piping, etc. Some of these equipment symbols are depicted in the schematic in Figure 8.4. Although a significant number of these symbols are used to describe some chemical processes, only a few are needed for simpler facilities. These symbols obviously reduce,

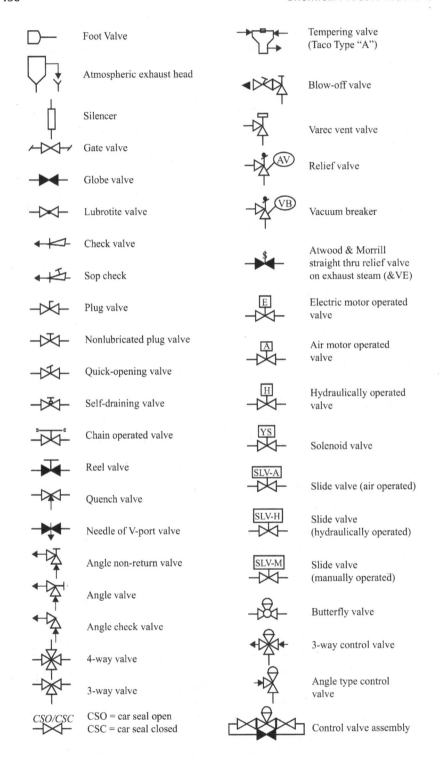

FIGURE 8.3 Commonly used valve symbols.

Chemical Process Equipment

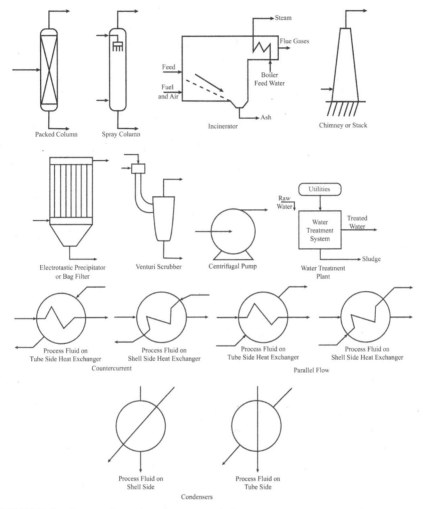

FIGURE 8.4 Commonly used equipment symbols.

and in some instances replace, detailed written descriptions of the process. Note that many of the symbols are pictorial, which helps in better describing process components, units, and equipment.

The degree of sophistication and details of a flowchart usually vary with both the preparer and time. It may initially consist of a simple freehand block diagram with limited information on the equipment; later versions may include line drawings with pertinent process data such as overall and componential flow rates, utility and energy requirements, environmental equipment, and instrumentation. During the later stages of the project, the flowchart will usually be a highly detailed P&ID.

8.9.4 Process P&IDs

The *P&ID*, which provides the basis for detailed design, offers a precise description of piping, instrumentation, and equipment. This key drawing defines the plant system, describes equipment, and shows all instrumentation, piping, and valving. It is used to train personnel and aids in troubleshooting during start-up and operation. The P&ID assigns item numbers to all equipment (e.g., towers, reactors, and tanks); gives dimensions of equipment and vessel elevations; and shows all piping, including line numbers, sizes, specifications, and all valves. All instrumentation is covered, giving numbers, function, types, and indicating whether the instrumentation is actuated electronically or pneumatically.

A general knowledge of the symbols for flow, level, pressure, and temperature controllers, as given in Figures 8.5–8.8, is needed to comprehend flow diagrams like the simple example presented in Figure 8.9 (from Theodore, Reynolds and

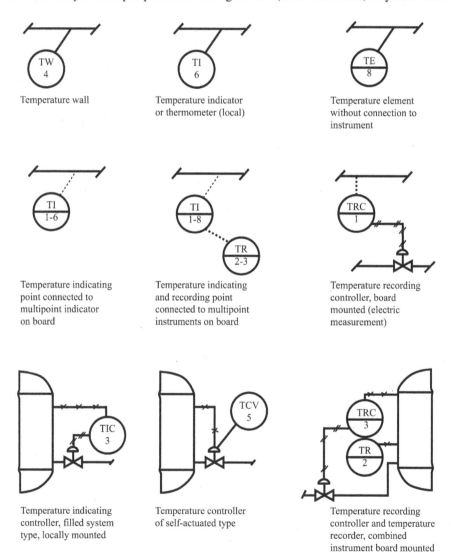

FIGURE 8.5 Typical instrumentation symbols for temperature.

Chemical Process Equipment

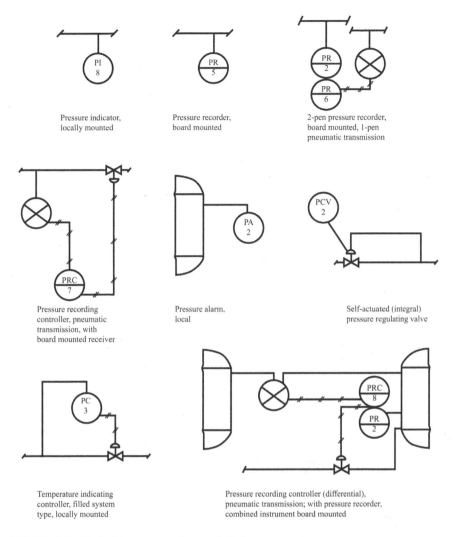

FIGURE 8.6 Typical instrumentation symbols for pressure.

Taylor 1989). In this vessel, with an inlet feed on top of the tank equipped with a flow controller, the level in the tank is maintained by a level-controlling device. When the level rises above the high-level point, the level controller sends a signal to a valve actuator and the valve is opened, dropping the liquid level within an acceptable range within the vessel. When the level approaches a specified lower limit value, the valve is closed. More complicated systems can be analyzed in a similar manner used in this simple example.

Certain essential information is often concisely provided via notations adjacent to the representation of each piece of process equipment. Experience has dictated the information required for common items such as pumps and vessels. For special equipment, overall dimensions and significant process and operating characteristics

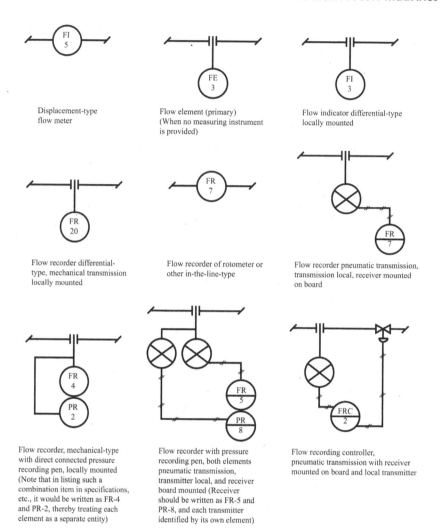

FIGURE 8.7 Typical instrumentation symbols for flow.

are often given. The notations provided in Table 8.1 are suggested for several common items. Any consistent system is satisfactory.

Process information that is often necessary includes the following:

Compressors

1. Service
2. Stages
3. Suction conditions
4. First stage suction
5. Second stage suction
6. Second stage discharge

Chemical Process Equipment

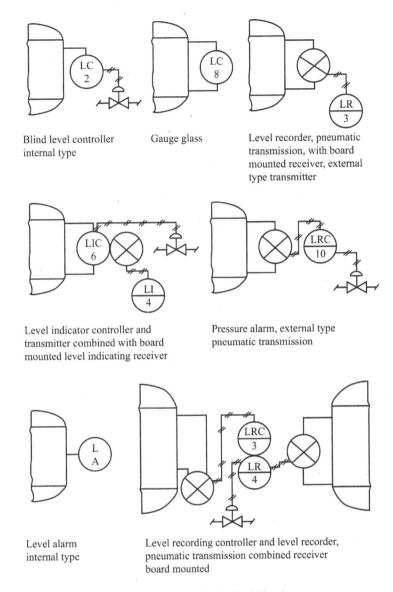

FIGURE 8.8 Typical instrumentation symbols for liquid level.

Heat Exchangers

1. Service
2. Differential pressure across shell and tubes
3. Heat transfer area
4. Duty (heat transfer rate)
5. Design conditions
6. Temperature and pressure at inlet and outlet

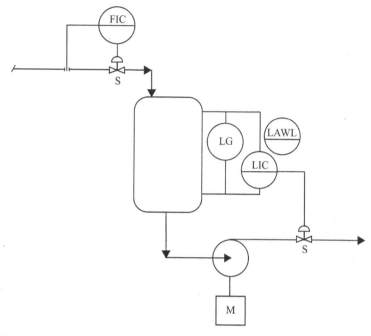

FIGURE 8.9 Vessel with level control on outlet. FIC, flow indicator controller; LG, level gauge; LIC, level indicator controller; LAWL, level alarm for water level; M, motor.

Pumps

1. Service
2. Size and type
3. Fluid being handled
4. Pump operating temperature
5. Fluid density at pump operating temperature
6. Design flow rate at pump operating temperature
7. Design operating service pressure and net positive suction head (NPSH)

TABLE 8.1
Typical Equipment Designations

Equipment Type	Designations
Compressors	K, C
Exchangers	E, C (for condensers), RB (for reboilers)
Heaters	H
Pumps	P or PU
Reactors	R
Storage tanks	ST
Towers	T
Vessels	V

Chemical Process Equipment

Vessels

1. Service
2. Diameter, height, wall thickness
3. Special features (lining, etc.)
4. Design conditions
5. Operating conditions

8.9.5 TREE DIAGRAMS

Another type of flow diagram applicable to the CPI is the *tree diagram*. Tree diagrams are used primarily in the study of hazards and/or chemical accidents and have become an integral part of any study or analysis involving risk and accident and emergency management (Theodore, Reynolds and Taylor 1989; Theodore and Dupont 2013). The following discussion details the two main types of analysis employed in the CPI, *fault tree analysis (FTA)*, and *event tree analysis (ETA)*.

8.9.5.1 Fault Tree Analysis

An FTA is carried out using a flow diagram that highlights conditions that cause system failure. FTA attempts to describe how and why an accident or other undesirable event has occurred or could occur in the future. FTA shows the relationship between the occurrence of the undesired event, the "top event," and one or more antecedent events, called "basic events." The top event may be, and usually is, related to the basic events via certain intermediate events. A fault tree diagram depicts the casual chain linking the basic events to the intermediate events and the latter to the top event. In this chain, the logical connection between events is indicated by so-called "logic gates." The principal logic gates are the AND gate, symbolized on the fault tree by $\triangle$, and the OR gate, symbolized by $\cap$.

8.9.5.2 Event Tree Analysis

An *ETA* uses a flow diagram that represents the possible steps leading to a failure or accident. The sequence of events begins with an initiating event and terminates with one or more undesirable consequences. In contrast to a fault tree, which works backward from an undesirable consequence to possible causes, an event tree works forward from the initiating event to possible undesirable consequences. The initiating event may be equipment failure, human error, power failure, or some other event that has the potential to adversely affect the environment or an ongoing process and/or equipment. The event tree may limit hazard analysis because it cannot quantify the probability of an event occurring. In addition, all the initial occurrences must be identified for a complete analysis. The event tree should be used to examine, rather than to evaluate, the possibilities and consequences of a failure. An FTA can be used to establish the probabilities of the event tree branches.

8.9.6 PREPARING FLOW DIAGRAMS

Before attempting to calculate the material or energy requirements of a process, it is desirable to develop a clear picture of the process. The best way to do this is to draw a flow diagram, i.e., a line diagram showing the successive steps of a process. As mentioned, flow diagrams are very important for saving time and eliminating mistakes. The beginner should learn how to draw them properly and cultivate the habit of sketching them on the slightest excuse. The following rules should be observed for the preparation of flow diagrams:

1. Show operating units by simple neat rectangles. Do not waste time in elaborate artwork since it is without advantage or even meaning (see Figure 8.10).

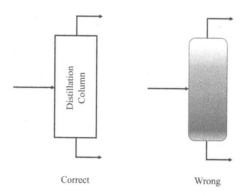

FIGURE 8.10 Simple process flow diagram symbols.

2. Make each material stream line represent an actual stream of material passing along a pipe, duct, chute, belt, or other conveying device. Show the gaseous mixture of carbon dioxide, oxygen, carbon monoxide, nitrogen, and water vapor obtained from the combustion of a hydrocarbon fuel and exiting from the stack of a furnace as the single material stream of flue gas, not as five separate streams. Refer to Figure 8.11 which shows the correct representation of a flue gas stream versus a sophisticated device that sorts components of the flue gas and delivers them as five separate products through five separate pipes.

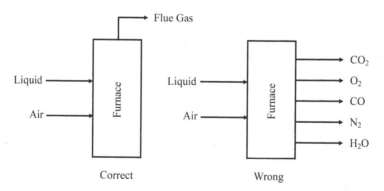

FIGURE 8.11 Correct and wrong representation of a flue gas containing CO_2, O_2, CO, N_2, and water vapor.

Chemical Process Equipment

3. Distinguish between "open" and "closed" material streams. In Figure 8.12a, open steam is being blown into the tank, mixing with the other materials in it. In Figure 8.12b, steam is passing through a coil, being kept separate from the tank contents, and heating them by transfer of heat through the walls of the coil.

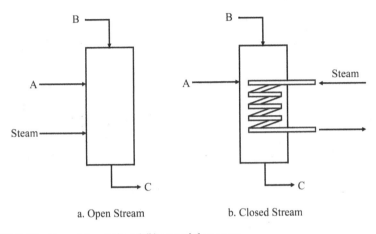

a. Open Stream   b. Closed Stream

FIGURE 8.12 Open (a) and closed (b) material streams.

4. Distinguish between a *continuous operation* in which a material is continuously introduced into the reactor versus *batch operations* in which a fixed charge of material is introduced to the reactor per cycle. Batch operations are often indicated by placing a double bar on the material stream lines of the entering materials as shown in Figure 8.13.

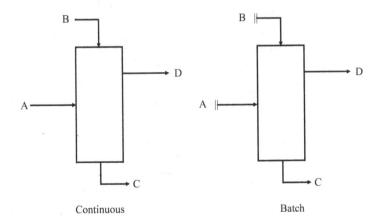

Continuous   Batch

FIGURE 8.13 Continuous versus batch operations.

5. Except for a few unusual situations, keep flow diagrams free of data regarding the material streams. An overburden of data clutters up the diagram and robs it of its main function, which is to present a clear picture of the materials as they move into, through, and out of the process. This is illustrated in Figure 8.14 with an air and coal stream entering a furnace.

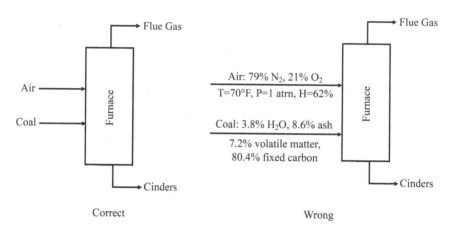

FIGURE 8.14 Inclusion versus exclusion of experimental data on flow diagrams.

8.10 ILLUSTRATIVE EXAMPLES

Five illustrative examples complement the material presented above.

8.10.1 Illustrative Example 1

As part of a plant's accident prevention program, flue gas from a process is mixed with recycled gas from an absorber (A), and the mixture passes through a waste heat boiler (H) which uses water as the heat transfer medium. It then passes through a water spray quencher (Q) in which the temperature of the mixture is further decreased and, finally, through an absorber (A) in which water is the absorbing agent (solvent) for one of the species in the flue gas stream. Prepare a simplified flow diagram for the process

Solution: Prepare a line diagram of the process. See Figure 8.15.

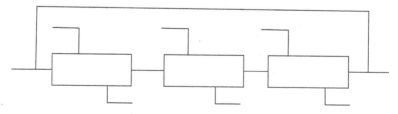

FIGURE 8.15 Line diagram for the process for Illustrative example 1.

Chemical Process Equipment

Label the equipment. See Figure 8.16.

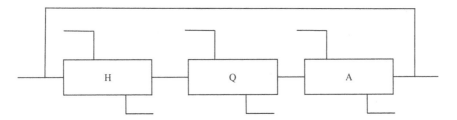

FIGURE 8.16 Equipment labels for the process for Illustrative example 1.

Label the flow streams. See Figure 8.17.

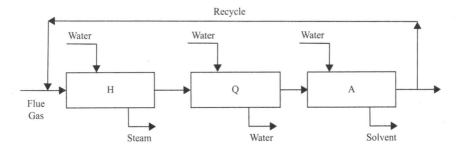

FIGURE 8.17 Flow stream labels for the process for Illustrative example 1.

8.10.2 Illustrative Example 2

The heat-generating unit in a coal fired power plant may be simply described as a continuous flow reactor into which fuel (mass flow rate F) and air (mass flow rate A) are fed, and from which effluents ("flue gas," mass flow rate E) are discharged.

  a. Draw a flow diagram representing this process. Show all flows into and out of the unit.
  b. Write a mass balance equation for this process.
  c. Suppose the fuel contains a mass fraction y of incombustible component C (for example, ash). Assume that all of the ash is carried out of the reactor with the flue gas (note that in reality, a fraction of the ash generated will remain within the heat-generating unit as bottom ash and must be removed periodically). Write a mass balance equation for component C.
  d. What is the mass fraction (z) of C in the exit stream E and what is the effect of increasing the combustion air flow upon z?

Solution:

a. Draw a flow diagram of the process. See Figure 8.18.

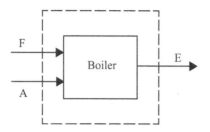

FIGURE 8.18 Flow diagram for the process for Illustrative example 2.

b. Write an overall mass balance equation for the system.

$$F + A = E \tag{8.1}$$

Component C is not combustible and is carried out with the exhaust, therefore $C_{in} = C_{out}$.

c. Write the component mass balance equation of C.

$$yF + (0)A = zE \tag{8.2}$$

d. Combining both equations to solve for z leads to:

$$E = F + A \tag{8.3}$$

$$yF = z(F + A) \tag{8.4}$$

$$z = y\left(\frac{F}{(F+A)}\right) \tag{8.5}$$

Therefore, increasing the combustion air flow (A) decreases the mass fraction of C in the exhaust (a dilution effect). For this reason, effluent concentrations are given at a specified excess air value (fraction or percent).

8.10.3 Illustrative Example 3

If a building fire occurs, a smoke alarm sounds with probability = 0.9 (i.e., it is expected to have a 10% failure rate). The sprinkler system functions with a probability = 0.7 whether or not the smoke alarm sounds. The consequences are: minor fire damage (alarm sounds, sprinkler works), moderate fire damage with few injuries (alarm sounds, sprinkler fails), moderate fire damage with many injuries (alarm fails, sprinkler works), or major fire damage with many injuries (alarm fails, sprinkler fails). Construct an event tree and indicate the probabilities for each of the four consequences

Solution: Determine the first consequence(s) of the building fire and list the probabilities of the first consequence. See Figure 8.19.

Chemical Process Equipment

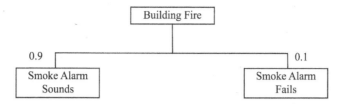

FIGURE 8.19 First consequences and their probabilities for events in Illustrative example 3.

Determine the second consequences of the building fire and list the probabilities of these consequences. See Figure 8.20.

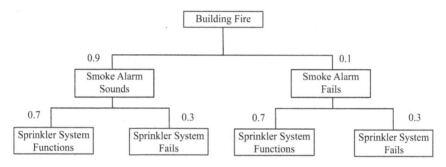

FIGURE 8.20 Second set of consequences and their probabilities for events in Illustrative example 3.

Determine the final consequences and calculate the probabilities of minor fire damage, moderate fire damage with few injuries, moderate fire damage with many injuries, and major fire damage with many injuries. See Figure 8.21.

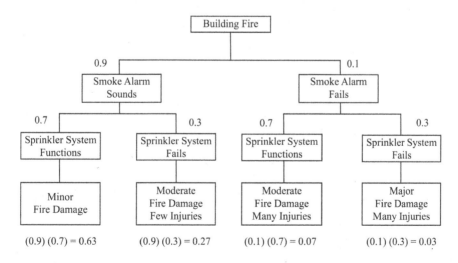

FIGURE 8.21 Final set of consequences and probabilities of damage outcomes for events in Illustrative example 3.

8.10.4 Illustrative Example 4

List the advantages and disadvantages of the three major classes of reactors – batch, CSTR, and fluidized bed – employed in industry.

Solution: The solution is provided in Table 8.2.

TABLE 8.2
Solution for Illustrative Example 4

Type of Reactor	Use in Industry	Advantages	Disadvantages
Batch	Small-scale production	High conversion per unit volume for one pass	High operating cost (labor)
	Intermediate or one-shot production	Flexibility of operation – same reactor can be used to produce one product one time and different product the next	Product quality more variable than with continuous operation
	Pharmaceutical Fermentation	Easy to clean	
CSTR	Liquid phase reactions	Continuous operation	Lowest conversion/unit volume
	Gas–liquid reactions	Easily adaptable to two-phase (gas–liquid) reactions	Bypassing and channeling possible with poor agitation
	Solid–liquid reactions	Good temperature control	
		Good operating control	
		Simplicity of construction	
		Low operating (labor) cost	
		Easy to clean	
Fluidized Bed	Gas–liquid reactions	Good mixing	Complex bed fluid mechanics
	Gas–solid reactions	Catalyst can be continuously regenerated with use of an auxiliary loop	Severe agitation can result in catalyst destruction, dust formation, and carry over in product streams
		Good uniformity of temperature	Uncertain scale-up

8.10.5 Illustrative Example 5

A 2,000-barrel (bbl) atmospheric pressure tank is frequently filled with cyclohexane. To avoid fire and explosion hazards, the vapor space is inerted with a nitrogen blanket at ambient temperature and pressure. Determine the amount of cyclohexane vapor vented if there is no vent recovery system installed on the tank. Also determine the mass of nitrogen lost during the tank filling process.

Given: The tank is at 80°F, cyclohexane molecular weight (MW) = 84.16 g/gmol, cyclohexane vapor pressure at 80°F, p'_{C1} = 2.051 psia, and at 85°F, p'_{C2} = 2.32 psia. Note: assume that the vapor in the tank is saturated with cyclohexane.

Solution: The total number of moles of cyclohexane, N_T, is given by:

$$N_T = PV/RT \qquad (8.6)$$

where V, volume = (2,000 bbl) (5.614 ft³/bbl) = 11,228 ft³; R, gas constant = 10.73 psia-ft³/lbmol-°R; T, temperature = 540°R; and P, pressure = 14.7 psia. Therefore,

$$N_T = PV/RT = (14.7\,\text{psia})(11,228\,\text{ft}^3)/\left[(10.73\,\text{psia}-\text{ft}^3/\text{lbmol}-°R)(540°R)\right]$$

$$= (16,5051.6\,\text{psia}-\text{ft}^3)/(5,794.2\,\text{psia}-\text{ft}^3/\text{lbmol}) = 28.49\,\text{lbmol}$$

Since the fraction of the volume occupied by saturated cyclohexane vapor is p'/P, the moles of cyclohexane released, N_{Ch}, are given by:

$$N_{Ch} = \frac{(N_T)(p'_{Ch})}{P} \qquad (8.7)$$

where p'_{Ch} = cyclohexane vapor pressure = 2.051 psia; and P, total tank pressure = 14.7 psia. Therefore:

$$N_{Ch} = (28.94\,\text{lbmol})(2.051\,\text{psia})/(14.7\,\text{psia}) = 3.98\,\text{lbmol}$$

Therefore, the mass "lost," M_{Ch}, is given by:

$$M_{Ch} = (N_{Ch})(\text{MW}) = (3.98\,\text{lbmol})(84.16\,\text{lb/lbmol}) = 335\,\text{lb}$$

The moles of nitrogen released, N_N, are then given by:

$$N_N = \frac{(N_T)(p_N)}{P} \qquad (8.8)$$

where p_N = nitrogen partial pressure = 14.7 psia − 2.051 psia = 12.65 psia. Therefore

$$N_N = \frac{(28.49\,\text{lbmol})(12.65\,\text{psia})}{14.7\,\text{psia}} = 24.51\,\text{lbmol}$$

$$M_N = (N_N)(\text{MW}) = (24.51\,\text{lbmol})(14\,\text{lb/lbmol}) = 343\,\text{lb}$$

PROBLEMS

8.1 Discuss the different types of heat exchangers that are employed in the CPI.
8.2 Discuss the major differences between gaseous adsorption and liquid adsorption.
8.3 Provide qualitative differences between gaseous absorption and gaseous stripping.
8.4 Every time an automobile gas tank is filled, the vapor space in the tank is displaced to the environment. Since gasoline is a hydrocarbon, it will eventually

be involved in the production of low-level ozone upon solar irradiation. Assume that the tank vapor space, the air, and the gasoline are all at 20°C. The vapor space of the tank is saturated with gasoline, and under these conditions, the gasoline has a vapor phase mole fraction = 0.4. The molecular weight of the vapor = 70 g/gmol, and the gasoline has a specific gravity = 0.62. A typical automobile averages 25 mpg, travels 12,500 mi/yr, and has a 15 gal capacity tank. Determine the annual volume of gasoline lost to the air when filling the gas tank (assume 2 gal reserve left in the tank).

8.5 Refer to the previous Illustrative example 5. Determine how much cyclohexane can be returned to the tank if the vented gas were compressed to 100 psig and cooled to 85°F as an alternative risk reduction measure.

8.6 A mixture of water and ethyl alcohol, 40% alcohol by mass, is fed to a flash unit where it is separated into two streams. The vapor and liquid output streams contain 80 and 28 mass percent alcohol, respectively. Draw a flow diagram of the process.

8.7 Potassium nitrate is obtained from an aqueous solution of 20% KNO_3. During the process, the aqueous solution of potassium nitrate is evaporated, leaving an outlet stream with a concentration of 50% KNO_3 which then enters a crystallization unit where the outlet product is 96% KNO_3 (anhydrous crystals) and 4% water. A residual aqueous solution which contains 0.55 g of KNO_3/g water also leaves the crystallization unit and is mixed with the fresh solution of KNO_3 at the evaporator inlet. Such recycling minimizes the release of potentially hazardous material to the environment and the loss of valuable raw materials.
  a. Draw a flow diagram of the process.
  b. Calculate the feed rate to the evaporator and the recycle rate to the evaporator in units of kg/hr when the feedstock flow rate is 5,000 kg/hr.

8.8 A runaway chemical reaction can occur if coolers fail (A) or there is a bad chemical batch (B). Coolers fail only if both Cooler #1 (C) AND Cooler #2 fail (D). A bad chemical batch occurs if there is a wrong mix (E) OR there is a process upset (F). A wrong mix occurs only if there is an operator error (G) AND instrument failure (H).
  a. Construct a fault tree.
  b. If the following annual probabilities are provided by the plant engineer, calculate the probability of a runaway chemical reaction occurring in a year's time.

$P(C) = 0.05 \quad P(D) = 0.08 \quad P(F) = 0.06 \quad P(G) = 0.03 \quad P(H) = 0.01$

REFERENCES

Coulson, J.M., Richardson, J.F., and Skinnott, R. 1983. *An Introduction to Chemical Engineering*. Elmsford, NY: Pergamon Press.

Green, D., and Southard, M.Z. (editors). 2019. *Perry's Chemical Engineers' Handbook*, 9th Edition. New York, NY: McGraw-Hill.

Liptak, B.G. 2006. *Instrument Engineers' Handbook*, 4th Edition. Boca Raton, FL: CRC Press, Taylor & Francis Group.
McCabe, W.L., Smith, J.C., and Harriot, P. 2014. *Unit Operations of Chemical Engineering*, 7th Edition. New York, NY: McGraw-Hill.
Sprague, B. 1999. *Manufacturing Assessment Planner (MAP) Toolkit*. Ann Arbor, MI: Michigan Manufacturing Technology Center (MMTC) and CAMP, Inc.
Theodore, L. 2008. *Air Pollution Control Equipment Calculations*. Hoboken, NJ: John Wiley & Sons.
Theodore, L. 2011. *Heat Transfer Operations for the Practicing Engineer*. Hoboken, NJ: John Wiley & Sons.
Theodore, L. 2012. *Chemical Reactor Analysis and Applications for the Practicing Engineer*. Hoboken, NJ: John Wiley & Sons.
Theodore, L. 2014. *Chemical Engineering: The Essential Reference*. New York, NY: McGraw-Hill.
Theodore, L., and Dupont, R.R. 2013. *Environmental Health Risk and Hazard Risk Assessment: Principles and Calculations*. Boca Raton, FL: CRC Press/Taylor & Francis Group.
Theodore, L., Reynolds, J., and Taylor, F. 1989. *Accident and Emergency Management*. New York, NY: Wiley-Interscience.
Theodore, L., and Ricci, F. 2011. *Mass Transfer Operations for the Practicing Engineer*. Hoboken, NJ: John Wiley & Sons.
Treybal, R.E. 1980. *Mass-Transfer Operations*, 3rd Edition. New York, NY: McGraw-Hill.

9 Chemical Processes
Fundamentals and Principles

9.1 INTRODUCTION

The term *chemical* process has been repeated referred to in a peripheral manner in the previous two chapters. This chapter is primarily concerned with chemical processes. As noted earlier, a process (plant) can regarded as a collection of individually operated process, i.e., *unit processes*. It is becoming increasingly evident that each separate unit of a plant *usually* influences all others in either a subtle or major way, justifying the unit process method of analysis. This approach is highlighted in this chapter. The key unit operations can ultimately be reduced to fluid flow, heat transfer, and mass transfer, subject areas that receive treatment in Chapter 8.

Many now believe that *chemistry* deals with the combination of atoms and with the forces between atoms described by *physics*. Atomic combination involves atomic forces, and it is one of the objects of *physical chemistry* to see how far the chemical interactions that occur between atoms and molecules can be interpreted by studying the forces existing within and between atoms. The study of atomic structure provides information of why atoms combine. Chemistry is the science upon which the chemical process rests. The responsibility of the practicing engineer is to apply the chemistry of a particular process using coordinated scientific and engineering fundamentals and processes.

A *chemical reaction* is a process by which atoms or groups of atoms are combined and/or redistributed, resulting in a change in the molecular composition and properties. The products obtained from reactants depend on the condition under which a chemical reaction occurs. The products are the end result of the chemical process and can be categorized into a host of areas, details of which are provided in Part III of this book.

This chapter presents information on the chemical process; the conservation laws of mass, energy, and momentum; stoichiometry; and optimal process design. Additional information and discussion is provided on general problem-solving techniques relevant to the chemical process industry (CPI). Seven illustrative examples are provided to demonstrate concepts presented throughout the chapter.

9.2 THE CHEMICAL PROCESS

The development of a process from its inception to operations requires the complete integration of a series of stages as listed below.

1. Process research
2. Research evaluation and process potential
3. Process development

4. Preliminary engineering studies
5. Pilot plant
6. Commercial plant

The conception of a process may be originated by an engineer, a scientist, or any other person. A successful process does not just happen; it is almost always founded on well-known, sound principles. Any process, whether for the production of gasoline, drugs, sulfuric acid, or rubber tires, may be visualized as a box into which raw materials and energy are fed and from which useful products, waste, and energy emerge. Nearly all chemical processes employ operations where the feed materials undergo chemical changes to produce a more valuable product.

When overall mass and energy balances have been established, individual components of a process can often be designed for optimum economic performance without reference to each other. This may require substantial simplification in calculations. Such calculations will often be of a similar nature regardless of process type, so that standard and/or traditional methods can be employed for most economic design.

As noted in Chapter 7, the overall operation of a chemical process may be divided into several small operations or steps such as flow of materials, filtration, extraction, distillation, and transfer of heat through which the materials in the process pass. The practitioner must understand each of the individual operations; however, one must never lose sight of the unified picture of the complete processes and not limit themselves to one small portion of the plant; in effect the idea of the complete process must be kept in mind.

To summarize, a process is built around the technology for the complete process rather than around the individual operations. The individual operations are important, they must be understood, and their principles must be mastered; but the practitioner must never lose sight of the final application(s) where all the smaller operations are combined into one smoothly operating process.

9.3 THE CONSERVATION LAW

To better understand the design, as well as the operation and performance of chemical processes, it is necessary to understand the fundamentals and principles underlying the conservation laws and stoichiometry. How can one predict what products will be emitted from effluent streams? At what temperature must a unit be operated to ensure the desired performance? How much energy in the form of heat is given off during a chemical reaction? Is it economically feasible to recover this heat? Is the design appropriate? The answers to these questions are rooted in the various theories of thermodynamics, chemistry, physics, and applied economics.

One of the keys necessary to answering the above questions is often obtained via the application of one or more of the conservation laws that are discussed in this section. Four important terms are defined below before proceeding to the conservation laws.

1. A *system* is any portion of the universe that is set aside for study.
2. Once a system has been chosen, the rest of the universe is referred to as the *surroundings*.

3. A system is described by specifying that it is in a certain *state*.
4. The *path*, or series of values certain variables assume in passing from one state to another, defines a process.

Mass, energy, and momentum are all conserved. As such, each quantity obeys the general conservation law below, as applied within a system.

$$\begin{Bmatrix} quantity \\ into \\ system \end{Bmatrix} - \begin{Bmatrix} quantity \\ out\ of \\ system \end{Bmatrix} + \begin{Bmatrix} quantity \\ generated\ in \\ system \end{Bmatrix} = \begin{Bmatrix} quantity \\ accumulated \\ in\ system \end{Bmatrix} \quad (9.1)$$

Equation 9.1 may also be written on a time rate basis:

$$\begin{Bmatrix} rate \\ into \\ system \end{Bmatrix} - \begin{Bmatrix} rate \\ out\ of \\ system \end{Bmatrix} + \begin{Bmatrix} rate \\ generated\ in \\ system \end{Bmatrix} = \begin{Bmatrix} rate \\ accumulated \\ in\ system \end{Bmatrix} \quad (9.2)$$

The conservation law may be applied at the macroscopic, microscopic, or molecular level. One can best illustrate the differences in these methods with an example. Consider a system in which a fluid is flowing through a cylindrical tube (see Figure 9.1) and define the system as the fluid contained within the tube between Points 1 and 2 at any time. If one is interested in determining changes occurring at the inlet and outlet of a system, the conservation law is applied on a "macroscopic" level to the entire system. The resultant equation (usually algebraic) describes the overall changes occurring *to* the system without regard for internal variations *within* the system. This approach is usually applied by the practicing engineer.

The microscopic approach is employed when detailed information concerning the behavior *within* the system is of interest. The conservation law is then applied to a *differential* element within the system that is large compared to an individual molecule, but small compared to the entire system. The resulting differential equation is then expanded via an integration to describe the behavior of the entire system. This is defined as the *transport phenomena approach* (Theodore 1970; Theodore, Dupont and Ganesan 2017).

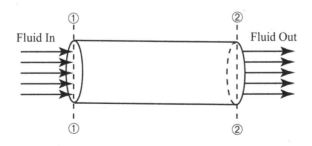

FIGURE 9.1 Fluid flow through a cylinder tube.

The molecular approach involves the application of the conservation laws to individual molecules. This leads to a study of statistical and quantum mechanics – both of which are beyond the scope of this text. In any case, the description at the molecular level is of little value to the practicing engineer. However, the statistical averaging of molecular quantities in either a differential or finite element within a system can lead to a more meaningful description of the behavior of a system.

Both the microscopic and molecular approaches shed light on the physical reasons for the observed macroscopic phenomena. Ultimately, however, for the practicing engineer, these approaches may be valid but are akin to attempting to kill a fly with a machine gun. Developing and solving these differential equations (in spite of the advent of computer software packages) is typically not worth the trouble.

Traditionally, the applied mathematician has developed differential equations describing the detailed behavior of systems by applying the appropriate conservation law to a differential element or shell within the system. Equations were derived with each new application. The engineer later removed the need for these tedious and error-prone derivations by developing a general set of equations that could be used to describe systems. These have come to be referred to by many as the *transport equations*. In recent years, the trend toward expressing these equations in vector form has gained momentum (no pun intended). However, the shell-balance approach has been retained in most texts where the equations are presented in componential form, i.e., in three coordinate systems – rectangular, cylindrical, and spherical. The componential terms can be "lumped" together to produce a more concise equation in vector form. The vector equation can be, in turn, re-expanded into other coordinate systems. This information is available in the literature (Theodore 1970; Bird, Stewart and Lightfoot 2007).

The macroscopic approach is primarily adopted and applied in this text, and little to no further reference to microscopic or molecular analyses will be made. This section's aim, then, is to express the laws of conservation for mass, energy, and momentum in algebraic or finite difference form.

9.4 CONSERVATION OF MASS, ENERGY, AND MOMENTUM

The conservation law for mass can be applied to any process, equipment, or system. The general form of this law is given by Equations 9.3 and 9.4:

$$\left\{\begin{array}{c}mass\\in\end{array}\right\} - \left\{\begin{array}{c}mass\\out\end{array}\right\} + \left\{\begin{array}{c}mass\\generated\end{array}\right\} - \left\{\begin{array}{c}mass\\consumed\end{array}\right\} = \left\{\begin{array}{c}mass\\accumulated\end{array}\right\} \quad (9.3)$$

$$I \quad - \quad O \quad + \quad G \quad - \quad C \quad = \quad A$$

or on a time rate basis by

$$\left\{\begin{array}{c}rate\ of\\mass\\in\end{array}\right\} - \left\{\begin{array}{c}rate\ of\\mass\\out\end{array}\right\} + \left\{\begin{array}{c}rate\ of\\mass\\generated\end{array}\right\} - \left\{\begin{array}{c}rate\ of\\mass\\consumed\end{array}\right\} = \left\{\begin{array}{c}rate\ of\\mass\\accumulated\end{array}\right\} \quad (9.4)$$

$$I \quad - \quad O \quad + \quad G \quad - \quad C \quad = \quad A$$

The law of conservation of mass states that mass can neither be created nor destroyed. Nuclear reactions, in which interchanges between mass and energy are known to occur, provide a notable exception to this law. Even in chemical reactions, a certain amount of mass-energy interchange takes place. However, in normal engineering applications, nuclear reactions do not occur and the mass-energy exchange in chemical reactions is so minuscule that it is not normally taken into account.

The law of conservation of energy, which, like the law of conservation of mass, applies for all processes that do not involve nuclear reactions, states that energy can neither be created nor destroyed. As a result, the energy level of a system can change only when energy crosses the system boundary, i.e.,

$$\Delta(\text{Energy level of system}) = \text{Energy crossing boundary} \qquad (9.5)$$

(Note: The symbol "Δ" means "change in.") Energy crossing the boundary can be classified into one of two different ways: heat, Q, or work, W. Heat is energy moving between the system and the surroundings by virtue of a temperature difference driving force, and heat flows from high temperature to low temperature. The entire system is not necessarily at the same temperature; neither are the surroundings. If a portion of the system is at a higher temperature than a portion of the surroundings and as a result, energy is transferred from the system to the surroundings, that energy is classified as heat. If part of the system is at a higher temperature than another part of the system and energy is transferred between the two parts, that energy is not classified as heat because it is not crossing the boundary. Work is also energy moving between the system and surroundings, but the driving force here is something other than temperature difference, e.g., a mechanical force, a pressure difference, gravity, a voltage difference, a magnetic field, etc. Note that the definition of work is a force acting through a distance.

The energy level of a system has three contributions: kinetic energy, potential energy, and internal energy. Any body in motion possesses kinetic energy. If the system is moving as a whole, its kinetic energy, E_k, is proportional to the mass of the system and the square of the velocity of its center of gravity. The phrase "as a whole" indicates that motion inside the system relative to the system's center of gravity does not contribute to the E_k term, but rather to the internal energy term. The terms external kinetic energy and internal kinetic energy are sometimes used here. An example would be a moving railroad tank car carrying liquid waste. (The liquid waste is the system.) The center of gravity of the waste is moving at the velocity of the train, and this constitutes the system's external kinetic energy. The liquid molecules are also moving in random directions relative to the center of gravity, and this constitutes the system's internal energy due to motion inside the system, i.e., internal kinetic energy. The potential energy, E_p, involves any energy the system as a whole possesses by virtue of its position (more precisely, the position of its center of gravity) in some force field, e.g., gravity, centrifugal, electrical, etc., that provides the system with the potential for accomplishing work. Again, the phase "as a whole" is used to differentiate between external potential energy, E_p, and internal potential energy. Internal potential energy refers to potential energy due to force fields inside the system. For example, the electrostatic force fields (bonding) between atoms and molecules provide these particles with the potential for work. The internal energy, U, is the sum of all internal kinetic and internal potential energy contributions (Theodore and Reynolds 1991).

The law of conservation of energy, which is also called the first law of thermodynamics, may now be written as:

$$\Delta(U + E_k + E_p) = Q + W \tag{9.6}$$

or equivalently as

$$\Delta U + \Delta E_k + \Delta E_p = Q + W \tag{9.7}$$

It is important to note the sign convention for Q and W adapted for the above equation. Since any term is always defined as the final minus the initial state, both the heat and work terms must be positive when they cause the system to gain energy, i.e., when they represent energy flowing from the surroundings to the system. Conversely, when the heat and work terms cause the system to lose energy, i.e., when energy flows from the system to the surroundings, they are negative in sign. This sign convention is not universal and the reader must take care to check what sign convention is being used by a particular author when referring to the literature. For example, work is often defined in some texts as positive when the system does work on the surroundings (Theodore and Reynolds 1991; Green and Perry 2008; Theodore 2014).

The general conservation law for momentum on a rate basis when applied to a volume element describes the rate of momentum and forces acting on a moving fluid in a volume element of concern at any time. Each rate of momentum or force term in the above equation can be expressed in terms of force to maintain dimensional consistency. The application of this conservation law finds extensive use in the field of fluid mechanics or fluid flow. It is suggested that the reader refers to the literature for more information (Theodore 1970; Theodore, Dupont and Ganesan 2017).

9.5 STOICHIOMETRY

When chemicals react, they do so according to a strict proportion. When oxygen and hydrogen combine to form water, the ratio of the amount of oxygen to the amount of hydrogen consumed is always 7.94 by mass and 0.500 by moles. The term stoichiometry refers to this phenomenon, which is sometimes called the chemical law of combining weights. The reaction equation for the combining of hydrogen and oxygen is:

$$2H_2 + O_2 = 2H_2O \tag{9.8}$$

In chemical reactions, atoms are neither generated nor consumed, merely rearranged with different bonding partners. The manipulation of the coefficients of a reaction equation so that the number of atoms of each element on the left of the equation is equal to that on the right is referred to as *balancing* the equation. Once an equation is balanced, the whole number molar ratio that must exist between any two components of the reaction can be determined simply by observation; these are known as *stoichiometric ratios*. There are three such ratios (not counting the reciprocals) in the above reaction. These are:

2 mol H_2 consumed/mol O_2 consumed
1 mol H_2O generated/mol H_2 consumed
2 mol H_2O generated/mol O_2 consumed

The unit mole represents either the gmol or the lbmol. Using molecular weights, these stoichiometric ratios (which are molar ratios) may easily be converted to mass ratios. For example, the first ratio above may be converted to a mass ratio by using the molecular weights of H_2 (2.016) and O_2 (31.999) as follows:

(2 gmol H_2 consumed) (2.016 g/gmol) = 4.032 g H_2 consumed
(1 gmol O_2 consumed) (31.999 g/gmol) = 31.999 g O_2 consumed

The mass ratio between the hydrogen and oxygen consumed is therefore:

4.032/31.999 = 0.126 g H_2 consumed/g O_2 consumed

These molar and mass ratios are used in material balances to determine the amounts or flow rates of components involved in chemical reactions.

Multiplying a balanced reaction equation through by a constant does nothing to alter its meaning. The reaction used as an example above is often written:

$$H_2 + \tfrac{1}{2}O_2 = H_2O \tag{9.9}$$

In effect, the stoichiometric coefficients of Equation 9.8 have been multiplied by 0.5. There are times, however, when care must be exercised because the solution to a problem may depend on the manner or form the reaction is written. This is the case with chemical equilibrium problems and problems involving thermochemical reaction equations.

There are two different types of material balances that may be written when a chemical reaction is involved: the molecular balance and the atomic balance. It is a matter of convenience which of the two types is used. Each is briefly discussed below.

The molecular balance is the same as that described earlier. Assuming a steady-state continuous reaction, the accumulation term, A, is zero for all components involved in the reaction, Equation 9.3 then becomes:

$$I + G = O + C \tag{9.10}$$

If a total material balance is performed, the above form of the balance equation must be used if the amounts or flow rates are expressed in terms of moles, e.g., lbmol or gmol/hr, since the total number of moles can change during a chemical reaction. If, however, the amounts or flow rates are given in terms of mass, e.g., kg or lb/hr, the G and C terms may be dropped since mass cannot be gained or lost in a chemical reaction. Thus,

$$I = O \tag{3.11}$$

In general, however, when a chemical reaction is involved, it is usually more convenient to express amounts and flow rates using moles rather than mass.

A material balance that is not based on the chemicals (or molecules), but rather on the atoms that make up the molecules, is referred to as an atomic balance. Since atoms are neither created nor destroyed in a chemical reaction, the G and C terms equal zero and the balance once again becomes:

$$I = O \tag{9.11}$$

As an example, consider once again the combination of hydrogen and oxygen to form water:

$$2H_2 + O_2 = 2H_2O \tag{9.8}$$

As the reaction progresses, O, and H, molecules (or moles) are consumed while H_2O molecules (or moles) are generated. On the other hand, the number of oxygen atoms (or moles of oxygen atoms) and the number of hydrogen atoms (or moles of hydrogen atoms) do not change. Care must also be taken to distinguish between molecular oxygen and atomic oxygen. If, in the above reaction, one starts out with 1,000 lbmol of O_2 (oxygen molecules), one may replace this with 2,000 lbmol of O (oxygen atoms).

Thus, a chemical equation provides a variety of qualitative and quantitative information essential for the calculation of the quantity of reactants reacted and products formed in a chemical process. As noted, a balanced chemical equation must have the same number of atoms of each type in the reactants on the left-hand side of the equation and in the products on the right-hand side of the equation. Thus, a balanced equation for butane combustion (reaction with oxygen to form oxidized end products CO_2 and H_2O) is:

$$C_4H_{10} + (13/2)O_2 = 4CO_2 + 5H_2O \tag{9.12}$$

Note that:

- Number of carbons in reactants = number of carbons in products = 4
- Number of oxygens in reactants = number of oxygens in products = 13
- Number of hydrogens in reactants = number of hydrogens in products = 10
- Number of moles of reactants is 1 mol C_4H_{10} + 6.5 mol O_2 = 7.5 mol total
- Number of moles of products is 4 mol CO_2 + 5 mol H_2O = 9 mol total

The reader should note that although the number of moles on both sides of the equation does not balance, the masses of reactants and products (in line with the conservation law for mass) must balance.

9.6 LIMITING AND EXCESS REACTANTS

Limiting and excess reactants involve an extension of the stoichiometric calculations provided above. Consider the following example.

When methane is combusted completely, the stoichiometric equation for the reaction is:

$$CH_4 + 2O_2 = CO_2 + 2H_2O \qquad (9.13)$$

The stoichiometric ratio of the oxygen to the methane is:

0.5 mol methane consumed/mol oxygen consumed

If one starts out with 1 mol of methane and 3 mol of oxygen in a reaction vessel, only 2 mol of oxygen would be used up, leaving an excess of 1 mol of oxygen in the vessel. In this case, the oxygen is called the *excess* reactant and methane is the *limiting* reactant. The limiting reactant is defined as the reactant that would be completely consumed if the reaction went to completion. All other reactants are excess reactants. The amount by which a reactant is present in excess of stoichiometric requirements (i.e., the exact number of moles needed to react completely with the limiting reactant) is defined as the percent excess and is given by Equation 3.14:

$$\% \, excess = \left(\frac{n - n_s}{n_s}\right) \times 100 \qquad (3.14)$$

where n = number of moles of the excess reactant at the start of the reaction; and n_s = the stoichiometric number of moles of the excess reactant.

In the example above, the stoichiometric amount of oxygen is 2 mol, since that is the amount that would react with the 1 mol of methane. The excess amount of oxygen is 1 mol, which is a percentage excess of 50% or a fractional excess of 0.50.

A detailed and expanded treatment of stoichiometry is available in references Green and Perry (2008) and Theodore (2014).

9.7 OPTIMUM PROCESS DESIGN

In almost every case encountered by an engineer, there are several alternative methods which can be used for any given process or operation. For example, formaldehyde can be produced by catalytic dehydrogenation of methanol, by controlled oxidation of natural gas, or by direct reaction between CO and H_2 under special conditions of catalyst, temperature, and pressure. Each of these processes contains many possible alternatives involving variables such as gas mixture composition, temperature, pressure, and choice of catalyst. It is the duty of the engineer to choose the best process and to incorporate into the design the equipment and methods which will yield the "best" or optimum results. The process engineer ordinarily prefers to replace the word "best" by "optimum" and then add a designating qualification such as "optimum economic process" or "optimum operation process."

If there are two or more methods for obtaining exactly equivalent final results, the preferred method would obviously be the one involving the least total cost. This is the basis of *optimum economic design*. One typical example of an optimum economic design is determining the pipe diameter to use when pumping a given amount of fluid from one point to another. Here the same final result (i.e., a set amount of

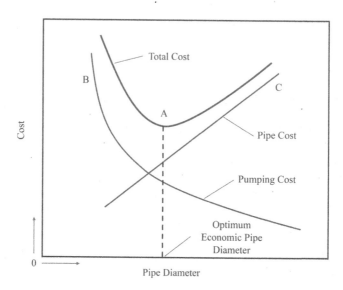

FIGURE 9.2 Optimal economic design for pipe diameter selection.

fluid pumped between two given points) can be accomplished by using a near infinite number of different pipe diameters. However, an economic analysis will show that one particular pipe diameter gives the least total cost. The total cost includes the cost for pumping the liquid and the cost (i.e., fixed charges) for the installed piping system.

A graphical representation showing the meaning of an optimum economic pipe diameter is presented in Figure 9.2. As shown in this figure, pumping cost increases with decreased size of pipe diameter because of frictional effects, while the fixed charges for the pipeline become lower when smaller pipe diameters are used because of the reduced capital investment. The optimum economic diameter is found where the sum of the pumping costs and fixed costs for the pipeline is at a minimum. This optimum point is shown in Figure 9.2 by the letter A, while the curves representing the cost of power required for pumping and the capital cost for pipe are shown by lines B and C, respectively.

From a chemical process perspective, the engineer is always concerned with the economic aspects of the chemistry of a process, particularly with *yields*, *conversions*, and *rates*. The operational efficiency of a chemical process usually is interpreted in terms of the yield and conversion. These terms are generally defined as follows:

$$\text{yield} = \frac{\text{moles of desired product formed}}{\text{moles of reactant consumed}} \qquad (9.14)$$

$$\% \text{ conversion} = \frac{\text{moles of reactant consumed}}{\text{moles of reactant fed}} \times 100 \qquad (9.15)$$

It is fair to say that the general approach in any process design involves a carefully balanced combination of theory, practice, and plain common sense. Early in any study the engineer must deal with many different types of experimental and empirical data

Chemical Process Industries: Fundamentals and Principles 187

and must be prepared to make many assumptions. Sometimes these assumptions are made because no absolutely accurate values or methods of calculation are available.

Another important factor in the approach to any design problem involves the previously mentioned economic conditions and limitations. The engineer must consider costs and probable profits constantly throughout all the design work. No matter how efficiently a process operates to produce a final product of high purity, the process is a failure if the product cannot be sold at a profit. Before a process becomes operational, a thorough market analysis must be made to determine how much of the product can be sold and at what price.

9.8 PROBLEM SOLVING

Engineers (and to a lesser extent scientists) are known for their problem-solving ability. It is probably this ability more than any other that has enabled many engineers to rise to positions of leadership and top management within their companies.

In problem solving, in academia and in industry, considerable importance is attached to a proper analysis of the problem, to a logical recording of the problem solution, and to overall professional appearance of the finished product calculations. Neatness and clarity of presentation should be distinguishing marks of the work. Engineering students should always strive to practice professional habits of problem-solving analysis and to make a conscious effort to improve the appearance of each document whether it is submitted for grading or is included in a design notebook.

The value of an engineer or scientist is determined by their ability to apply basic principles, facts, and methods in order to accomplish some useful purpose. In this modern age of industrial competition, the ultimate definition of a useful purpose is usually based on a tangible profit of monetary value. It is not sufficient, therefore, to have a knowledge and understanding of physics, chemistry, mathematics, mechanics, stoichiometry, thermodynamics, the unit operations, chemical technology, and other related engineering and scientific subjects; they must also have the ability to apply this knowledge to practical situations and in making these applications recognize the importance of costs and profits.

Engineers and scientists who have mastered the method of problem solving are considerably more successful in their work than are people not trained in this technique. In the past, many problems were of such a routine nature that a resort to deductive reasoning would suffice, and premises of deduction could be taken from handbooks. However, many of the engineering problems of today cannot be solved by mere "handbook techniques." Experimentation, research, and development have indeed become significant activities in today's world.

9.8.1 Generic Problem-solving Techniques

Certain methods of logic and techniques of calculation are fundamental to the solution of many problems, and there is a near infinite number of these approaches of solution. Words such as creative, ingenuity, original, etc., appear in all these approaches. What do they all have in common? They provide a systematic, logical approach to solving problems, and what follows is the authors' definition of a generic approach.

The methodology of solving problems has been discussed by most mathematicians and logicians since the days of Aristotle. *Heuristic* ("serving to discover") is the term often given to this study of the methods and rules of solving problems.

Nearly always, a stepwise approach to the solution is desirable. The basic steps in all problem-solving techniques include:

1. Understanding the problem
2. Devising a plan for solution
3. Carrying out the plan
4. Looking back to review and check the solution

9.8.2 A Specific Problem-solving Approach

Here is an approach drawn from the files of the lead author of this book that involves six steps to solving problems:

1. The first step, and perhaps the most important one, is that of a problem definition. It is not possible to establish rules for problem definition that are sufficiently general to be useful. Environmental problems related to the CPI are so diverse that it is up to the analyst to clearly state the nature of their problem. This will establish a definite objective for the analysis and is invaluable in outlining a path from the problem to the solution.
2. The next step is a definition of the theory that governs the phenomena of the problem. This theory is usually available from a variety of sources, both published and unpublished, but for those isolated cases where there is no theory available, it is worthwhile to postulate one (or several) and to test its validity later by comparing (if available) the solution of the mathematical model with experimental results. One of the advantages of the computerized approach is the ability to rapidly obtain solutions to various cases; this makes comparisons among alternate theories practically possible.
3. In the third step, the theory, as applied to the problem, is written in mathematical symbols. This necessary step forces the analyst to develop a clear, unambiguous definition of the problem. The physical system may be described by a set of simultaneous algebraic and differential equations. These equations must be written in the most direct form possible, and no manipulations are required at this stage. It is, however, worthwhile at this point to simplify the equations (where possible) by omitting insignificant terms. Care is needed to ensure that any terms omitted are indeed insignificant to the solution of the problem. It is often possible to eliminate entire equations by merely neglecting minor fluctuations in certain intermediate variables. For example, if the heat capacity of a component mixture (to be purified) required for a heat balance varies by only 1% of its value due to expected variations in composition, an average constant number could be substituted rather than include an equation in the model to continuously compute its value.
4. Having assembled the equations, a procedural method for the solution is required. A consideration of the solution required from the model is a

necessary preliminary step to the computation phase. A list of the various cases to be studied and the information that is expected in each case will reveal possible redundant situations as well as assist the programming of the computation phase.
5. The computation phase that follows may offer several alternate routes to the solution. The method selected will depend on the complexity of the equations to be solved. There are three general levels of computation complexity:
   a. The most elementary is common sense, i.e., the solutions desired can be obtained from the model by inspection if the equations or the solutions required are sufficiently simple. It should be realized that this technique cannot be extrapolated to more complex cases without requiring increasing amounts of pure guesswork.
   b. The next level, again restricted to systems of modest complexity, solves the equation by analytical or numeric techniques. A considerable amount of skill is required, however, to solve even some of the simplest problems. Selecting the most fruitful path may be the only feasible method for problems of even fair complexity.
   c. The solution may require an optimizing method.
6. The sixth and last phase is the study and verification of the solutions obtained from the mathematical model. Any unexpected solutions should be rationalized to ensure that no errors have occurred in the computations; also, some of the computer or logic should be specifically designed to check the validity of the results of the mathematical model.

9.8.3 Some General Comments

Here is what one of the authors has stressed to his students in terms of developing problem-solving skills and other creative thinking (Abulencia and Theodore 2015).

1. Carefully define the problem at hand.
2. Obtain all pertinent data and information.
3. Initially, generate an answer or solution.
4. Examine and evaluate as many alternatives as possible, employing "what if" scenarios.
5. Reflect on the above over time.
6. Consider returning to Step 1 and repeat/expand the process as necessary to reach a satisfactory final solution.

Many now believe creative thinking should be part of every student's education. Here are some ways that have proven to nudge the creative process along:

1. Break out of the one-and-only answer rut.
2. Use creative thinking techniques and games.
3. Foster creativity with assignments and projects.
4. Be careful not to punish creativity.

The above-suggested activities will ultimately help develop a critical thinker that:

1. Raises important questions and problems, formulating them clearly and precisely.
2. Gathers and assesses relevant information, using abstract ideas to interpret it effectively.
3. Comes to well-rounded conclusions and solutions and tests them against relevant criteria and standards.
4. Thinks open-mindedly with alternative systems of thought, recognizing and assessing, as need be, their assumptions, implications, and practical consequences.
5. Communicates effectively with others in figuring out solutions to complex problems.

The analysis aspect of a problem remains. It essentially has not changed. The analysis of a new problem in engineering and science can still be divided into four basic steps:

1. Consideration of the process in question
2. Mathematical description of the process, if applicable
3. Solution of any mathematical relationships to provide a solution
4. Verification of the solution

Summarizing, knowledge alone is not enough for the efficient solution of engineering problems, skill is also required. This skill is acquired not only through diligent practice, but is facilitated by application of a methodology such as presented above.

9.9 ILLUSTRATIVE EXAMPLES

Seven illustrative examples complement the material presented above.

9.9.1 Illustrative Example 1

The following information is provided for an ideal gas: pressure = 1.0 atm, temperature = 60°F, molecular weight of gas = 29 lb/lbmol. Determine the density of the gas in lb/ft³.

Solution: Rewrite the ideal gas law in terms of the density, ρ.

$$PV = nRT \tag{9.16}$$

$$n = m/MW \tag{9.17}$$

where m = mass of gas and MW = molecular weight. Thus,

$$PV = (m/MW)RT \tag{9.18}$$

and

$$m/V = \rho = P(MW)/RT \tag{9.19}$$

Substituting in the previous equation gives

$$\rho = P(MW)/RT = (1\,\text{atm})(29\,\text{lb/lbmol})/\left[(0.73\,\text{atm}-\text{ft}^3/\text{lbmol}-°R)(60+460°R)\right]$$
$$= (29\,\text{atm}-\text{lb/lbmol})/(379.6\,\text{atm}-\text{ft}^3/\text{lbmol}) = 0.0764\,\text{lb/ft}^3$$

Note: The choice of R is arbitrary, provided consistent units are employed. Since the molecular weight of the given gas is 29 lb/lbmol, the calculated density can be assumed to apply to air. The effect of pressure, temperature, and molecular weight on density can also be obtained directly from the ideal gas law. Increasing the pressure and molecular weight increases the density, while increasing the temperature decreases the density.

9.9.2 ILLUSTRATIVE EXAMPLE 2

An external gas stream is fed into an air pollution control device at a rate of 10,000 lb/hr in the presence of 20,000 lb/hr of air. Due to environmental constraints, 1,250 lb/hr of a conditioning agent are added to assist the treatment of the gas stream. Determine the rate of product gases exiting the unit in lb/hr.

Solution: First, write the conservation law for mass to the control device on a rate basis.

$$(\text{rate of mass in}) - (\text{rate of mass out}) + (\text{rate of mass generated})$$
$$= (\text{rate of mass accumulated})$$

Apply this equation subject to the conditions in the problem statement. Remember that mass is not generated and steady-state conditions apply, therefore:

$$(\text{rate of mass in}) = (\text{rate of mass out})$$

or simply

$$m_{in} = m_{out}$$

Substituting yields:

$$m_{in} = 10,000 + 20,000 + 1250 = 31,250\,\text{lb/hr}$$

Since $m_{in} = m_{out}$, the product gas flow rate is then:

$$m_{out} = 31,250\,\text{lb/hr}$$

9.9.3 ILLUSTRATIVE EXAMPLE 3

The reaction equation for the combustion of propane is as follows:

$$C_3H_8 + O_2 \rightarrow CO_2 + H_2O \quad (9.20)$$

1. Balance this equation and determine the ratio of reactants to products.
2. Using the balanced equation, determine the standard cubic feet (scf) of air required for stoichiometric combustion of 1.0 scf propane (C_3H_8) and calculate both the stoichiometric air requirements and flue gas produced from the complete combustion of propane.

Solution: The balanced stoichiometric equation is:

$$C_3H_8 + 5O_2 \rightarrow 3CO_2 + 4H_2O \quad (9.21)$$

Although the number of moles on both sides of the equation do not balance, the masses of reactants and products (in accordance with the conservation law for mass) must balance.

The scf of O_2 required for the complete combustion of 1 scf of propane can now be determined. From the balanced equation, there are 5 moles of O_2 required per mole of propane. Since the moles of ideal gas are directly proportional to their volume,

$$\text{scf of } O_2 = 5(1 \text{scf of propane}) = 5 \text{scf}$$

From a mass balance, 44 lb of propane has reacted with 160 lb of O_2 to form 132 lb of CO_2 and 72 lb of H_2O, with an initial and final total mass of 204 lb.

The amount of N_2 in a quantity of air that contains 5 scf of O_2 is given by the following knowing that the nitrogen to oxygen volume (or mole) ratio in air is 79/21:

$$\text{scf of } N_2 = (79/21)(\text{scf of } O_2) = (79/21)(5) = 18.81 \text{scf}$$

The stoichiometric amount of air is then

$$\text{scf of air @ stoichiometric} = \text{scf of } N_2 + \text{scf of } O_2 = 18.81 + 5 = 23.81 \text{scf}$$

In order to calculate the amount of flue gas produced, one must include the nitrogen from the air in addition to the products of combustion, that is, carbon dioxide and water vapor.

$$\text{scf of flue gas} = \text{scf of } N_2 + \text{scf of } CO_2 + \text{scf of } H_2O = 18.81 + 3.0 + 4.0 = 25.81 \text{scf}$$

9.9.4 ILLUSTRATIVE EXAMPLE 4

Describe the difference between thermodynamics and kinetics.

Chemical Process Industries: Fundamentals and Principles

Solution: Thermodynamics deals with the transformation of energy from one form to another. The energy balance equation is an expression of the *first law* of thermodynamics. The *second law* essentially states that in a process involving heat transfer alone, energy may be transferred only from a higher temperature to a lower one. Thermodynamics is also useful in determining the composition and distribution of phases in equilibrium as well as predicting the distribution of chemical species in reaction *equilibrium*.

Kinetics considers the rate at which chemicals react, not the equilibrium distribution that might result. Data on the rate of reactions are necessary in the design of industrial chemical reactors.

Additional details of thermodynamics and kinetics are available in the literature (Theodore, Ricci and VanVliet 2009; Theodore 2014).

9.9.5 Illustrative Example 5

Briefly describe the difference between unit operations and chemical reactors.

Solution: Any chemical process can be broken down into a series of changes steps involving *physical* or are *chemical* changes. The physical changes involve a combination of one or more of the transfer processes described in this chapter, i.e., fluid flow, heat transfer, and mass transfer. The chemical changes are chemical reactions that are carried out in various types of industrial chemical reactors.

9.9.6 Illustrative Example 6

Describe the difference between humidification and dehumidification.

Solution: In *humidification*, water or another liquid is vaporized, and the heat of vaporization must be transferred to the liquid. *Dehumidification* is the condensation of water vapor from air, or in general, the condensation of any vapor from a gas. Additional details are available in the literature (Theodore 2014).

9.9.7 Illustrative Example 7

Calculate the composition of air on a mass basis.

Solution: Since combustion calculations assume dry air to contain 21% oxygen and 79% nitrogen on a mole or volume basis, 1 mol of air consists of 0.21 mol of oxygen and 0.79 mol of nitrogen. The 0.21 mol of O_2 equals 6.72 lb; while the 0.79 mol of N_2 equals 22.12 lb. Therefore, on a weight basis, dry air contains 23.3% O_2 by weight (6.72/28.84) and 76.7% N_2 by weight (22.12/28.84).

PROBLEMS

9.1 Describe the role of *Understanding the Problem* in the problem solution process.
9.2 Describe the role *Looking Back* plays in the problem solution process.
9.3 Discuss the importance of pilot plans to the chemical design process.

9.4 Describe the role of materials of construction, particularly as it applies to corrosion, in chemical process design.

9.5 The reaction equation for the combustion of chlorobenzene is as follows:

$$C_6H_5Cl + O_2 \rightarrow CO_2 + H_2O + HCl \tag{9.22}$$

a. Balance this equation and determine the ratio of reactants to products.
b. Using the balanced equation, determine the scf of air required for stoichiometric combustion of 1.0 scf chlorobenzene and calculate both the stoichiometric air requirements and flue gas produced from its complete combustion.

9.6 Using the calculation approach from Illustrative example 1 and the problem statement of Illustrative example 2, determine the density of the product gas mixture exiting the unit, as well as its volumetric flow rate using the following additional data, assuming isobaric conditions:

The pressure of all gas streams = 1 atm

The temperature of the external gas stream fed into the air pollution control device = 250°F

The temperature of the added air and vapor conditioning agent = 75°F

REFERENCES

Abulencia, P., and Theodore, L. 2015. *Open-Ended Problems: A Future Engineering Approach*. Salem, MA: Scrivener-Wiley Publishing.

Bird, R., Stewart, W., and Lightfoot, E. 2007. *Transport Phenomena*, Revised 2nd Edition. Hoboken, NJ: John Wiley & Sons.

Green, D., and Perry, R. (Ed.). 2008. *Perry's Chemical Engineers' Handbook*, 8th Edition. New York, NY: McGraw-Hill.

Theodore, L. 1970. *Introduction to Transport Phenomena*. Scranton, PA: International Textbook Co.

Theodore, L. 2014. *Chemical Engineering: The Essential Reference*. New York, NY: McGraw-Hill.

Theodore, L., Dupont, R.R., and Ganesan, K. 2017. *Unit Operations in Environmental Engineering*. Beverly, MA: Scrivener Publishing.

Theodore, L., and Reynolds, J. 1991. *Thermodynamics*. A Theodore Tutorial, East Williston, NY, originally published by the USEPA/APTI, RTP, NC.

Theodore, L., Ricci, F., and VanVliet, T. 2009. *Thermodynamics for the Practicing Engineer*. Hoboken, NJ: John Wiley & Sons.

10 Industry-Specific Processes

10.1 INTRODUCTION

By now, the reader is no doubt aware that the term *chemical process* in this book is loosely defined to represent all industrial processes. Even the so-called "non-chemical" processes, e.g., manufacturing, architecture and urban planning, energy, the military, terrorism, etc., all involve some form of chemical activity. All of these processes are addressed in this chapter as well as in Part III since the practicing engineer usually has contact with more than one industry. This chapter presents information on the early chemical process industry (CPI) and its evolution over time. Four illustrative examples are provided to demonstrate concepts presented throughout the chapter.

10.2 THE EARLY CPI

A view of the range of the CPI at the middle of the 20th century is summarized in Table 10.1. This table includes both the chemical categories prominent at the time along with one of the main products associated with each industry. Some of the major chemical companies at this time are listed in Table 10.2.

TABLE 10.1
Chemical Process Industries at the Middle of the 20th Century in the United States

Industry	Typical Product
Inorganic Chemicals	Sulfuric acid
Organic Chemicals'	Methanol
Petroleum & Petrochemicals	Gasoline
Pulp & Paper	Paper
Pigments & Paint	Zinc oxide
Rubber	Synthetic rubber
Plastics	Polyesters
Synthetic Fibers	Nylon
Minerals	Glass and ceramics
Cleansing Agents	Synthetic detergents
Biochemicals	Pharmaceuticals and drugs
Metals	Steel

TABLE 10.2
Early Chemical and Chemical Process Companies at the Middle of the 20th Century in the United States

Industry Sector	Company Name	Industry Sector	Company Name
Chemical	Air Reduction Company	Chemical Process (cont.)	Sun Chemical Corporation
	Allied Chemical Corporation		United Carbon Company
	American Cyanamid Company	Drug and Pharmaceutical	Abbott Laboratories
	American Potash & Chemical Corporation		Bristol-Myers Company
	Atlas Powder Company		Eli Lilly & Company
	Diamond Alkali Company		Merck & Company
	Dow Chemical Company		Parke, Davis & Company
	E. I. du Pont de Nemours & Company		Chas. Pfizer & Company
	W.R. Grace & Company		G. D. Searle & Company
	Hercules Powder Company		Upjohn Company
	Hooker Chemical Corporation		Warner Lambert Pharmaceutical Company
	Monsanto Chemical Corporation	Extractive and Fertilizers	American Agricultural Chemical Company
	Olin Mathieson Chemical Corporation		Texas Gulf Sulphur Company
	Reichold Chemicals, Inc.		U.S. Borax & Chemical Corporation
	Rohm and Haas Company	Petroleum	Atlantic Refining Company
	Stauffer Chemical Company		Cities Service Company
	Thiokol Chemical Corporation		Gulf Oil Corporation
	Union Carbide Corporation		Phillips Petroleum Company
	Witco Chemical Company		Shell Oil Company
	Wyandotte Chemical Corporation		Sinclair Oil Corporation
Chemical Process	Colgate-Palmolive Company		Socony Mobil Oil Company
	Corning Glass Works		Standard Oil Companies
	Eastman Kodak Company	Pulp & Paper	Kimberly-Clark Corporation
	Food Machinery & Chemical Corporation		St. Regis Paper Company
	Minnesota Mining & Mfg. Company		Scott Paper Company
	National Distillers & Chemical Corporation	Rubber	Firestone Tire & Rubber Company
	National Starch & Chemical Corporation		General Tire & Rubber Company
	Owens-Illinois Glass Company		B. F. Goodrich Company
	Pittsburgh Plate Glass Company		Goodyear Tire & Rubber Company
	Procter and Gamble Company		U.S. Rubber Company
	Sherwin-Williams Company	Synthetic Fibers	Celanese Corporation of America

10.3 THE SHREVE CPI

More recently the Shreve CPI summary (Basta 1997) provides an insight into more contemporary chemical industries in the United States. The "Shreve List" is provided in Table 10.3. Note that the list also includes a product/process normally associated with each industry.

TABLE 10.3
The Shreve CPI List

Industry	Typical Product/Process
Water Conditioning & Environmental Protection	Water conditioning chemicals
Energy, Fuels, Air Conditioning, & Refrigeration	Energy
Coal Chemicals	Distillation of coal tar
Fuel Gases	Liquefied natural gas
Industrial Gases	Hydrogen
Industrial Carbon	Activated carbon
Ceramics Industries	Refractories
Portland Cements, Calcium, & Magnesium Compounds	Gypsum
Glass Industries	Specialty glasses
Salt & Miscellaneous Sodium Compounds	Sodium chloride
Chlor-Alkali Industries	Soda ash
Electrolytic Industries	Aluminum
Electrothermal Industries	Calcium carbide
Phosphorus Industries	Baking powder
Potassium Industries	Potassium chloride
Nitrogen Industries	Ammonia
Sulfur & Sulfuric Acid	Sulfuric acid
Hydrochloric Acid & Miscellaneous Inorganic Chemicals	Hydrochloric acid
Nuclear Industries	Nuclear reactors
Explosives, Toxic Chemical Agents, & Propellants	Rocket propellants
Photographic Products Industries	Color photography
Surface-Coating Industries	Paints
Food & Food By-Product Processing Industries	Meat & Poultry
Agricultural Industries	Fertilizers
Fragrances, Flavors, & Food Additives	Perfume formulation
Oil, Fats, & Waxes	Vegetable oils
Soap & Detergents	Glycerin
Sugar & Starch Industries	Sugar
Fermentation Industries	Industrial alcohol
Wood Chemicals	Distillation of hardwood
Pulp & Paper Industries	Paper Stock
Plastics Industries	Chemical intermediates for resins
Synthetic Fiber & Film Industries	Polyester
Rubber Industries	Synthetic rubber
Petroleum Refining	Gasoline
Petrochemicals	Alkylation
Intermediates, Dyes & their Application	Nitration
Pharmaceutical Industry	Vitamins

10.4 THE THEODORE-DUPONT CPI

The CPIs presented in detail in Part III of this book is an attempt by the authors to condense the two previous lists of CPIs into one that contains sectors where health and environmental risks are significant. This Theodore-Dupont CPI list is presented in Table 10.4 and reflects the contents of Part III where each of these industrial sectors receives extensive treatment including applications and illustrative examples relevant to health and environmental risks concerns.

TABLE 10.4
The Theodore-Dupont CPI List

Industry
Inorganic Chemicals
Organic Chemicals
Petroleum Refining
Energy and Power
Pharmaceuticals
Food Products
Nanotechnology
Military & Terrorism
Travel & Weather
Architecture & Urban Planning
Environmental Sector

10.5 ILLUSTRATIVE EXAMPLES

Four illustrative examples complement the material presented above.

10.5.1 ILLUSTRATIVE EXAMPLE 1

Discuss some end uses of products associated with the pulp and paper industry.

Solution: Typical products from this industrial process sector include: paper, cardboard, and fiberboard. Uses of these products include: books, newspaper, boxes, and some building materials.

10.5.2 ILLUSTRATIVE EXAMPLE 2

Indicate whether esterification, humidification, and hydrolysis are unit operations.

Solution: Esterification and hydrolysis involve chemical conversions. Humidification is a unit operation.

10.5.3 Illustrative Example 3

Discuss some of the end products generated by the minerals industry.

Solution: Typical products from the minerals industry include: glass, ceramics, and cement. End uses include: windows, containers, bricks, pipe, and concrete.

10.5.4 Illustrative Example 4

Discuss the biochemical industry in general terms.

Solution: The biochemical process industry primarily includes pharmaceuticals, fermentation, products, and food. These industries either use processes involving biological action (such as fermentation) or produce products that are biologically active (such as penicillin in its medicinal uses). Pharmaceuticals are produced by controlled natural biological processes and by synthetic organic chemistry. Although much of the food industry is not generally considered part of the CPI those products which have been highly processed, such as sugar, are often considered part of the industry. Fermentation processes produce industrial and beverage alcohols, acetone, and acetic acid by biological action on various sugars.

PROBLEMS

10.1 Discuss the principle uses of synthetic fibers.
10.2 Which of the following are unit operations: distillation, pyrolysis, alkylation, and heat transfer?
10.3 Discuss the metals industry in general terms.
10.4 Discuss some of the end uses associated with the metals industry.

REFERENCE

Basta, N. 1997. *Shreve's Chemical Process Industries Handbook*, 6th Edition. New York, NY: McGraw-Hill.

11 Emergency Planning and Response

11.1 INTRODUCTION

Nothing is so destructive to a plant as fires, explosions, and uncontrolled emissions. Precautions to prevent these events must be taken into consideration in the design of any chemical process. Likewise, employees must be protected from events that compromise their health and safety on an ongoing basis. Safety measures not only keep an employee safely and consistently on the job, and keep equipment working, but also actually save money by reducing premiums paid by employers for liability and fire insurance. Too frequently familiarity with chemicals breeds carelessness; hence, well-run plants implement plans to prevent accidents and provide ongoing programs for detecting and alerting employees when accidents do happen to minimize property loss and health impacts. For certain, adequate safety and accident prevention measures require expert guidance.

This chapter addresses planning for emergencies and how to respond appropriately when they occur. Although much of the material in this chapter may appear to be dated, the response procedures in place still apply. The reader should note that the presentation is geared primarily for local and state personnel. However, the same basic approach is applied to planning and response activities for industrial applications.

Section 11.2 explains some reasons for planning ahead and discusses laws that require community groups to develop emergency response plans. Regardless of the existence of such laws, however, it makes good sense to plan ahead. For example, once an explosion has occurred, it is probably too late for analysis. The topic of Section 11.3 is the Planning Committee; this group should be composed of people who can make the planning effort successful, that is, government leaders, industry specialists, police, firefighters, health specialists, and local residents. Section 11.4 describes the hazards survey. Before a plan can be developed, an inventory of the potential hazards in a community must be gathered; then the risks associated with each hazard must/can be assessed and prioritized.

The main section of this chapter can be found in Section 11.5. It details the items that should be included in an emergency plan, which specifies the actions to be taken during an emergency and identifies the critical personnel and their responsibilities. A clear, concise, stepwise approach is the goal of the emergency plan. The next section discusses training. Service groups need to be trained for emergencies before such events occur. Public officials should be apprised of their roles in emergencies. Communications during an emergency (Section 11.7) are critical. Notification to the proper government agencies is required by law; a clearly understood and well-publicized notice is important to control the public's response in cases of serious accidents and emergencies. The manner in which information regarding an emergency is communicated can be just as crucial as the information itself. Many injuries and deaths during a disaster are the result of panic, and panic is often triggered by misinformation or a lack of relevant and timely information. The implementation of the plan (Section 11.8) includes keeping the plan current. An occasional audit is imperative to keep the plan from becoming obsolete (US EPA 1987a). This chapter concludes with seven illustrative examples and problems provided to elaborate on concepts presented throughout the chapter.

11.2 THE NEED FOR EMERGENCY RESPONSE PLANNING

Emergencies have occurred in the past and will continue to occur in the future. A few of the many commonsense reasons to plan ahead are provided in the following (Krikorian 1982):

1. Emergencies will happen; it is only a question of time.
2. When emergencies occur, the minimization of loss and the protection of people, property, and the environment can be achieved through the proper implementation of an appropriate emergency response plan.
3. Minimizing the losses caused by an emergency requires planned procedures, clear lines of responsibility, designated authority, accepted accountability, and trained, experienced people. With a fully implemented plan, these goals can be achieved.
4. If an emergency occurs, it may be too late to plan. Lack of preplanning can turn an emergency into a disaster.

A particularly timely reason to plan ahead is to ease the "chemophobia," or fear of chemicals, which is so prevalent in society today. So much of the recent attention to emergency planning and so many newly promulgated laws are a reaction to the tragedy at Bhopal. The probable causes of "chemophobia" are lack of information and misinformation. Fire is hazardous, and yet it is used regularly at home. Most adults have understood the hazards associated with fire since the time of the cavemen. By the same token, hazardous chemicals, necessary and useful in modern society, are not something to fear. Chemicals need to be carefully used and their hazards understood by the general public. A well-designed emergency plan that is understood by the individuals responsible for action, as well as by the public, can ease concerns over emergencies and reduce "chemophobia." People will react during an emergency; how they react can be managed through education. When ignorance is pervasive, the likely behavior during an emergency is panic.

An emergency plan can minimize loss of life and property by helping to assure the proper responses to the hazards at hand. "Accidents become crises when subsequent events, and the actions of people and organizations with a stake in the outcome, combine in unpredictable ways to threaten the social structures involved" (Shrivastava 1987). The wrong response, as easily as no response, can turn an accident into a disaster. For example, if a chemical fire is doused with water, which causes the emission of toxic fumes, it would have been better to let the fire burn itself out. For another

Emergency Planning and Response

example, suppose people are evacuated from a building into the path of a toxic vapor cloud; they might well have been safer staying indoors with closed windows. Still another example is offered by members of a rescue team who become victims because they were not wearing proper breathing protection. The proper response to an emergency requires an understanding of the hazards. A plan can provide the right people with the information they need to respond properly during an emergency.

In addition to the commonsense reasons mentioned above, there are legal reasons to plan. Recognizing the need for better preparation to deal with chemical emergencies, Congress enacted the Superfund Amendments and Reauthorization Act of 1986 (SARA). One part of SARA is the stand-alone act called *Title III* (the Emergency Planning and Community Right-to-Know Act [EPCRA] of 1986). This act requires federal, state, and local governments and industry to work together to develop emergency plans and "community right-to-know" reporting on hazardous chemicals used by industry. These new requirements build on US Environmental Protection Agency's (EPA) Chemical Emergency Preparedness Program and numerous state and local programs that are aimed at helping communities deal with potential chemical emergencies (US EPA 1987a).

Most large industries have had emergency plans designed for on-site personnel for quite some time. The protection of people, property, and, thus, profits has made emergency plans and prevention methods common in industry. On-site emergency plans are also often required by insurance companies. One way to minimize the effort required for emergency planning is to expand existing industry plans to include all significant hazards and all people in an adjacent community.

11.3 THE PLANNING COMMITTEE

Emergency planning should grow out of a team process coordinated by a leader. The team may be the best vehicle for including people representing various areas of expertise in the planning process, thus producing a more meaningful and complete plan. The team approach also encourages planning that will reflect a consensus of the entire community and be sensitive to environmental justice concerns that may be relevant in a particular community setting. Some individual communities and areas that included several communities had formed advisory councils before SARA requirements came into effect. These councils served as an excellent resource for the planning team recommended for SARA planning activities (Beranek et al. 1987).

The following considerations are important when selecting the members of a team that will bear overall responsibility for emergency planning:

1. The group must possess, or have ready access to, a wide range of expertise relating to the community, its industrial facilities, its transportation systems, and the mechanics of emergency response and response planning.
2. The members of the group must agree on their purpose and be able to work cooperatively.
3. The group must be representative of all the elements of the community that have substantial interest in reducing the risks posed by emergencies.

While many individuals have an interest in reducing the risks posed by hazards, their differing economic, political, and social perspectives may cause them to favor different means of promoting safety. For example, people who live near an industrial facility that manufactures, uses, or emits hazardous materials are likely to be greatly concerned about avoiding threats to their lives. They are likely to be less concerned about the costs of developing accident prevention and response measures than some of the other team members. Others in the community, for example, those representing industry or the budgeting group, are likely to be more sensitive to costs. They may be more anxious to avoid expenditures for unnecessarily elaborate prevention and response measures. Also, industry facility managers, although concerned with reducing risks posed by hazards, may be reluctant, for proprietary reasons, to disclose materials and process information beyond what is required by law. These differences can be balanced by a well-coordinated team that is responsive to the needs of the entire community.

Agencies and organizations bearing emergency response responsibilities may have differing views about the role they should play in case of an incident. The local fire department, an emergency management agency, and a public health agency are all likely to have some responsibilities during an emergency. However, each of these organizations might envision a very different set of actions at the emergency site. An emergency management plan will serve to detail the actions of each response group during an emergency and establish an agreed upon chain of command prior to any incident occurring.

In organizing the community to address the problems associated with emergency planning, it is important to bear in mind that all affected parties have a legitimate interest in the choices among planning alternatives. Therefore, strong efforts should be made to ensure that relevant stakeholder groups are included in the planning process. The need for unity of the committee during both the planning and implementation stages increases for larger numbers of different community groups. Each group has a right to participate in the planning, and a well-structured, well-organized planning committee should serve the entire community.

By law, the planning committee should include the following (US EPA 1987b):

1. Elected and state officials
2. Civil defense personnel
3. First aid personnel
4. Local environmental personnel
5. Transportation personnel
6. Owners and operators of facilities subject to the SARA
7. Law enforcement personnel
8. Firefighting personnel
9. Public personnel
10. Hospital personnel
11. Broadcast and print media
12. Community groups

Other individuals who could also serve the community well and should be a part of the committee include technical professionals, city planners, academic and university researchers, and local volunteer help organizations (Schulze 1987).

The local government has a great share of the responsibility for emergency response within its community. The official who has the power to order evacuation, fund fire and emergency units, and educate the public is a key person to emergency planning and the resulting response effort. For example, an entire plan might fail if a necessary evacuation were not ordered in time. Although politics should be disassociated from technical decisions, such linkage is inevitable in emergency planning. Distasteful options that require political courage are often necessary. In a given situation, for example, one may need to decide whether to evacuate a section of town where there is some doubt about the necessity of evacuation, but the worst-case consequence of not evacuating would be deadly. A public official can build support for future candidacy by using the issue of chemical safety as a bandwagon, but mistakes in handling emergencies are measured by a strong instrument, e.g., the election, and a failed emergency plan can be fatal to a political career. Politics is a social feedback device which, when used properly, can aid government leaders in making correct decisions. A political career can also be destroyed by an error in reading the social feedback. An effective plan can save elected officials' hours of media criticism after a crisis because the details of a response were organized by someone on the team before events occurred. Because of the power elected officials have locally, they are likely to take the leadership roles on such committees.

The independence of fire and police units from political control is somewhat traditional. Recognition of the freedom necessary to conduct public safety work gives police and fire units the option of rejecting outside control. However, to be successful, community emergency response planning must be universally developed and implemented. The fire and police departments, which are likely to be first on the emergency scene, will be required to act immediately. Their knowledge of the hazards and the plan is important to the plan's effectiveness as well as to their own safety. The police are best suited for evacuation and crowd control or protection of evacuated areas. They will need to understand all such assigned roles. The fire service groups will likely be on the scene to control the effects of the accident. The fire service people must bring to the committee their expertise in managing emergencies, a real asset. The firefighters must also learn from the committee the special considerations to be given to emergencies other than structural fires.

Environmental agencies may be among the best suited for evaluating risk. Their expertise usually includes a sound knowledge of the particular features of the local environment, such as location of flood plains and water resources, and the hazards of certain chemicals. They should be used to support the risk evaluation effort. The local or state environmental agencies are also a source of inventories of hazards on industrial sites, information necessary to guide the committee's response planning.

The state or local health agencies will help the committee to understand adverse health effects. For example, the risks associated with different chemicals can be evaluated by health agency personnel or by their contacts at universities and research institutions. These experts can assist in evaluating the health risks of various hazards. The owners and operators of facilities handling hazardous chemicals can be an asset in emergency planning because of their knowledge of the safety features already in place in their facilities. The representatives of industry also have access to information about the hazards of each chemical, either from the supplier or from the company's research department. Knowledge of on-site prevention features at an industrial site can help to sort through the potential hazards listed and to focus on the significant ones.

The local planners in a city or community may also be equipped to assist in emergency planning. The agendas of these groups typically include developing the community, creating jobs, and establishing economic stability; thus, the planners have their own reasons for wanting a community to be viewed as safe. The planners are also likely to have detailed information about the community, that not only includes road maps and transportation routes, but also locations of highly populated areas and industrial sectors. Understanding the locations of people and places in a community is important to planning and assessing significant risks posed to the community. The local planners can serve the committee because of their knowledge of these demographic and industry-related features of their community.

Toxicologists, meteorologists, chemists, plus environmental and chemical engineers are among the technical professionals who have experience and knowledge about chemicals, hazards, and preventive designs. The committee needs individuals with such expertise to assist in the preparation of plans that are technically and scientifically sound as well as safe. The evaluation of hazards and the design of appropriate emergency responses should reflect rational choices, not political options.

The management or control of the committee during planning, and especially during implementation, is essential. As suggested earlier, it is a given that the emergency plan will be generated by different individuals with different priorities. The different groups will have their own legitimate interests, and each interest will have to be weighed against its value to the plan. The committee leader must demonstrate respect for the interests of each of the individuals, as well as for each member's contributions. The committee leader is likely to be chosen for several reasons; among these should be:

1. The degree of respect held for the person by groups and individuals with an interest in the emergency plan
2. The time and resources the person will be able to devote to the work of the committee
3. The person's history of working relationships with concerned community agencies and organizations
4. The person's management and communication skills
5. The person's present responsibilities and background related to emergency planning, prevention, and response

Personal considerations, as well as institutional ones, should be weighed when selecting a committee leader. If one candidate has all the right resources to address the issues of emergency planning and implementation but is unable to interact with local officials, someone else may be a better choice for committee chair. Since the committee leader must coordinate this large group of people with different priorities and areas of expertise, the choice of the leader is critical to the success of the committee (O'Reilly 1987).

Emergency Planning and Response

11.4 HAZARD SURVEYS

A survey of hazards or foreseeable threats in the community must be performed and evaluated to characterize potential disasters by type and extent. Without such information, an appropriate plan cannot be developed. An inventory of the community protection assets, hazard sources, and risks must be completed before an effective plan can be written.

Although a plan for a city divided by a river may not be applicable to a desert city on a seismic fault, duplication can be an enemy of cost efficiency. Thus, wherever possible, any emergency plans that already exist in the community should be used as a starting point. Community groups that may have developed such plans include civil defense organizations, fire departments, the Red Cross, public health agencies, and local industry councils. Existing plans should be studied and their applicability to the proposed community plan evaluated.

Local government departments, such as those dealing with transportation, water, power, and sewer, may have valuable response resources. These should be listed and then compared to the needs of the plan. Some examples of valuable resources for emergency response include the following:

1. Trucks
2. Construction equipment (e.g., backhoes and flatbeds)
3. Laboratory services (e.g., water department)
4. Fire vehicles
5. Police vehicles
6. Emergency personal protection equipment (PPE)
7. Breathing apparatus
8. Gas masks
9. Buses or cars
10. Communication equipment (e.g., ham radios)
11. Local TV and radio stations
12. Ambulances
13. Burn treatment equipment
14. Stocks of medicines
15. Fallout shelters
16. Trained medical technicians and first aid personnel
17. Trained emergency personnel
18. Trained volunteer personnel (e.g., Red Cross)

The potential sources of hazards should be listed for risk assessment. SARA requires certain industries to provide information to the planning committee. Information about small as well as large industries is necessary to permit the committee to evaluate *all* significant hazard risks to the community. The information required by the SARA (some of which was provided in Chapter 3) includes:

1. The chemical name
2. The quantity stored over time

3. The type of chemical hazard (e.g., toxicity, flammability, ignitability, and corrosivity)
4. Chemical properties and characteristics (e.g., liquid at certain temperatures, gas at certain pressures, reacts violently with water)
5. Storage description and storage location on the site
6. Safeguards or prevention measures associated with the hazardous chemical storage or handling design, such as dikes, isolation of incompatible substances, fire resistant equipment, etc.
7. Control features for prevention such as temperature and pressure controllers and fail-safe design devices, if included in the process design
8. Recycle control loops intended for accident prevention
9. Emergency shutdown features

The planning committee should designate hazard sources on a community map. This information probably already exists and can be obtained locally from the transportation department, environmental protection agency, city planning department, community groups, and industry sources. Some of the data to be represented on the community map are:

1. Industrial and other sites of possible chemical accidents
2. Wastewater and water treatment plants at which chlorine is stored
3. Potable and surface water
4. Drainage and runoff patterns
5. Population location and density in different areas
6. Transportation routes for children
7. Commuter routes
8. Truck transport roads
9. Railroad lines, yards, and crossings
10. Major highways, noting merges and downhill curves
11. Hospitals, nursing homes, and schools with sensitive populations
12. Fallout shelters

The potential for natural disasters, based on the history and knowledge of the region and local geology, should also be indicated in the plan. Items such as seismic fault zones and flood plains and potentials for hurricanes, tornadoes, and winter storms should be noted.

The risk inventory or risk evaluation is the next part of the hazard survey. It is not usually practical to expect the plan to cover every potential accident. When the hazards have been evaluated, the plan should be focused on the most significant ones. This risk assessment stage requires the technical expertise of many people in order to compare the data and determine their relevance. Among the important factors to be considered in performing the risk evaluation are the following:

1. The routes of transport of hazardous substances to determine where a release could occur.
2. The proximity of all hazards to people and other sensitive environmental receptors should be examined.
3. The toxicology of different exposure levels should be reviewed.

Emergency Planning and Response

When the significant risks have been listed, the hazard survey is complete and the plan can be developed.

11.5 PLANNING FOR EMERGENCIES

Successful emergency planning begins with a thorough understanding of the event or potential disaster being planned for. The impacts on public health and the environment must also be estimated. Some of the emergencies that should be included in the plan are (Michael, Bell and Wilson 1986):

1. Earthquakes
2. Explosions
3. Fires
4. Floods and tsunamis
5. Hazardous chemical leaks, both, gas or liquid
6. Power or utility failures
7. Radiation incidents
8. Tornadoes or hurricanes
9. Transportation accidents

The likely emergency zone must be studied to estimate the potential impact on the public or the environment of accidents of different types. For example, a hazardous gas leak, fire, or explosion may cause a toxic cloud to spread over a great distance. The minimum affected area, and thus the area to be evacuated, should be estimated based on an atmospheric dispersion model of hazardous or explosive vapors that might be accidentally released from a facility. Various models can be used; the more complex models often produce more realistic results, but the simple and faster models may provide adequate data for planning and response purposes (US EPA 1987c).

In formulating the plan, some general assumptions may be made:

1. Organizations do a good job when they have specific assignments.
2. Various resources will need to be carefully coordinated
3. Most of the necessary resources are likely to be already available in the community (in industrial facilities or city departments).
4. People react more rationally when they have been apprised of a situation.
5. Coordination is basically a social process, not a legal one.
6. Disorganization and reorganization are common in a large group.
7. Flexibility and adaptability are basic requirements for a coordinated team.

The objective of the plan should be a procedure that uses the combined resources of the community in a way that will:

1. Safeguard people during emergencies
2. Minimize damage to property and the environment
3. Initially contain the incident and ultimately bring it under control
4. Effect the rescue and treatment of casualties

5. Provide authoritative information to the news media (for transmission to the public)
6. Secure the safe rehabilitation of the affected area
7. Preserve relevant records and equipment for subsequent inquiry into causes and consequences and for improvements of implemented responses

During the development of the plan, these assumptions and objectives should be kept in mind. Although prevention is an important goal in accident and emergency management, it is not really the objective of the emergency response plan. The plan should focus on minimizing damage when emergencies do occur (Krikorian 1982). Key components of the emergency action plan include the following (US EPA 1987a):

1. Emergency actions other than evacuation
2. Escape procedures when necessary
3. Escape routes clearly marked on a site map and perhaps also on the roads
4. A method of accounting for people after evacuation
5. Description and assignment of rescue and medical duties
6. A system for reporting emergencies to the proper regulatory agencies
7. A means of notification of the public by a reliable and meaningful alarm system
8. Responsibilities of contact and coordination personnel

SARA originally called for each community group, as designated by the governor, to have a plan by 1988. Specific requirements included the following:

1. The identification of all facilities as well as transportation routes for extremely hazardous substances (EHSs)
2. The establishment of emergency response procedures, both on and off plant sites (facility owner and operator actions, as well as the actions of local emergency and medical personnel)
3. The establishment of methods of determining when releases occur and what areas and populations may be affected by them
4. A listing of community and industry emergency equipment and facilities, along with the names of those responsible for this equipment and its upkeep
5. The description and scheduling of a training program to teach methods for responding to chemical emergencies
6. The establishment of methods and schedules for exercises or drills to practice and test emergency response plans
7. The designation of a community coordinator and a facility coordinator to implement the plan
8. The designation of facilities (e.g., hospitals, schools, nursing homes, natural gas plants, etc.) that are subject to added risk and provision for their protection

A standard format that could be followed might incorporate the following:

1. A statement promulgating the plan
2. A purpose for the plan

Emergency Planning and Response

3. Assumptions made in developing the plan
4. A discussion of the plan's weaknesses and vulnerabilities
5. A clear statement of when the plan will be executed
6. A stepwise narrative explanation of how the plan works (for those who will direct or coordinate the plan)
7. A chart of the major disaster functional groups, including the departments and volunteers who are responsible for coordinating or supporting each function
8. A description of the responsibilities of each functional group (e.g., duties and actions of police, etc.)
9. A list of the necessary equipment, its location, and contact persons for obtaining each item or unit
10. A method for communicating each type of emergency to the public, the functional groups, and the responsible agencies
11. A list of the emergency coordinator's tools and/or resources
12. Training details and schedules
13. The plan implementation schedule, including slots for routine audits and updates

Different emergencies are likely to require different response actions. Specific steps for coping with four types of emergency situations are outlined as follows:

1. Volatile toxic releases
   a. The release should be deluged with water.
   b. The people who will possibly be affected by the toxic cloud should be warned to close their windows or, if necessary, evacuate.
   c. Police with protective equipment should check the homes that have been affected.
2. Flammable chemical fire
   a. Access to the area should be controlled.
   b. The fire should be prevented from spreading.
   c. The fire should be extinguished by professionals using proper personal protective gear and modern firefighting equipment.
3. Chemical spill
   a. The spilled substance should be contained.
   b. Medical personnel with protective equipment should be available to administer to those affected.
   c. A rescue team with protective equipment should collect the spilled material in appropriate containers.
   d. The spill residue should be transported properly to a permitted treatment and disposal facility.
4. Tornado
   a. Emergency warnings should be issued to people to move to shelters.
   b. Equipment in factories should be shut down.
   c. Squads of search and rescue teams and medical personnel should be rushed to the affected areas after the tornado has passed to aid those affected.

The details of the plan will be different from community to community, and the appropriate responses will differ according to the events anticipated. Obviously, each community must develop a plan tailored to its own needs.

11.6 TRAINING OF PERSONNEL

The education of the public is critical to securing public support for the emergency plan; the real hazards in the community must be made known, as well as what to do in an emergency. Most people are not aware of the reality of hazards in their communities. The common perception is that hazards exist elsewhere, as do the resulting emergencies (CMA 1987). The education of the populace about the true hazards associated with routine discharges from plants in the neighborhood and preparing that populace for emergencies may be a real challenge for the planning committee. People must be taught how to react to an emergency, i.e., how to recognize and report an incident, how to react to alarms, and what other actions to take. A possible initial result of SARA Title III may be a fear of industrial discharges on the part of the public (Cathcart 1985). News stories can be misleading if based on hazardous chemical inventories, accidental release data, or annual emissions reports of questionable accuracy or if taken out of context. It should be possible to put such information into perspective through training programs.

The personnel at an industrial plant who are trained in the operation of the facility are critical to proper emergency response. They must be taught to recognize abnormalities and excursions in operations and to report them immediately. Plant operators should be taught how to respond to various types of accidents. Internal emergency response teams can also be trained to contain the emergency until outside help arrives or, if possible, to resolve the emergency all together. It is especially important to train plant personnel in shutdown, notification, and evacuation procedures.

Training is important for the emergency teams to assure that their roles are clearly understood and that accidents can be responded to safely and properly without delay. As discussed earlier, the emergency teams include the police, fire, and medical personnel, and the volunteers who will be required to take action during an emergency (Cathcart 1985). These people must be knowledgeable about the potential hazards. For example, specific antidotes for different health-related conditions must be known by medical personnel prior to any potential accident. The entire emergency team must also be taught the use of personal protective equipment. Local government officials also need training. Since these officials have the power to order an evacuation, they must be aware of the circumstances under which such action is necessary, and they must understand that the timing of an evacuation is critical before an emergency occurs. Local officials also control the use of city equipment and therefore must understand what is needed for an appropriate response to a given emergency.

Media personnel, such as print and broadcast reporters, editors, etc., must also be involved in the training program since it is important that the public receive accurate and timely information. If incorrect or distorted information about an emergency is disseminated, panic can easily result. For this reason, it is important for print and broadcast journalists to be somewhat knowledgeable about the potential hazards and the details of emergency responses so they can report to the public appropriately.

Emergency Planning and Response

Training for emergencies should be done routinely:

1. When a new member is added to the group
2. When someone is assigned a new responsibility within the community
3. When new equipment or materials are acquired for use in emergency response
4. When emergency procedures are revised
5. When a practice drill shows inadequacies in performance of duties
6. At least once annually

Any training program should address five questions:

1. How can potential hazards be recognized? (This can be determined by periodic review of hazards and prevention measures.)
2. To whom should a hazard be reported?
3. What actions constitute proper responses to special alarms or signals?
4. Where are the evacuation routes?
5. What precautions (e.g., donning appropriate personal protective equipment) are to be taken when responding to an emergency?

It is important for emergency procedures to be performed as planned. This requires regular training to ensure that people understand and remember how to react. The best plan on paper is likely to fail if the persons involved are reading it for the first time as an emergency is occurring. People must be trained *before* an emergency happens.

11.7 NOTIFICATION OF PUBLIC AND REGULATORY OFFICIALS

Notifying the public of an emergency is a task that must be accomplished with caution. People will react in different ways when receiving notification of an emergency. Many will simply not know what to do, some will not take the warning seriously, and others will panic. Proper training in each community, as discussed in the previous section, can help minimize panic and can condition the public to make the correct response in a time of stress.

Methods of communicating an emergency will differ from community to community, depending on its size and resources. Some techniques for notifying the public include:

1. The sounding of fire department alarms in different ways to indicate emergencies of various kinds
2. Chain phone calls (this method usually works well in small towns)
3. Announcements made through loudspeakers from police cars or the vehicles of volunteer teams traveling throughout the community

Once the emergency has been communicated, an appropriate response by the public must be evoked. For this to occur, an accepted plan that people know and understand must be put into effect. Since an emergency can quickly become a disaster if panic ensues, the plan should include the appropriate countermeasures to bring the situation back under control.

Information reported to the emergency coordinator must be carefully screened. A suspected "crank call" should be checked out before an alarm is sounded. By taking no immediate action, however, the team runs the obvious risk that the plan will not be implemented in time. Therefore, if a call cannot be verified as bogus, a response must begin and local police should be dispatched quickly to the scene of the reported emergency to provide firsthand information of the actual situation.

The print and broadcast media can be a major resource for communication, and one job of the emergency coordinator is to prepare information for reporters. The emergency plan should include a procedure to pass along information to the media promptly and accurately.

Certain types of emergencies must be reported to government agencies; it is not always sufficient to notify just the response team. For example, state and federal laws require the reporting of hazardous releases and nuclear power plant problems. There are also more specific requirements under SARA Title III for reporting chemical releases. Facilities that produce, store, or use a listed hazardous substance must immediately notify the Local Emergency Planning Committee (LEPC) and the State Emergency Response Commission if there is a release of one or more substances specifically listed in SARA. These substances include 402 extremely hazardous chemicals on the list prepared by the Chemical Emergency Preparedness Program and chemicals subject to the reportable quantities requirements of the original Superfund (US EPA 1987a). The initial notification can be made by telephone, radio, or in person. Emergency notification requirements involving transportation incidents can be satisfied by dialing 911. The emergency planning committee should provide a means of reporting information on transportation accidents quickly to the emergency coordinator.

SARA requires that the notification of an industrial emergency includes:

1. The name of the chemical released
2. Whether it is known to be acutely toxic
3. An estimate of the quantity of the chemical released into the environment
4. The time and duration of the release
5. Where the chemical was released (e.g., air, water, and land)
6. Known health risks and necessary medical attention that will be required
7. Proper precautions, such as evacuation
8. The name and telephone number of the contact person at the plant or facility at which the release occurred

As soon as is practical after the release, there must be a written follow-up emergency notice, updating the initial information and giving additional information on response actions already taken, known or anticipated health risks, and advice on medical attention. Law has required the reporting and written notices since October 1986.

Emergency Planning and Response

11.8 PLAN IMPLEMENTATION

Once an emergency plan has been developed, its successful implementation can be assured only through constant review and revision. Helpful ongoing procedures include:

1. Routine checks of equipment inventory, status of personnel, status of hazards, and population densities
2. Auditing of the emergency procedures
3. Routine training exercises
4. Practice drills

The emergency coordinator must assure that the emergency equipment is always in readiness. Siting the control center and locating its equipment is also the coordinator's responsibility. There should be both a main control center and an alternate, both in carefully chosen locations. The following items should be present at the control center:

1. Copies of the current emergency plan
2. Maps and diagrams of the area
3. Names and addresses of key functional personnel
4. Means to initiate alarm signals in the event of a power outage
5. Communication equipment (e.g., phones, radios, TV, and two-way radios)
6. Emergency generators and lights
7. Evacuation routes identified on area maps
8. Self-contained breathing equipment for possible use by the control center crew
9. Miscellaneous furniture, including cots

Inspection of emergency equipment such as fire trucks, police cars, medical vehicles, personal safety equipment, and alarms should be routinely performed.

The plan should be audited on a regular basis, at least annually, to ensure that it is current. Items to be updated include the list of potential hazards and emergency procedures (adapted to any newly developed technology). A guideline for auditing the emergency response plan, adapted from literature published by the Chemical Manufacturers Association, is available in the literature (Cathcart 1985).

Certain operational aspects of the plan should be practiced to assure that the proper response will be realized if and when an actual emergency occurs. The drill scenario should be prepared almost as carefully as the emergency plan itself. Both preannounced and surprise drills should be held, observed, and evaluated to pinpoint deficiencies in the plan and to determine whether new training is required. The following questions should be used in evaluating drills:

1. What types of drills are performed?
2. What aspects are tested?
3. How often are the drills held?
4. Are there both announced and unannounced drills?

5. Comparing response times for announced and unannounced drills, are times for unannounced drills much longer?
6. What time of the day are the drills held?
7. Who is responsible for evaluating the drill?

Once deficiencies have been identified, the plan should be revised to correct them. Such testing and revision should be done regularly; the interval between tests and revisions should not exceed 1 year.

For the interested reader, further information on emergency planning and response is available in the literature (FEMA 1993, 1996; US EPA 2017; Emergency Management BC 2021; Ready.gov 2021).

11.9 ILLUSTRATIVE EXAMPLES

Seven illustrative examples complement the material presented above.

11.9.1 ILLUSTRATIVE EXAMPLE 1

Although this chapter addresses emergency response planning from an industrial perspective, explain why it would be advantageous to explore emergency response planning at the home or office.

Solution: There are many accident and safety concerns in the home or workplace. Preparing to address and prevent these potential accident risks ensures protection against injury and death in these everyday situations. The following are some potential accident areas that should be considered in the home or office.

1. Keep stairs clear of debris. The same concern for tripping and falling hazards applies at an industrial plant.
2. Install fire/smoke detectors to detect fires in the home or office. The same concern for fire detection and response applies to an industrial plant.
3. Keep the house or office well ventilated. Build-up of irritants and/or hazardous air constituents is a concern for industrial plants as well.

11.9.2 ILLUSTRATIVE EXAMPLE 2

Provide an overview of the EPCRA.

Solution: EPCRA was passed in response to concerns regarding the environmental and safety hazards posed by the storage and handling of toxic chemicals. These concerns were triggered by the disaster in Bhopal, India, in which more than 2,000 people died or suffered serious injury from the accidental release of methyl isocyanate. To reduce the likelihood of such a disaster in the United States, Congress imposed requirements on both states and regulated facilities.

EPCRA established requirements for federal, state, and local governments, Native American Tribes, and industry regarding emergency planning and "Community Right-to-Know" reporting on hazardous and toxic chemicals. The Community

Emergency Planning and Response

Right-to-Know provisions helped increase the public's knowledge and provided access to information on chemicals at individual facilities, their uses, and releases into the environment. States and communities, working with facilities, can use the information to improve chemical safety and protect public health and the environment. EPCRA has four major provisions, the details of which are codified in 40 CFR Part 370:

1. Emergency planning (Sections 301–303)
2. Emergency release notification (Section 304)
3. Hazardous chemical storage reporting requirements (Sections 311–312)
4. Toxic chemical release inventory (Section 313)

11.9.3 Illustrative Example 3

Describe the Toxics Release Inventory (TRI) program.

Solution: EPCRA's primary purpose is to inform communities and citizens of chemical hazards in their community. Sections 311 and 312 of EPCRA require businesses to report to state and local governments on the locations and quantities of chemicals stored on-site to help communities prepare to respond to chemical spills and similar emergencies. EPCRA Section 313 requires EPA and the states to annually collect data on releases and transfers of certain toxic chemicals from industrial facilities and make the data available to the public in the TRI. In 1990, Congress passed the Pollution Prevention Act that required that additional data on waste management and source reduction activities also be reported under TRI. The goal of TRI is to empower citizens, through information, to hold companies and local governments accountable in terms of how toxic chemicals are managed.

The EPA compiles the TRI data each year and makes it available through several data access tools, including the TRI Explorer (https://enviro.epa.gov/triexplorer/tri_release.chemical) and Envirofacts (https://enviro.epa.gov). There are other organizations that also make the data available to the public through their own data access tools, including Unison Institute that puts out a tool called "RTKNet" (rtknet.org) and the Environmental Defense Fund that has developed a tool called "Scorecard" (www.Scorecard.org).

The TRI program has expanded significantly since its inception in 1987. The EPA has issued rules to roughly double the number of chemicals included in the TRI to approximately 650. Seven new industry sectors have also been added to expand coverage significantly beyond the original covered industries, that is, manufacturing industries. Most recently, the EPA has reduced the reporting thresholds for certain persistent, bioaccumulative, and toxic (PBT) chemicals to provide additional information to the public on these chemicals.

11.9.4 Illustrative Example 4

Provide information on PPE.

Solution: PPE may be viewed as the workers' last line of defense against injury in the workplace. When the work environment cannot be made safe by incorporating sound engineering, tried and tested work practices, and administrative controls, PPE is employed as a "last resort." PPE effectively creates a barrier between the worker and the health or hazard problem but does not reduce or eliminate the problem. PPE includes safety goggles, helmets, face shields, gloves, safety shoes, hearing protection, full-body protective wear, and respirators.

11.9.5 Illustrative Example 5

Describe safety requirements for pressure vessels.

Solution: Pressure vessels of any type require safeguards to protect personnel against their accidental failure. The American Society of Mechanical Engineers (ASME) gives specific standard requirements for safety devices, instrumentation, and controls. Chemical process equipment should include overpressure relief devices, including safety relief valves and rupture discs, where necessary. Provisions should be made to ensure that the contents of the vessel are not released into areas where personnel are put at risk and placed in danger.

11.9.6 Illustrative Example 6

In 1947, two ships docked in Texas City, TX, with tons of ammonium nitrate fertilizer and other cargo aboard. These ships caught fire, burned, and exploded over a period of more than 16 hr. The explosions were so powerful that almost 600 people were killed and more than 3,500 people were injured in the surrounding community. The dock area and much of the city were destroyed. One of the ship's anchors was thrown approximately 2 miles inland where it still lies today as a memorial to the incident.

As described earlier, the EPCRA, among other things, requires any facility that produces, uses, or stores any chemical on a published list in excess of the "threshold planning quantity" to notify local emergency response entities (such as the fire department, police department, hospitals, etc.) of the quantity, identity, and nature of these chemicals; to cooperate with an LEPC; and to develop an emergency plan to be used in the event of an accidental release.

While the regulatory definition of a "facility" includes transportation vessels and port authorities for release reporting, these entities are exempt from notification and emergency planning requirements. As a result, emergency response planning against another Texas City disaster is not a requirement of the EPCRA legislation.

Prepare a list of areas of concern that would have to be addressed if the notification and emergency planning requirements were applied to port areas. Among other things, you may wish to address matters such as the short residence time of in-transit materials and the political (as opposed to legal) ramifications of applying regulation of this kind to foreign flag carriers. *Note*: This is an open-ended question with many correct answers. Additional examples/problems of this type are available in Flynn and Theodore (2002).

Emergency Planning and Response

Solution: The following areas should be addressed in the notification and emergency planning for port areas:

1. Who shall be responsible for notification and/or emergency planning – the shipper, transporter, or port operator?
2. How shall the inventory of the materials flowing in and out of the area be maintained?
3. Should there be a minimum storage time that triggers notification and emergency planning?
4. Should an emergency plan be developed for the release of every chemical that ever flowed through the port even though some of those chemicals may never be present in the area again?
5. What notification and emergency planning criteria should be adopted for large quantities of listed materials that frequently flow through the port but are present for only short periods of time?
6. Should limits be placed on quantities of some materials being stored in the port area at a given time?
7. Should ports be classified as to what materials are allowed to enter them?
8. Should segregation of cargo by compatibility groups be required for materials waiting to be loaded or transshipped?
9. Should port areas be rezoned to reduce the potential risk to the surrounding population?
10. Are evacuation plans possible for port areas in large cities?
11. Is there sufficient authority under current law to accomplish this task or is new legislation required?
12. What would be the political consequences of requiring foreign ships to adhere to these regulations?
13. What would be the cost of applying these regulations to port areas?

11.9.7 Illustrative Example 7

One of the first basic steps in an Occupational Safety and Health Administration (OSHA) study is to estimate the concentrations of any pollutants that are being generated in the process. A large laboratory with a volume of 1,100 m^3, at 22°C and 1 atm contains a reactor which may emit as much as 1.5 gmol of hydrocarbons into the room if a seal ruptures. If the hydrocarbon mole fraction in the room air becomes greater than 850 ppb$_v$ it constitutes a health hazard.

  a. Calculate the total number of gmol of air in the room.
  b. Suppose the reactor seal ruptures and the maximum amount of hydrocarbons is emitted almost instantaneously. Assume that the air flow in the room is sufficient to make the room behave as a completely mixed reactor, i.e., the air composition is spatially uniform. Calculate the concentration of hydrocarbon in the room in units of ppb$_v$ and compare it to the health risk concentration provided above.

Solution: First calculate the gmol of air in the room employing Charles' law, noting that at 0°C 1 gmol of an ideal gas occupies 22 L (1,000 gmol occupy 22.4 m³).

$$n_{air} = (1,100 \text{ m}^3)(1,000 \text{ gmol}/22.4 \text{ m}^3)[273 \text{ K}/(22+273 \text{ K})] = 45,440 \text{ gmol air}$$

Calculate the mole fraction of hydrocarbons in the room air in units of ppm$_v$ and ppb$_v$:

$$y_{HC} = (1.5 \text{ gmol hydrocarbons})/(45,440 \text{ gmol air}) = 3.3 \text{ ppm}_v = 3,300 \text{ ppb}_v$$

Since 3,300 ppb$_v$ ≫ 850 ppb$_v$, there is a definite health risk that needs to be addressed in emergency planning and response.

PROBLEMS

11.1 What four steps should be taken routinely to ensure that an emergency plan is viable?

11.2 Describe the "correct" way to write an emergency response plan.

11.3 List some of the precautions that should be taken when wearing personal protective equipment.

11.4 In an emergency, what methods can be used to notify the public? What methods could be used if the emergency includes a power failure?

11.5 What information should a chemical Safety Data Sheet (SDS) contain?

11.6 A chemical reactor at a plant site has exploded. Provide specific steps for coping with this accident/emergency.

11.7 A large office building has collapsed to a pile of rubble and debris after an explosion and a subsequent fire. The building was originally constructed in 1966 and was fully occupied since that time. Search, rescue, and cleanup has been ordered. What potentially hazardous materials can be expected in the ash, rubble, and debris? And what protective measures should be put in place to protect search, rescue, and cleanup personnel?

11.8 A train has collided with a truck at an intersection in the industrial area of a major city. A tank car and flatbed car filled with containers have derailed. The tank car is lying in a ditch alongside the tracks, surrounded by some containers that have broken loose from the flatbed car. The following information is known:
   a. The tank car is labeled "hydrogen fluoride."
   b. The UNNA number shown on all the containers is 1806. (The UNNA number is the United Nations North America number. This numbering system was developed by the US Department of Transportation and has since become the UN standard system for classifying hazardous materials.)

If you were responding to this incident, what additional information would you want to know?

REFERENCES

Beranek, W., McCullough, J.P., Pine, S.H., and Soulen, R.L. 1987. Getting Involved in Community Right-to-Know. *Chemical and Engineering News* 65(43): 62.

Cathcart, C. 1985. *Community Awareness & Emergency Response, Program Handbook.* Washington, DC: Chemical Manufacturers Association.

Chemical Manufacturers Associations. 1987. Title III: The Right to Know, the Need to Plan. *ChemEcology* (13): 2.

Emergency Management BC. 2021. *Emergency Management Planning Guide for Local Authorities and First Nations*, 2nd Edition. Victoria, BC, Canada. https://www2.gov.bc.ca/assets/gov/public-safety-and-emergency-services/emergency-preparedness-response-recovery/local-government/em_planning_guide_for_la_fn.pdf

Federal Emergency Management Agency. 1993. *Emergency Management Guide for Business and Industry*, FEMA 141. Washington, DC. https://www.fema.gov/pdf/library/bizindst.pdf.

Federal Emergency Management Agency. 1996. *SLG 101: Guide for All-Hazard Emergency Operations Planning.* Washington, DC. https://www.fema.gov/pdf/plan/slg101.pdf.

Flynn, A., and Theodore, L. 2002. *Accident and Emergency Management for the Chemical Process Industries.* Boca Raton, FL: CRC Press (originally published by Marcel Decker).

Krikorian, M. 1982. *Disaster and Emergency Planning.* Loganville, AL: Institute Press.

Michael, E., Bell, O., and Wilson, J. 1986. *Emergency Planning Considerations for Specialty Chemical Plants.* Boston, MA: Stone and Webster Engineering Corporation.

O'Reilly, J. 1987. *Emergency Response to Chemical Accidents. Planning and Coordinating Solutions.* New York, NY: McGraw-Hill.

Ready.gov. 2021. *Emergency Response Plan.* FEMA/Ready.gov, Washington, DC. https://www.ready.gov/business/implementation/emergency.

Schulze, R. 1987. *Superfund Amendments and Reauthorization Act of 1986 (SARA Title III).* Richardson, TX: Trinity Consultants Incorporated.

Shrivastava, P. 1987. *Anatomy of a Crisis.* Cambridge, MA: Ballinger Publishing Company.

US Environmental Protection Agency. 1987a. Other Statutory Authorities: Title III: Emergency Planning and Community Right-to-Know. *EPA Journal* 13(1): 28–30.

US Environmental Protection Agency. 1987b. *Hazardous Materials Emergency Planning Guide.* Washington, DC: National Response Team.

US Environmental Protection Agency. 1987c. *Title III Fact Sheet. Emergency Planning Community Right-to-Know.* Washington, DC: Office of Solid Waste and Emergency Management.

US Environmental Protection Agency. 2017. *How to Better Prepare Your Community for a Chemical Emergency. A Guide for State, Tribal and Local Agencies.* Washington, DC: Office of Land and Emergency Management. https://www.epa.gov/epcra/how-better-prepare-your-community-chemical-emergency-guide-state-tribal-and-local-agencies.

Part III

Health Risk Calculations for Specific Chemical Process Industries

This part is an introduction to a number of health and hazard risk calculations for a dozen chemical process industry types. Chemical process industry here refers to an industry that uses processes primarily involved in chemical transformations, but a physical and/or mechanical change may also be involved. The industries reviewed in this part include the more conventional industries of: inorganic chemicals (Chapter 12), organic chemicals (Chapters 13), petroleum refining (Chapters 14), energy and power (Chapter 15), pharmaceuticals (Chapter 16), food products (Chapter 17), and the perhaps non-traditional industries that are involved with nanotechnology (Chapter 18), military and terrorism (Chapter 19), weather and climate (Chapter 20), architecture and urban planning (Chapter 21), and environmental considerations (Chapter 22). Each chapter contains information on industry-specific processes, followed by a number of illustrative examples highlighting health and hazard risk calculations based on information and procedures detailed earlier in the text.

It should also be noted that because of space considerations, only the 11 process industries listed are considered in Part III. There are nearly 100 process industries that could justifiably be classified as part of the chemical process industry. The reader is referred to Basta (1997) for some of the "other" processes. Hopefully the material presented for the industries selected to be included in this part will allow the reader to extend these calculations and approaches for risk determination to those industries that were not explicitly reviewed here.

12 Inorganic Chemicals

12.1 INTRODUCTION

Inorganic chemistry is that field of chemistry in which chemical reactions and properties of all chemical elements and their compounds, with the exception of hydrocarbons (compounds composed of carbon and hydrogen) and their derivatives, are studied. The chemistry of carbon-hydrogen compounds is termed organic chemistry and is addressed in the next chapter. Historically, inorganic chemistry arose with the ancient study of minerals and the search for ways to extract metals or other substances from their ores. Increased understanding of the chemical behavior of the elements and of inorganic compounds has led to the discovery of many new classes of inorganic substances. Modern inorganic chemistry overlaps parts of many other scientific fields including biochemistry, metallurgy, minerology, organic chemistry, physical chemistry, and solid-state physics. Other branches of inorganic chemistry include solid-state chemistry, which is concerned with the chemistry of semiconductors; high-temperature and high-pressure chemistry; geochemistry; and the chemistry of the trans-uranium elements, the elements of the actinide series, and the rare earth elements.

As indicated in Chapter 7, inorganic chemical processes have existed for thousands of years. Many early processes involved simple treatment of natural materials. For example, limestone was heated by fire to produce lime for mortar; clay was fired to produce water-resistant pottery, etc. Slowly more complex processes evolved. These may have consisted of several steps involving the reaction of several natural raw materials to produce the desired end product. With the development of the scientific chemical industry, many more inorganic chemicals were produced. This chapter describes a number of these important inorganic chemical products including sulfuric and hydrochloric acid, ammonium nitrate, sodium chloride, cement, and glass. This chapter concludes with four illustrative examples and problems provided to elaborate on concepts presented throughout the chapter.

12.2 GENERAL COMMENTS

Inorganic processes are often more easily developed than are organic processes. Inorganic reactions often proceed rapidly and go nearly to completion with few if any side reactions resulting in undesirable by-products. In addition, the problem of thermal decomposition is unimportant for most inorganic materials compared to the heat-sensitivity of many organic compounds.

Some of the inorganic processes operating today have developed over many years; others are of recent development. In some cases, a newer, more economical process has displaced an older process for making the same chemical. For example, the contact process for the manufacture of sulfuric acid has essentially replaced the older chamber process for its manufacture.

A number of questions arise in the analysis of an inorganic chemical process:

1. From what raw material can the product be made, and where are the raw materials available?
2. What alternative manufacturing processes may be used?
3. What are the markets for the product?
4. Considering the raw materials, processes, and markets, is the production of the material economically viable? This economic analysis may consider the best location for a plant. Location may be a compromise between raw material source and market locations. Transportation, power, and labor costs are important factors in this economic analysis.

Table 12.1 lists a few of the major inorganic chemicals produced today. The balance of this chapter examines the processes involved in the manufacture of several of these key inorganic chemicals.

TABLE 12.1
Major Inorganic Chemicals Currently Produced

Aluminum chloride
Ammonia
Ammonium nitrate
Ammonium sulfate
Calcium carbide
Calcium phosphate
Carbon dioxide
Chlorine
Hydrochloric acid
Hydrofluoric acid
Hydrogen
Nitric acid
Oxygen
Phosphoric acid
Phosphorus
Potassium hydroxide
Sodium
Sodium bicarbonate
Sodium bichromate
Sodium carbonate
Sodium chromate
Sodium chlorate
Sodium hydroxide
Sodium phosphate
Sodium silicate
Sodium sulfate
Sulfuric acid

12.3 SULFURIC ACID

Sulfuric acid is the largest tonnage inorganic chemical produced. Although very little of it reaches the consumer in the form of acid, large quantities are used in the production of many consumer items. The three raw materials needed to manufacture sulfuric acid are air, water, and sulfur. The first two ingredients are readily available and essentially free, while the last ingredient may be obtained in a number of forms. Most sulfur that is produced is used in the manufacture of sulfuric acid, but a significant quantity is also used in the production of rubber, carbon disulfide, and other chemical products. About 80% of sulfuric acid is made from elemental sulfur. The Frasch process (Basta 1997) accounts for most of the elemental sulfur production. This raw material exists in nature both in the free state and combined in ores such as pyrite (FeS_2). It is also an important constituent of petroleum and natural gas as H_2S.

Various industrial and academic research groups are developing new uses for sulfur. Among some of these are: as an additive to asphalt, sulfur concretes and mortars, plant and soil treatment, sulfur-alkali batteries, and foamed sulfur insulation.

There are two main processes for producing sulfuric acid, the chamber process and the contact process. The older chamber process employs nitric acid and nitrogen oxides to oxidize SO_2 to SO_3. The contact process catalytically oxidizes SO_2 to SO_3 and is used in essentially all new plants built today. Nearly 90% of all sulfuric is produced by the contact process because the chamber method produced acid of an uneconomically low concentration. The three chemical reactions in the contact process are:

$$\text{Combustion: } S + O_2 \rightarrow SO_2 \qquad (12.1)$$

$$\text{Conversion: } SO_2 + 1/2 O_2 \rightarrow SO_3 \qquad (12.2)$$

$$\text{Absorption: } SO_3 + H_2O \rightarrow H_2SO_4 \qquad (12.3)$$

Eight major uses of sulfuric acid are listed below:

1. Phosphate fertilizer
2. Inorganic pigments
3. Ammonium sulfate
4. Aluminum sulfate
5. Iron and steel pickling
6. Petroleum products
7. Rayon
8. Nonferrous metal processing

12.4 HYDROCHLORIC ACID

Hydrogen chloride is a colorless, corrosive, non-flammable gas with the chemical formula HCl. HCl dissolves readily in water forming a solution, hydrochloric acid, that contains 40.3% by weight HCl and has a specific gravity of 1.20. This solution fumes strongly in moist air but dilution stops the fuming. As with other gases, HCl becomes less soluble as the water temperature increases and is less soluble in alcohol, ether, and in other organic liquids.

HCl was discovered in the 15th century by Basilius Valentinius. Commercial production of hydrochloric acid began in England when legislation was passed prohibiting the indiscriminate discharge of HCl into the atmosphere. This legislation forced manufacturers, using the Leblanc process for soda ash, to absorb the waste HCl in water. As more uses for hydrochloric acid were discovered, plants were built solely for its production.

In solution in water, the molecules of HCl ionize, becoming positively charged hydrogen ions and negatively charged chloride ions. Because it ionizes easily, hydrochloric acid is a good conductor of electricity. The hydrogen ions give hydrochloric acid its acidic properties, so that all solutions of HCl and water have a sour taste, corrode active metals forming metal chlorides and hydrogen, turn litmus red, neutralize alkalies, and react with salts of weak acids forming chlorides and the weak acids. It is also typical of the hydrogen compounds formed by all the elements of the halogen family, which includes fluorine, bromine, and iodine.

Commercial grade hydrochloric acid is called muriatic acid and is available in anhydrous form in steel cylinders at a very considerable increase in cost due to the expense of the cylinders used for its containment. The largest users of hydrochloric acid are the metal, chemical, pharmaceutical, food, and petroleum industries. It is estimated that metal industries consume approximately half of the hydrochloric acid sold, while chemical and pharmaceutical manufacturing and processing account for another third of hydrochloric acid consumption. The major use of hydrochloric acid is in steel pickling (surface treatment to remove mill scale).

12.5 AMMONIUM NITRATE

The Stengel process for the manufacture of ammonium nitrate is based upon the simple reaction of ammonia and aqueous nitric acid, with subsequent drying, crystallization, and grinding of the product. To conserve materials and to produce a more uniform product, several recycle streams are employed. Ammonium nitrate is used primarily as a fertilizer and as an explosive.

A plant would typically use the following raw materials and process train to manufacture ammonia nitrate:

1. Liquid anhydrous ammonia and aqueous nitric acid would be used as the raw ingredients.
2. Ammonia would be produced from the reaction of hydrogen from the reforming of natural gas and high-purity nitrogen recovered from the air. Ammonia would be stored below ambient temperature, approximately 10°F.
3. The nitric acid solution (approximately 59% HNO_3) would be manufactured by the catalytic oxidation of ammonia and would be stored in large stainless steel storage tanks.
4. The basic reaction to form ammonium nitrate is as follows:

$$NH_3 + HNO_3 \rightarrow NH_4NO_3 \qquad (12.4)$$

The ammonia would be vaporized and heated to 255°F by steam before being fed to the reactor.
5. Aqueous nitric acid is heated to 300°F and fed to an acid charge tank where it is mixed with two recycle streams containing ammonium nitrate. The first acid heater would be stainless steel, but a second would be made of tantalum to avoid corrosion by the high temperature acid.
6. The ammonium nitrate is formed as liquid droplets and the water in the acid is vaporized. This reaction is fast and goes essentially to completion. A slight excess of ammonia is used.
7. The nitrate solution passes through an air heater into a partial condenser where all the nitrate and ammonia and some water are condensed. The separated ammonium nitrate is then packaged for shipment to purchasers.

Because ammonium nitrate is classified as explosive, its manufacture, transportation, and use are regulated by Federal law, and appropriate safety precautions must be followed at each step in the process.

12.6 SODIUM CHLORIDE

Salt, also known as sodium chloride, has a chemical formula NaCl. The term *salt* is also applied to substances produced by the reaction of an acid with a base, known as a neutralization reaction. Salts are characterized by ionic bonds, relatively high melting points, high electrical conductivity when melted or in solution, and a crystalline structure when in the solid state.

Common table salt is a white solid, soluble in hot or cold water, slightly soluble in alcohol, but insoluble in concentrated hydrochloric acid. In the crystalline form NaCl is transparent and colorless, shining with an ice-like luster. NaCl often contains trace quantities of magnesium chloride, $MgCl_2$, magnesium sulfate, $MgSO_2$, calcium sulfate, $CaSO_4$, potassium chloride, KCl, and magnesium bromide, $MgBr_2$.

Salt is widely distributed in nature. It is found in solution in all ocean water in concentrations of approximately 30 g/L (0.255 lb/gal). NaCl is also distributed throughout many rivers and inland lakes and seas, the concentration varying from 0.002% in the Mississippi River to 12% in the Great Salt Lake and 8% in the Dead Sea.

Salt may occur in the form of surface crust or layer in swamps and dry lake bottoms, especially in extremely arid regions. The mineral halite, more commonly known as rock salt or massive salt, occurs in beds deposited by the dehydration of ancient bodies of salt water. NaCl is constantly being formed by the action of rivers and streams on rocks containing chlorides and compounds of sodium. Salt melts at 804°C (1,479°F) and volatilizes at temperatures just slightly above this. Its specific gravity is 2.17.

Important since prehistoric times as a seasoning agent and to preserve foods, salt was a symbol of enduring faith to peoples of the ancient world and was commonly used in the religious rites of the Greeks, Romans, Hebrews, and Christians.

The simplest method of obtaining salt from areas near oceans or seas is by the evaporation of salt water. This process is expensive, however, and is used only when cheaper methods are unavailable. Several evaporation methods may be used, the most important of which is solar evaporation in which the heat of vaporization is derived from the sun. Steam evaporation in vacuum pans and covered kettles, and direct-heat evaporation in open kettles and pans (Theodore and Dupont 2021) can also be used for NaCl production. Most commercial salt is produced by steam or direct-heat evaporation of rock salt brine.

Salt has more than a thousand uses. Most familiar is its use as a seasoning and as such it is an essential constituent of the diet of human beings and other warm-blooded animals. Common table salt marketed for consumption in inland areas often has small quantities of iodides added to prevent the occurrence of goiters. NaCl is also the basic raw material for numerous chemical compounds such as sodium hydroxide, sodium carbonate, sodium sulfate, hydrochloric acid, sodium phosphates, and sodium chloride and chlorine and is the source of many other compounds through its derivatives. Practically, all the chlorine produced in the world is manufactured by the electrolysis of NaCl, and the production of chlorine and sodium hydroxide accounts for nearly 50% of salt usage in the United States. Salt is also used in the regeneration of sodium zeolite water softeners and has many applications in the manufacture of organic chemicals.

Other commercially important chlorine and sodium compounds produced from NaCl include chloroform, carbon tetrachloride, bleaching powder, washing soda, and baking soda. NaCl is widely used as a preservative for meats, in dyeing, and in the manufacture of soap and glass. Because of its transparency to infrared radiation, salt crystals are used for making prisms and lenses of instruments used to measure infrared radiation.

12.7 CEMENT

Cement is defined as any material that hardens and becomes strongly adhesive after application in plastic form. The term cement is often used interchangeably with glue and adhesive. In engineering and building construction the term usually refers to a finely powdered, manufactured substance consisting of gypsum plaster or portland cement that hardens and adheres after being mixed with water and fine and coarse aggregates.

Cements are used for various purposes such as in the binding of fine (sand) and coarse (gravel) aggregates to form concrete, for uniting the surfaces of various materials, or for coating surfaces to protect them from chemical attack. Cements are made in a wide variety of composition for a wide variety of uses. They may be named for their principal constituents such as calcareous cements which contain silica and epoxy cement which contains epoxy resins. They can also be named for the materials they join, i.e., glass or vinyl cement; for objects to which they are applied, i.e., boiler cement; or for their characteristic properties, i.e., hydraulic cement which hardens under water or acid-resistant or quick-setting cement. Cements that resist high temperatures are called refractory cements.

Although various types of mineral-based cement are of ancient origin, modern cements have their origin in the middle 18th century. The term portland cement was first used in 1834 by Joseph Aspdin, a British cement maker, because of the resemblance between concrete made from his cement and Portland stone which was used widely as a building material at the time in England. The first modern portland cement, made from lime and clay or shale material heated until they form cinders (called clinkers in the industry) and then ground to a fine powder, was produced in England in 1845. At the time, cement was usually made in upright kilns where the raw materials were spread between layers of coke, which was then burned. The first rotary kilns used in cement production were introduced around 1880. Portland cement is now almost universally used for structural concrete.

Cements set or harden by the evaporation of the plasticizing liquid such as water, alcohol, or oil; by internal chemical change; by hydration; or by the growth of interlacing sets of crystals. Other cements harden as a result of chemical reaction with the oxygen or carbon dioxide in the atmosphere. Typical portland cements are mixtures of tricalcium silicate (3 $CaO \cdot SiO_2$), tricalcium aluminate (3 $CaO \cdot Al_2O_3$), and dicalcium silicate (2 $CaO \cdot SiO_2$) in varying proportions, together with small amounts of magnesium and iron compounds. Gypsum is often added to slow the hardening process. Portland cement is manufactured from lime-bearing materials, usually limestone, together with clays, shales, or blast furnace slag containing alumina and silica. By varying the percentage of its normal components or adding others, portland cement can be given various desirable characteristics such as rapid hardening, low heat during hydration, and resistance to alkalis. Rapid hardening cements, sometimes called high-early strength cements, are made by increasing the proportion of tricalcium silicate or by finer grinding of the cement itself.

12.8 GLASS

Glass is generally defined as an amorphous, artificial substance made primarily of silica fused at high temperatures with borates or phosphates. Glass is neither a solid nor a liquid but exists in a vitreous, or glassy state in which molecular units have disordered arrangement but sufficient cohesion to produce mechanical rigidity. Glass is cooled to a rigid state without the occurrence of crystallization and heat can reconvert glass to a liquid form. Usually transparent, glass can also be translucent or opaque. Color varies with the ingredients added during the production of the glass.

Molten glass is plastic and can be shaped by means of several techniques. When cold, glass can be carved. At low temperatures glass is brittle and breaks with a shell-like fracture on the broken face. Such natural materials as obsidian and tektites (from meteors) have compositions and properties similar to those of man-made glass.

The basic ingredient of glass is silica derived from sand, fling, or quartz. Silica can be melted at very high temperatures to form fused silica glass. Because this glass has a high melting point and does not shrink or expand greatly with changing temperatures, it is suitable for laboratory apparatus and for such objects subject to heat shock as telescope mirrors. For most glass, silica is combined with other raw materials in various proportions to achieve an end product with the desired properties. More than 800 different glass compositions are produced to yield a range of desired properties and functionality. Alkali fluxes, commonly the carbonates of sodium or potassium, lower the fusion temperature and viscosity of silica. Limestone or dolomite (calcium and magnesium carbonates) act as stabilizers for the glass. Other ingredients such as lead and borax give certain physical properties to the produced glass. The wide range of uses for the material has resulted in the development of a wide range of glass types.

Window glass, in use since the 1st century AD, was originally made by casting, or by blowing hollow cylinders that were slit and flattened into sheets. The crown process was a later technique in which glass was blown and shaped into a flattened globe or crown that was spun to form a large circular sheet. Today nearly all window glass is produced by the float glass process in which glass is drawn upward from a molten pool fed from a tank furnace. The Fourcault process draws this molten glass through a submerged, slotted refractory block into a vertical annealing furnace that produces flat glass that is cut into sheets of the final product.

Bottles, jars, and other glass containers are produced by automatic processes that combine pressing to form the open end of the container and blowing to form the hollow body of the container. Most lenses used in eyeglasses, microscopes, telescopes, cameras, and certain other optical instruments are made from optical glass. Optical glass differs from other glass in the way in which it bends or refracts light. The manufacture of optical glass is a delicate and exacting operation. Photosensitive glass is similar to photographic film in that gold or silver ions in the material will respond to light. This glass is used in printing and the reproduction process. Heat treatment following an exposure to light produces permanent changes in photosensitive glass.

Glass contains certain metals which form a localized crystallization when exposed to ultraviolet radiation. If heated to high temperatures, the class will convert to crystalline ceramics with mechanical strength and electrical insulating properties greater than that of ordinary glass. Such ceramics are now manufactured for such diverse uses as in cooking ware and missile radomes. Other metallic glasses can be magnetized and demagnetized, are strong and flexible, and show great potential for use in transformers.

By drawing out molten glass to diameters of a few ten-thousands of an inch, it is possible to produce fibers that can be woven or felted like textile fiber. Woven into textile fabrics, glass fibers make excellent drapery and upholstery materials because of their chemical stability, strength, and resistance to fire and water. Glass fabrics alone, or in combination with resins, make excellent electrical insulation. By impregnating glass fibers with plastics a composite fiberglass is formed that combines the strength and inertness of glass with the impact resistance of the plastic. Interestingly, this composite fiberglass has found application in the air pollution control of particulates (Theodore 2007).

12.9 ILLUSTRATIVE EXAMPLES

Four illustrative examples complement the material presented above.

12.9.1 Illustrative Example 1

The number of daily inorganic chemical batches at a plant has a Poisson distribution with a mean of 5. Find the probability of at least eight batches produced per day.

Inorganic Chemicals

Solution: The Poisson distribution was described in Chapter 5. The probability of at least eight batches per day is determined using Equation 5.24, noting that the standard deviation is equal to the mean for a Poisson distribution:

$$f(x) = \frac{e^{-\mu}\mu^x}{x!}, x = 0, 1, 2, \ldots \tag{5.24}$$

$$P(X \geq 8) = e^{-5}\left[P(X=8) + P(X=9) + P(X=10) + P(X=11) + P(X=12) + \cdots\right]$$

$$P(X \geq 8) = e^{-5}\left[5^8/8! + 5^9/9! + 5^{10}/10! + 5^{11}/11! + 5^{12}/12! + \cdots\right]$$

$$P(X \geq 8) = (0.00674)[9.69 + 5.38 + 2.69 + 1.22 + 0.51 + \cdots]$$

$$P(X \geq 8) = (0.065 + 0.036 + 0.018 + 0.008 + 0.003 + \cdots) = 0.131 = 13.1\%$$

Thus, the probability of producing at least eight batches of this inorganic chemical is approximately 13%. How else could this example be solved?

12.9.2 ILLUSTRATIVE EXAMPLE 2

Assuming an average lifetime of an inorganic chemical plant is 10 years, determine the reliability of the plant lasting at least 20 years.

Solution: As indicated in Chapter 5, reliability of systems is often described using an exponential pdf. In this example, the average failure rate, λ, is 1/average lifetime = 1/10 years = 0.1/yr.

Therefore, using the exponential distribution function, the probability that the plant will last for 10 years is:

$$P(T > 20) = \int_{20}^{\infty} 0.1e^{-0.1t} dt = -e^{-\infty} + e^{-0.1(20)} = 0 + 0.135 = 13.5\%$$

12.9.3 ILLUSTRATIVE EXAMPLE 3

The dioxin concentration measurements ($\mu g/m^3$) in an inorganic process slip stream are approximately normally distributed with a mean $\mu = 102$ and a standard deviation $\sigma = 3.75$. Find the percentage of readings that are between 99 and 106.5 $\mu g/m^3$.

Solution: Using Equation 5.35 to generate standard normal variable, Z, for this example yields:

$$P(T_1 < T < T_2) = P\left(\frac{T_1 - \mu}{\sigma} < \frac{T - \mu}{\sigma} < \frac{T_2 - \mu}{\sigma}\right) \tag{5.35}$$

$$P(99 < T < 106.5) = P\left(\frac{99-102}{3.75} < \frac{T-102}{3.75} < \frac{106.5-102}{3.75}\right) = P(-0.80 < Z < 1.20)$$

Using the values in Table 5.4 for the standard normal distribution, the following can be determined:

$$P(-0.90 < Z < 1.20) = (0.5 - 0.212) + (0.5 - 0.115) = 0.288 + 0.385 = 0.673 = 67.3\%$$

In effect, approximately 2/3 of the dioxin readings will be between 99 and 106.5 µg/m³.

12.9.4 Illustrative Example 4

Consider the following equation, $y = \sqrt[x]{x} = x^{1/x}$. Several applications in later chapters will involve determining the maximum value of this function. For this problem, determine the maximum value of this function and the corresponding value of x.

Solution: To begin the solution, first take the natural log of both sides of the function y as:

$$\ln(y) = \ln(\sqrt[x]{x}) = \frac{1}{x}\ln(x) \tag{12.5}$$

Next, take the derivative with respect to x of both sides and set this derivative to 0:

$$\frac{d(\ln(y))}{dx} = -\frac{1}{x^2}\ln(x) + \left(\frac{1}{x}\right)\left(\frac{1}{x}\right); \left(\frac{1}{y}\right)\frac{d(y)}{dx} = \frac{-\ln(x)+1}{x^2}; \frac{d(y)}{dx} = y\left[\frac{-\ln(x)+1}{x^2}\right]$$

$$\frac{d(y)}{dx} = \sqrt[x]{x}\left[\frac{-\ln(x)+1}{x^2}\right] = x^{1/x}\left[\frac{-\ln(x)+1}{x^2}\right] = x^{1/x-2}\left[-\ln(x)+1\right] = 0 \tag{12.6}$$

Solving for x yields:

$$x^{1/x-2}\left[-\ln(x)+1\right] = 0; \left[-\ln(x)+1\right] = 0; \ln(x) = 1; x = e^1 = 2.718$$

To determine the maximum value of the function, substitute this value of x back into the function as:

$$\text{Maximum } y = (2.718)^{1/2.718} = (2.718)^{0.368} = 1.445$$

PROBLEMS

12.1 A random variable X denoting the useful life in decades of a distillation column processing liquid waste from an inorganic processing facility in Southeast Asia has a probability density function (pdf) of:

$$f(x) = \frac{3}{8}x^2; 0 < x < 2; f(x) = 0; \text{elsewhere}$$

Determine the cumulative distribution function (cdf) of X.

Inorganic Chemicals

12.2 Refer to Illustrative example 2. Determine the reliability of the plant lasting at least 8 years.

12.3 Refer to Illustrative example 3. Find the percentage of readings that are at least 108 μg/m^3.

12.4 The failure rate per year, Y, of a coolant recycle pump in an inorganic plant's wastewater treatment plant has a log-normal distribution. If ln Y has a mean of 2.0 and a variance of 1.5, find $P(0.175 < Y < 1)$.

REFERENCES

Basta, N. 1997. *Shreve's Chemical Process Industries Handbook*, 6th Edition. New York, NY: McGraw-Hill.

Theodore, L. 2007. *Air Pollution Control Equipment Calculations*. Hoboken, NJ: John Wiley & Sons.

Theodore, L., and Dupont, R.R. 2021. *Introduction to Desalination. Principles and Calculations*. New York, NY: McGraw-Hill.

13 Organic Chemicals

13.1 INTRODUCTION

Organic chemistry is that branch of chemistry in which carbon compounds and their reactions are studied. A wide variety of classes of substances such as drugs, vitamins, plastics, natural and synthetic fibers, as well as carbohydrates, proteins, and fats are composed of organic molecules. Organic chemists determine the structures of organic molecules, study their various reactions, and develop processes for the synthesis of organic compounds. Organic chemistry has had a profound effect on life since the mid-20th century. It has synthesized natural and artificial materials that have improved health, increased comfort, and added to the convenience of nearly every product manufactured today.

Basic chemical principles indicate that the molecular formula of a compound indicates the number of each kind of atom in a molecule of that substance. Fructose, or grape sugar ($C_6H_{12}O_6$), consists of molecules containing 6 carbon atoms, 12 hydrogen atoms, and 6 oxygen atoms. Nearly 20 other compounds, however, have this same molecular formula. Even an analysis that gives the percentage of carbon, hydrogen, and oxygen cannot distinguish $C_6H_{12}O_6$ from ribose, $C_5H_{10}O_5$, another sugar in which the ratios of elements are exactly the same, i.e., 1:2:1.

The ability of carbon to form covalent bonding with other carbon atoms in long chains and rings, however, does distinguish carbon from other elements. Other elements are not known to form chains of greater than eight like atoms. This property of carbon, and the fact that carbon nearly always forms four bonds to other atoms, accounts for the large number of known organic compounds. At least 80% of the nearly 10 million known chemical compounds contain carbon.

This chapter presents information on the organic chemical process industry, highlighting five key organic chemicals, i.e., phenol, ethyl acetate, plastic, paper, and rubber. It should be noted that some overlap exists between this chapter on organic chemicals and the next chapter on refining. Four illustrative examples are provided to demonstrate concepts presented throughout the chapter.

13.2 GENERAL COMMENTS

The chemical industry produces a wide variety of synthetic organic chemicals. Major categories include dyes, pigments, flavors and perfume materials, medicinals, plasticizers, plastics and synthetic resins, synthetic fibers, synthetic rubbers, rubber-processing chemicals, surface-active agents (detergents), pesticides and other agricultural chemicals, and intermediates.

Intermediates, which are chemicals used in the manufacture of other organic chemicals, have the largest annual production of any of the categories listed. A few of the more important intermediates are acetic anhydride, aniline, formaldehyde, phenol, phthalic anhydride, and styrene. Organic chemicals have a wide variety of end uses, and in some cases, they can be produced by several alternate processes.

TABLE 13.1
Major Organic Chemicals in Production

Chemical
Acetanilide
Acetic acid
Acetic anhydride
Acetone
Acetylsalicylic acid
Amyl acetates
Aniline
Barbituric acid
Butyl alcohol
Carbon disulfide
Carbon tetrachloride
Chlorobenzene
Cresols
Cresylic acid
Dibutyl phthalate
Dichlorodiphenyltrichloroethane (DDT)
Ethyl acetate
Ethyl alcohol
Ethylene glycol
Formaldehyde
Methanol
Phenol
Phthalic anhydride
Pyridine
Styrene

Because many synthetic organic chemicals are produced from raw materials obtained from petroleum or natural gas, they may also be classified as petrochemicals (see the next chapter). For example, phenol is produced primarily from benzene, much of which is produced from petroleum. Therefore, phenol may also be considered a petrochemical. A list of 25 of the major organic chemicals manufactured in the United States is provided in Table 13.1.

13.3 PHENOL

Phenol (formerly called carbolic acid) is an aromatic organic compound, C_6H_5OH. It is weakly acidic and resembles alcohols in structure. The colorless, needlelike crystals of purified phenol melt at 43°C (109°F) and boil at 182°C (360°F). During storage the crystals become pink and finally reddish brown. Phenol is soluble in organic solvents and slightly soluble in water at room temperature, but infinitely soluble above 66°C (150.8°F). It is a constituent of coal tar.

Organic Chemicals

The term *phenol* is also used for any of a group of related acidic compounds that are hydroxyl derivatives of aromatic hydrocarbons, such as cresols and resorcinol. An example of a phenol derivative is pH indicator phenolphthalein ($C_{20}H_{14}O_4$), which is a chemical compound prepared by a reaction of phenol and phthalic anhydride in the presence of sulfuric acid. Phenol was first used as a disinfectant in 1867 by the British surgeon Joseph Lister for sterilizing wounds, surgical dressings, and instruments. Dilute solutions are useful antiseptics, but strong solutions are caustic and scarring to tissue. Less irritating and more efficient germicides have replaced phenol, but it is widely used in the manufacture of resins, plastics, insecticides, explosives, dyes, and detergents, and as raw material for the production of medicinal drugs such as aspirin.

Phenol is a very large volume intermediate. There are five alternative processes for producing it. Originally, the only source of phenol was from coal tar produced in the destructive distillation of coal in the manufacture of coke for blast furnaces. The phenol present in the coal tar was simply separated and purified. After World War I, sulfonation, chlorination, and regenerative processes were introduced. These processes used benzene as a raw material. Until World War II, most benzene was produced as a by-product of the destructive distillation of coal to yield coke. Since then, benzene is produced primarily from petroleum. The cumene process for phenol is a more recent development which produces acetone as an important coproduct.

Cumene is manufactured by the reaction of benzene and propylene. Both the benzene and propylene are derived from petroleum. Air is used for oxidation, and dilute sulfuric acid is supplied to the reactor. Fresh cumene feed and a recycle of unreacted cumene are fed to a reactor where the small quantity of methylstyrene from the recycle is reacted with hydrogen at 210°F and 1 atm to convert it to cumene as shown in Equation 13.1.

α-Methylstyrene + H_2 $\xrightarrow{\text{Nickel Catalyst}}$ Cumene (13.1)

The nickel catalyst is finely divided and is suspended in the reaction mixture by mechanical agitation. The cumene product is decanted from the reactor, filtered to remove catalyst, and sent to a stirred reactor for oxidation.

The air oxidation of pure cumene that follows give low yields, but the yield may be increased by emulsifying the cumene in a buffered aqueous sodium carbonate solution. This reactor is operated at 265°F, and the reaction is shown in Equation 13.2. Oxygen is supplied by bubbling air through the solution. The organic phase is separated from the reaction mixture and passed to another reactor where the cumene hydroperoxide is split in an acid medium to give phenol and acetone (Equation 13.3).

$$\text{Cumene} + O_2 \rightarrow \text{Cumene hydroperoxide} \quad (13.2)$$

$$\text{Cumene hydroperoxide} \xrightarrow[\text{Catalyst}]{H_2SO_4} \text{Phenol} + \text{Acetone} \quad (13.3)$$

Phenol produced from cumene must compete with phenol produced by the other processes. The sale of the coproduct acetone is equally important in determining the economics of the process. Acetone is used as a solvent and as an intermediate for a wide variety of chemicals such as acetic anhydride, methyl isobutyl ketone, and methyl methacrylate. Acetone produced by the cumene/phenol process must also compete with acetone produced by other processes.

13.4 ETHYL ACETATE

Ethyl acetate is produced by a simple reaction between ethyl alcohol and acetic acid. Typical of organic processes, its production is concerned largely with the separation of the product from unreacted reactants and impurities. The reaction between an organic acid and an alcohol is called *esterification*. In the esterification of ethyl alcohol and acetic acid, a small amount of concentrated sulfuric acid is used as a catalyst. The reaction is reversible and gives a 67% yield at equilibrium with equal moles of reactants. To obtain higher overall yields, an excess of alcohol is used, and the water formed in the reaction must be removed.

The three starting materials are common industrial chemicals. Ethyl alcohol is produced from ethylene obtained from petroleum or natural gas. Acetic acid may be used in any concentration in aqueous solution. It is produced by a variety of processes. The largest volume method of manufacturer is the reaction of methane with carbon monoxide. Acetic acid can also be produced by the oxidation of acetaldehyde or ethyl alcohol. The basic reaction is shown in Equation 13.4:

$$CH_3CH_2OH + CH_3COOH \xrightarrow[\text{Catalyst}]{H_2SO_4} CH_3COOCH_2CH_3 + H_2O \quad (13.4)$$

Ethyl alcohol Acetic Acid   Ethyl acetate

Approximately stoichiometric quantities (Theodore 2014) of ethanol and acetic acid are fed to a mixer-preheater, along with the sulfuric acid catalyst. The reaction is fast and reaches the near-equilibrium conversion of 65 percent of the acetic acid.

Organic Chemicals

The resulting mixture is fed to a distillation column. To force the equilibrium toward complete conversion of the acetic acid, an ethanol-rich recycle stream is supplied to the column, and the ethyl acetate and excess ethyl alcohol are distilled off as crude product. Water and sulfuric acid are withdrawn from the bottom of the column. The crude product is sent to a second distillation column (Theodore and Ricci 2011) for further processing.

13.5 PLASTIC

Plastics is a term applied to organic polymeric materials (those consisting of large organic molecules) that are formed into desired shapes by extruding, molding, casting, or spinning. The molecules can be either natural; including cellulose, wax, and natural rubber; or synthetic including polyethylene and nylon. The starting materials are resins in the form of pellets, powders, or solutions, from which finished products are formed.

Plastics are characterized by high strength-to-density ratios, excellent thermal and electrical insulation properties, and good resistance to acids, alkalies, and solvents. These large molecules of which they consist may be linear, branched, or cross-linked, depending on the plastic. Linear or branched molecules are thermoplastic (soften when heated), whereas cross-linked molecules are thermosetting (permanently hardened when heated). Plastics can be categorized in several ways: by the polymerization process that forms them, by their processibility, by their chemical nature, and by their end use.

In terms of early history, the development of plastics began about 1860, when the decimation of elephant herds resulted in an enormous increase in the price of ivory. Pheelan and Collander, a US firm manufacturing billiard and pool balls, offered a prize of S10,000 for a satisfactory substitute for ivory. In an attempt to win this prize, the US inventor John Wesley Hyatt (1837–1920) developed a method of pressure-working pyroxylin, a cellulose nitrate of low nitration that had been plasticized with camphor and a minimum of alcohol solvent. Although Hyatt did not win the prize, his product, patented under the trademark Celluloid, was used in the manufacture of objects ranging from dental plates to men's collars. Despite its flammability and deterioration under prolonged exposure to light, Celluloid achieved a notable commercial success.

Other plastics were introduced gradually over the next few decades. Among them were the first totally synthetic plastics, the family of phenolformaldehyde resins sold under the trademark Bakelite. Other plastics introduced during this period includes modified natural polymers such as rayon made from cellulose products.

The manufacture of plastic products involves forming a resin to a desired final shape. The resins, which are basically organic compounds, are produced from monomers synthesized in the petrochemical industry. In fact, the starting raw material for plastics is petroleum. In many instances, various chemical additives are used in plastics to produce a desired characteristic. Typical additives include antioxidants, ultraviolet stabilizers, plasticizers, flame retarders, antistatics, lubricants, pigments, and fillers. Antioxidants protect a polymer from chemical degradation by oxygen or ozone. The same is true for ultraviolet stabilizers, which protect against natural weathering. Plasticizers make plastic more flexible; a good example is polyvinyl chloride, which, when plasticized, can be used as vinyl furniture coverings.

Plastics lend themselves to an exceedingly wide range of applications due to their toughness, water resistance, ease of fabrication, and remarkable color range. Plastics have a wide range of industrial applications. Engineering plastics are used for structural and industrial purposes, such as in gears and cams. Typical plastics used in these cases are polycarbonates, nylons, acetals, and phenolics. Plastics such as nylons, acetals, and Teflon are also used for their wear resistance.

Acrylics, celluloses, phenolics, polyethylene, polystyrene, and other plastics are used in such light-duty or decorative applications as handles, cases, and moldings. Large, hollow shapes and housings use polyester, polycarbonate, polystyrene, or cellulose butyrate. Composite materials involve a reinforcing system (usually a fiber made of glass or carbon) in a plastic resin matrix. This reinforcement greatly enhances certain properties of the materials, such as dimensional stability and flexural strength.

Still another plastic variation is the production of foams, which can be either rigid or flexible. Foams are made by generating a gas that is dispersed through the plastic during processing. They have a wide variety of uses because they have reasonably good strength and stability combined with exceptionally low density.

Plastics and their manufacture involve a number of health hazards (Theodore and Dupont 2012), principally in relation to the manufacture of the monomer used for polymerization. Vinyl chloride, for example, is a poison that has been cited in a number of serious cases of industrial pollution. This topic will be revisited in Chapter 22.

13.6 PAPER

Paper is defined as material manufactured by the webbing of vegetable cellulose fibers. Cellulose is not only the most abundant substance available in nature, but it is probably the most versatile, yet the simplest, renewable material known. Its conversion to paper products is the everyday function of the pulp and paper industry, and the resulting multitude of different kinds of useful items is the proof that pulp has become indispensable to modern civilization. Paper is used for writing and printing, for wrapping and packaging, and for a variety of special purposes ranging from the filtration of precipitates from solutions to the manufacture of certain types of building materials. In the 21st-century civilization, paper is a basic material, and the development of machinery for its high-speed production has been largely responsible for the increase in literacy and the raising of educational levels of people throughout the world.

Paper was first made in AD 105. The material used to produce this first paper was probably the bark of the mulberry tree and was made on a mold of bamboo strips. The earliest known paper still in existence was made from rags about AD 150. For approximately 500 years the art of papermaking was confined to China, but in 610 AD it was introduced to Japan, and into Central Asia about 750 AD. Paper made its appearance in Egypt about 800 AD but was not manufactured there until 900 AD.

The use of paper was introduced into Europe by the Moors, and the first papermaking mill was established in Spain about 1150 AD. In succeeding centuries, the craft spread to most of the European countries. The first paper mill in England was established in 1495, and the first such mill in America was built in 1690. The

Organic Chemicals

increasing use of paper in the 17th and 18th centuries created shortages of rags, which were the only satisfactory raw material known to European papermakers. As a result, many attempts were made at the time to devise substitute raw materials but none were commercially viable. The solution to the limited raw material problem was the introduction of the groundwood process of pulp making about 1840 and the first of the chemical pulp processes approximately 10 years later.

13.7 RUBBER

Rubber is defined as a natural or synthetic substance characterized by elasticity, water repellence, and electrical resistance. Natural rubber is obtained from the milky white fluid called latex found in many plants; synthetic rubbers are produced from unsaturated hydrocarbons. In its natural state, rubber exists as a colloidal suspension in the latex of rubber-producing plants. Crude rubber from some plant sources is generally contaminated by an admixture of resins that must be removed before the rubber is suitable for use. Such crude rubbers include guttapercha and balata, which are products of various tropical plants in the sapodilla family.

Any artificially produced substance that resembles natural rubber in essential chemical and physical properties may defined as synthetic rubber. Such substances, also called elastomers, are produced by condensation or polymerization, of certain unsaturated hydrocarbons. The basic materials of synthetic rubber are monomers: compounds of relatively low molecular weight that form the building blocks of huge molecules called polymers. After fabrication, the synthetic rubber is cured by vulcanization.

Synthetic rubber research initiated in the United States during World War II led the synthesis of a polymer of isoprene identical in chemical composition to natural rubber. Various types of synthetic rubber are in production, with their composition determined by the specific use for which they are intended. Annual consumption of synthetic rubber in the United States averages three to four times the average annual consumption of natural rubber.

Rubber has become a material of tremendous economic and strategic importance and is an excellent barometer of the industrialization of nations. The rubber industry currently involves the production of monomers or raw materials for natural and synthetic rubbers, the production of various rubbers themselves, the importation of natural rubber, the production of rubber chemicals, and the fabrication of rubber products.

13.8 ILLUSTRATIVE EXAMPLES

Four illustrative examples complement the material presented above regarding the organic chemical industry.

13.8.1 Illustrative Example 1

Calculate the probability that a pump in an organic chemical production facility will last at least five times its expected life.

Solution: The describing equation for this problem is an exponential distribution as shown in Equation 13.5:

$$P(T) = e^{-\lambda t} \tag{13.5}$$

where $\lambda = 1/a$; a = the expected life of the pump; and t = time to failure. For this problem, $t = 5a$, so that:

$$P(T > 5a) = e^{-(1/a)(5a)} = e^{-5} = 0.0067 = 0.67\%$$

13.8.2 ILLUSTRATIVE EXAMPLE 2

The probability of a contaminated batch of organic chemical is 0.01. Find the mean and standard deviation for this failure if described by a binomial distribution (Section 5.2.1) of a total of 400 batches.

Solution: The mean or expected value for the binomial failure distribution is:

$$Np = 400(0.01) = 4$$

Thus, on average four batches *can be expected* to be contaminated within a 400-batch work flow.

The *variance* of the binomial distribution is given as:

$$Npq = 400(0.01)(0.99) = 3.96$$

Thus, the standard deviation of the distribution is the square root of the variance or:

$$\text{Standard Deviation} = \sqrt{3.96} = 1.99$$

13.8.3 ILLUSTRATIVE EXAMPLE 3

Consider the organic synthesis process described in Figure 13.1. Equipment Components A, B, C, and D have their respective reliabilities of 0.90, 0.90, 0.80, and 0.90. The process fails if Component A fails, if both Components B and C fail, or if Component D fails. Determine the overall reliability of this organic synthesis process.

Solution: Using Equations 5.41 and 5.42 for the description of the reliability of series and parallel systems, respectively, this problem can be solved as follows:

Components B and C constitute a parallel subsystem connected in series to Components A and D. The reliability of the parallel subsystem yields:

$$R_P = 1 - (1 - 0.80)(1 - 0.90) = 1 - 0.2(0.1) = 0.98$$

The reliability of the overall system, then is:

$$R_S = (0.90)(0.98)(0.90) = 0.79 = 79\%$$

Organic Chemicals

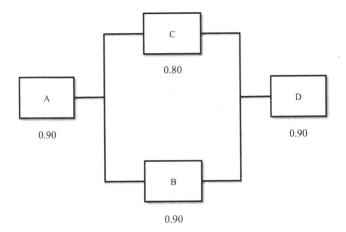

FIGURE 13.1 Organic synthesis process with corresponding equipment reliability values.

13.8.4 ILLUSTRATIVE EXAMPLE 4

The downwind concentration, C (μg/L), of a hazardous organic chemical generated from a chemical synthesis process was recently described by RAT Associates as:

$$C = a\sqrt[x]{x} = ax^{1/x} \qquad (13.6)$$

where a for this chemical is 130. Calculate the downwind distance, x (km), where the concentration of this organic chemical is a maximum and calculate the maximum concentration that would occur at this location.

Solution: Refer to Illustrative example 4 to determine the maximum value of x for this function. From this example, it was found that the maximum value for this function is at $x = e = 2.718$ km. The maximum concentration found at this location is determined as:

$$C = 130\sqrt[2.718]{2.7183} = 130(1.445) = 187.8\,\mu g/L$$

PROBLEMS

13.1 Refer to Illustrative example 1. Calculate the probability that a pump will survive at least 10 times its expected life.

13.2 Suppose that X has a log-normal distribution with $\alpha = 2$ and $\beta = 0.1$. Calculate the probability that the value of X will be between 6 and 8.

13.3 The lifetime to failure of an organic chemical synthesis process can be described by a Weibull distribution with $\alpha = 2$ and $\beta = 0.5$. Outline how to determine the probability that the process lasts a given period of time without failure if the failure rate is described by the function $t^{-1/2}$ over its entire time domain.

13.4 A recent study by FAT Consultants has concluded that the concentration of a toxic organic contaminant above approximately 150 μg/L would create an immediately dangerous to life and health (IDLH) situation for any human receptor. If the concentration of this chemical from an emergency release vent from a chemical synthesis unit can be described by Equation 13.9, where $a = 110$ for this chemical, calculate what the downwind distance is to the maximum concentration, what this maximum concentration would be, and the maximum downwind distance beyond which an IDLH concentration does not occur.

REFERENCES

Theodore, L. 2014. *Chemical Engineering: The Essential Reference.* New York, NY: McGraw-Hill.

Theodore, L., and Dupont, R.R. 2012. *Environmental Health Risk and Hazard Risk Assessment: Principles and Calculations.* Boca Raton, FL: CRC Press, Taylor & Francis Group.

Theodore, L., and Ricci, F. 2011. *Mass Transfer Operations for the Practicing Engineer.* Hoboken, NJ: John Wiley & Sons.

14 Petroleum Refining

14.1 INTRODUCTION

The terms oil and petroleum have been used interchangeably by industry since the early 1900s. However, there are some who refer to petroleum as the crude oil prior to any treatment or refining. Both terms are employed interchangeably in this chapter as a matter of convenience. Typical petroleum reservoirs are mostly sandstone or limestone formations containing oil. The viscosity of the oil in these formations may be as thin as gasoline or as thick as tar. The oil may be almost clear or black. Petroleum was created by the decay of biological materials similar to the processes creating coal and is considered a nonrenewable energy source as it takes millions of years to form.

Petroleum occurs naturally as an oily, bituminous liquid composed of various organic chemicals. It is found in large quantities below the surface of the Earth and is used as a fuel and as a raw material in the chemical industry. Modem industrial societies use it primarily to achieve a degree of mobility – on land, at sea, and in the air – that was barely imaginable 100 years ago. In addition, petroleum and its derivatives are used in the manufacture of medicines and fertilizers, foodstuffs, plasticware, building materials, paints, cloth, various fuels and to generate electricity.

The chemical composition of all petroleum is principally hydrocarbons, although a few sulfur-containing and oxygen-containing compounds are usually present; the sulfur content varies from about 0.1% to 5%. Petroleum contains gaseous, liquid, and solid elements. As noted above, the consistency of petroleum varies from liquid as thin as gasoline to liquid so thick that it will barely pour. Small quantities of gaseous compounds are usually dissolved in the liquid. When larger quantities of these compounds are present, the petroleum deposit is associated with a deposit of natural gas.

Three broad classes of crude petroleum exist: the paraffin types, the asphaltic types, and the mixed-base types. The paraffin types are composed of molecules in which the number of hydrogen atoms is always two more than twice the number of carbon atoms. The characteristic molecules in the asphaltic types are naphthenes, composed of twice as many hydrogen atoms as carbon atoms. In the mixed-base group are both paraffin hydrocarbons and naphthenes.

Crude petroleum composition varies substantially with its location, but all oils are primarily hydrocarbon compounds, including paraffins (such as butane or octane), cycloparaffins (such as cyclohexane), and aromatics (such as benzene). Oxygen as organic acids may reach 3% in some crude oils. Sulfur compounds are present in many crude oils. For example, Wyoming crudes may have as much as 5% sulfur. Nitrogen compounds, metals, etc., are also sometimes found in low concentrations in crude oils.

These compounds are formed under the Earth's surface by the decomposition of marine organisms. The remains of tiny organisms that live in the sea, and to a lesser extent, those of land organisms that are carried down to the sea in rivers and of plants that grow on the ocean bottoms, are enmeshed with the fine sands and silts that settle to the bottom in quiescent sea basins. Such deposits, which are rich in organic materials, become the source rocks for the generation of crude oil. The process began many millions of years ago with the development of abundant life, and it continues to this day. The sediments grow thicker and sink into the seafloor under their own weight. As additional deposits pile up, the pressure on the ones below increases several thousand times, and the temperature rises by several hundred degrees. The mud and sands harden into shale and sandstone; carbonate precipitates and skeletal shells harden into limestone; and the remains of the dead organisms are transformed into crude oil and natural gas.

Once the petroleum forms, it flows upward in the Earth's crust because it has a lower density than the brines that saturate the interstices of the shales, sands, and carbonate rocks that constitute the Earth's crust. The crude oil and natural gas rise into the pores of the coarser sediments lying above. Frequently, the rising material encounters an impermeable shale or dense layer of rock that prevents further migration, and the oil has become trapped and a reservoir of petroleum or natural gas is formed. A significant amount of the upward-migrating oil, however, does not encounter impermeable rock but instead flows out at the surface of the Earth or onto the ocean floor.

This chapter presents general information about the petroleum refining industry, highlighting drilling processes, refining operations, petrochemicals and transportation and transmission characteristics of the industry. It should be noted that some overlap exists between this chapter on refining and the previous chapter on organic chemicals because of the similar nature of the chemicals involved in each. Six illustrative examples and six problems are provided to demonstrate concepts presented throughout the chapter.

14.2 DRILLING

Drilling operations include the activities necessary to bore through the Earth's crust to access crude oil and natural gas reservoirs. During drilling operations, specially formulated muds are circulated through a bore hole to remove cuttings from around the drill bit, to provide lubrication for the drill string, to protect the walls of the bore hole, and to control downhole pressure. Internal combustion engines often driven by diesel and natural gas provide energy to rotate the drill bit, hoist the drill string, and circulate the mud, and these engines are sources of hydrocarbon air emissions and criteria pollutants (carbon monoxide, CO; particulate matter; nitrogen oxides, NO_x; and sulfur oxides, SO_x).

Cuttings are separated from the mud at the well surface as the mud is passed through shale shakers, desanders, desilters, and degasses. The mud flows to a tank for recycling, and the cuttings, which may be contaminated with hydrocarbons, are pumped to a waste pit for disposal. Because the waste pits may be open to the atmosphere, they are a potential source of hydrocarbon emissions. One other potential air emission source during drilling is blowouts.

Another method to increase oil-field production, and one of the more exciting engineering achievements in the last half century, has been the construction and operation of offshore drilling rigs. These drilling rigs are installed, operated, and serviced on

Petroleum Refining

an offshore platform in water up to a depth of several hundred meters. The platforms may either float or sit on legs attached to the ocean floor, where it is capable of resisting waves, wind, and in Arctic regions, ice floes.

As in traditional rigs, a *derrick* is basically a device for suspending and rotating drill pipe, to the end of which is attached a drill bit. Additional lengths of drill pipe are added to the drill string as the bit penetrates farther and farther into the Earth's crust. The force required for cutting into the Earth comes from the weight of the drill pipe itself. To facilitate the removal of the cuttings, mud is constantly circulated down through the drill pipe, out through nozzles in the drill bit, and then up to the surface through the space between the drill pipe and the bore hole. Successful bore holes have been drilled right on target to depths of more than 6.4 km (more than 4 mi) below the surface of the ocean. Offshore drilling has resulted in the development of a significant additional petroleum reserve, which in the United States amounts to 5% of the total reserves.

During the last few decades, new techniques have been developed to tackle oil exploration at sea. One of the earliest was the *jack-up rig*. It gets its name from a set of steel legs that rest on the seabed and can be extended to jack up the drilling platform above the reach of the waves. The legs can also be raised through the deck so that the platform can be towed to another location. Jack-up rigs can also operate in deep water.

Another drilling rig has been specially designed for exploration in deeper water. This has huge buoyancy tanks that enable it to float out to a drilling site. There, the tanks are partially filled with water, making the rig sink lower in the sea and giving it more stability in stormy weather. Anchors are also used to keep the rig in place above the well. This type of rig, known as a *semisubmersible rig*, allows exploration in water depths of more than a thousand feet. For exploration even deeper water, oil companies use a specially equipped drill ship.

These oil fields that lie below the sea may cover an area of some 30 mi^2, the size of a modern city, and many wells need to be driven to access these widely distributed reservoirs. The discovery of vast oil fields beneath the frozen areas of Alaska confronted technicians and engineers with a spectacular challenge. Oil could hardly be found in a more remote and inhospitable place. These supplies are thousands of miles from industrial centers where fuel is needed and the Artic seas are frozen for most of the year making it impossible to transport cargoes of oil by sea.

14.3 REFINING/PROCESSING

Once oil has been produced from an oil field, it is treated with chemicals and heat to remove water and solids, and the natural gas is separated. The oil is then stored in a tank, or battery of tanks, and later transported to a refinery by truck, railroad tank car, barge, or pipeline. Large oil fields all have direct outlets to major·common-carrier pipelines that lead to refineries for crude oil processing.

Refining of petroleum was initially confined to distillation of the crude oil into several fractions. Products initially of interest were kerosene for lighting, lubricating oil, and paraffin wax. The more volatile fractions, including gasoline, were too dangerous to use for lighting and were frequently discarded. In 1900, kerosene was the most important petroleum product. With the advent of the automobile at the turn of the century, gasoline for internal combustion engines rapidly became the major product. Only a limited quantity of rather low-grade gasoline could be obtained by distillation of crude oil; so it was necessary to develop processes to convert other components of the crude oil to materials that could be used as gasoline. Much of the complexity of a modern oil refinery is due to the numerous processes for making higher-grade gasoline. More recently, the growing interest in the production of chemicals and plastics from certain components of petroleum has led to many new processes for the recovery of these components from the refinery streams.

Distillation (Theodore and Ricci 2011; Theodore 2014) may be defined as the separation of the components of a liquid feed mixture by a process involving partial vaporization through the application of heat. In general, the vapor evolved is recovered in liquid form by condensation. The more volatile (lighter) components of the liquid mixture are obtained in the vapor discharge at a higher concentration. The extent of the separation is governed by two important factors: the properties of the components involved and by the physical arrangement of the unit used for the distillation.

Continuous distillation is carried out in a series of large distillation columns. One column separates the most volatile components from the bulk of the oil. A second column further separates the oil into several fractions, and a third column, operating under a vacuum, separates the least volatile (heavy) components obtained from the bottom of the second column into several fractions. Auxiliary equipment includes heat exchangers, furnaces for heating the feed, and steam strippers for purifying the product streams. All the streams leaving the crude distillation unit are complex mixtures. The mixtures are usually characterized by their boiling range. The most volatile components vaporize at the lowest temperature, and the heaviest components vaporize at the highest temperature at a given pressure. To prevent thermal decomposition, the heaviest components are distilled under vacuum to lower their boiling temperatures.

Catalytic cracking (Theodore 2012) is a major process for converting a heavy fraction, gas oil, into high-octane gasoline. As the name implies, the process splits large molecules into smaller ones. A typical catalytic reaction is shown in Equation 14.1:

$$\underset{\text{Nonane}}{C_9H_{20}} \xrightarrow[\text{Catalyst}]{900°F} \underset{\text{Propylene}}{CH_3CH=CH_2} + \underset{\text{Hexane}}{C_6H_{14}} \qquad (14.1)$$

Because the feed is a complex mixture, many reactions occur. The major products from catalytic cracking are high-octane gasoline (93 octane unleaded), heating oil, and gases (such as propylene, butane, and butylene) which are used in various other refinery processes.

Thermal cracking is usually used to split molecules of the heavy components in the residuum (reduced crude). It may be used for producing gasoline and a heavy residue; or it may produce a catalytic cracker feed and a fuel oil. The process involves heating the reduced crude to approximately 950°F under high pressure above 200 psi. The molecules are split, and the new species are separated by distillation.

In *catalytic reforming*, the carbon atom chains in the petroleum molecules are rearranged, usually into branched chain or cyclic patterns. Any sulfur is usually removed as H_2S from the reforming feed and other refinery streams may be recovered as elemental sulfur or converted to sulfuric acid. Sulfur must frequently be removed from refinery streams to prevent poisoning of catalysts and to improve the properties of products.

Light-gas recovery consists of a series of distillation columns where various dissolved gases present in petroleum are separated. These gases are primarily hydrocarbons with one to four carbon atoms. Methane (CH_4) is burned as a fuel. Ethane (C_2H_6) may be burned with the methane, or it may be separated and cracked to give ethylene (C_2H_4), a starting material for many petrochemicals. The propane (C_3H_8)-propylene (C_3H_6) fraction is separated in the light-gas recovery unit and sent to catalytic polymerization. The butane (C_4H_{10})-butylene (C_4H_8) fraction is separated and sent to an alkylation plant for further processing.

Catalytic polymerization involves the combination of two or more olefinic hydrocarbons to produce a larger molecule. A typical polymerization reaction involving propylene is shown in Equation 14.2. The produced propylene dimer is used in gasoline blending:

$$2C_3H_6 \rightarrow C_6H_{12} \qquad (14.2)$$
Propylene   Propylene dimer

Alkylation may involve the combination of an isoparaffin with an olefin to produce a larger molecule of lower volatility and higher octane for gasoline. For example, isobutane and butylene are shown in Equation 14.3 reacting by alkylation to form isooctane:

```
       CH3                                    CH3   H  H
        |                          Heat        |    |  |
 CH3 - C -H  +  CH3CH=CHCH3  ----------->  CH3-C -- C--C - CH3
        |                        Catalyst     |    |  |
       CH3                                   CH3   H  CH3
```
(14.3)

Isobutane   *n*-Butylene   Isooctane
                              (2,2,4-Trimethylpentane)

Various other reactions to give different isomers of octane may also occur. An alkylation reaction of particular interest is the reaction of benzene with propylene to product cumene as shown in Equation 14.4:

```
     H    H                                    H    H
      \  /                                      \  /
       C=C                                       C=C           CH3
      /   \                                     /   \           |
  H--C     C--H  + CH2=CHCH3  →            H--C     C-----C-H
      \   /                                     \   /          |
       C=C                                       C=C          CH3
      /  \                                      /  \
     H    H                                    H    H
```
(14.4)

  Benzene   Propylene   Cumene

Isomerization involves the alteration of the arrangement of the atoms in a molecule without changing the number of atoms in the molecule as shown in Equation 14.5:

$$\begin{array}{c}\text{H H H H}\\\text{H-C-C-C-C-H}\\\text{H H H H}\end{array} \rightarrow \begin{array}{c}\text{H}\\\text{H-C-H}\\\text{H H}\\\text{H-C-C-C-H}\\\text{H H H}\end{array} \qquad (14.5)$$

Straight chain   Branched chain

The most important ketone produced is acetone. Its main supply comes from catalytic dehydrogenation of isopropyl alcohol, by-products of the cumene phenol process. The conversion of isopropyl alcohol to acetone is shown in Equation 14.6. Acetone from this reaction is recovered by condensation from the hydrogen and is purified by simple distillation:

$$(CH_3)_2(CHOH)(g) \rightarrow (CH_3)_2 CO(g) + H_2(g) \qquad (14.6)$$

Major uses of petroleum produced in the United States are for industry and transportation. In 2020, transportation consumed approximately 24 quads of energy, which represents approximately one-third of total energy consumption. Approximately 50% of this transportation energy demand was supplied by petroleum products, and in 2020 the United States was a net exporter of 0.63 MMb/d of petroleum across the globe (US Energy Information Administration 2021).

14.4 PETROCHEMICALS

Petrochemical production is the most rapidly growing segment of the chemical industry today. Development of new processes and products is expected to continue in the future. It is estimated that petrochemicals represent approximately 50% of the total chemical production in the United States. Although the petrochemical industry is large compared with chemical manufacturing, it is small when the entire petroleum industry is considered. Less than 2% of all petroleum and natural gas produced is used for petrochemical production. By definition, a petrochemical is produced from raw materials derived from petroleum or natural gas. Often several refinery streams are potential sources of raw materials leading to a given finished product. In other cases, a refinery stream may be useful for several purposes, and economic considerations determine for which purpose the stream should be employed.

Petrochemicals are used in many areas of everyday life. They have such diverse applications as fertilizer (ammonia), explosives (toluene for TNT), synthetic rubber (butadiene, styrene, etc.), etc. Petrochemicals may be aliphatic, aromatic, or inorganic. The largest tonnage petrochemical is ammonia, an inorganic chemical utilizing hydrogen from various petroleum sources. It is considered a petrochemical even though less than 20% of its mass (the hydrogen) is from a petroleum source. The principal petrochemicals produced in the United States are listed in Table 14.1.

TABLE 14.1
Major Petrochemicals in Production

Chemical
Acetaldehyde
Acetic acid
Acetic anhydride
Acetone
Acetylene
Ammonia
Butadiene
Carbon black
Carbon tetrachloride
Chloroethanes and ethylenes
Ethanolamines
Ethyl alcohol
Ethyl chloride
Ethylene dichloride
Ethylene glycol
Ethylene oxide
Formaldehyde
Glycerol
Hydrazine
Hydrogen peroxide
Isopropyl alcohol
Maleic anhydride
Methanol
Methyl ethyl ketone
Perchloroethylene
Phenol
Phthalic anhydride
Polyethylene
Propyl alcohol
Styrene
Vinyl chloride

It should be noted that some of these chemicals are intermediates in the production of other chemicals.

14.5 TRANSPORTATION/TRANSMISSION

Pipelines are the safest and cheapest way to move large quantities of either crude oil or refined petroleum across land. About 100,000 miles of small gathering lines and large trunk lines move crude oil from wells to refineries. Other transport methods include ships (and barges) and trains. Details follow (Skipka and Theodore 2014).

14.5.1 Pipelines

Oil (and natural gas) can be transported in their natural states through pipelines buried underground or even on the seabed. Hundreds of miles of underground pipeline have been laid to transport oil (and gas) ashore from important offshore producing areas such as the North Sea and the Gulf of Mexico. There are also land pipelines carrying both oil (and gas) products over thousands of miles between producing areas and centers of population in North America, as well as in the Middle East.

Some of the earliest pipelines used to move fuel were made from wood. In the United States during the 19th century, holes were drilled through the center of tree trunks to provide tubes for gas distribution with inside diameters of up to 3 in. Main distribution lines for crude oil (and gas) are now almost always made of steel. The largest one, crossing Alaska from Prudhoe Bay to Valdez, has a diameter of 48 in and would be able to carry 100 million tons of oil a year if used to its full capacity.

Most land pipelines are buried at least 3 ft underground, often under fields where crops are growing undisturbed by the fuel passing beneath. Underwater pipelines may also be buried in a trench on the seabed. All oil and gas pipelines are coated with a layer of bitumen or fiberglass to prevent corrosion while underwater pipelines have an additional coating of concrete for extra protection against the effects of seawater. Long pipelines usually need pumping stations every 50–150 mi to give the oil or gas an additional pressure boost along the way, and in remote areas these stations sometimes get their energy supply by using a minute quantity of the fuel being carried in the line. The flow velocity is approximately 5 mi/hr. Although pipelines are a very efficient way of transporting oil and gas, they need to be cleaned regularly to remove waste and other deposits that the fuels leave on the inside walls, particularly where there are bends in the line. Cleaning is carried out by using the oil or gas flow to push through a device known as a pig. This odd name comes from the initials of pipeline inspection gadget, which describes another of its uses. The pig has a diameter that exactly fits the inside dimensions of the pipe and will scrape away blockages with its outer edge as it moves along the pipe section being cleaned.

14.5.2 Ships

Much of the world's oil comes from areas such as the Middle East that are too distant from main markets to make transport by pipeline either economic or practical. Oil from these regions is shipped to North America, Europe, and Japan in specially built tankers. The first oil tanker, the Glückauf, was launched in 1866 and could carry just 300 tons of oil. Modern vessels can carry up to 500,000 tons. These supertankers are more than 1,300 ft long and hide their bulk beneath the surface like icebergs.

14.5.3 Trains

Oil is transported by train in small quantities, usually across short distances. This mode is employed when the receiver is not near pipelines or major terminals. Jobbers

Petroleum Refining

handle the wholesale distribution of most petroleum products. The retailer, i.e., a gasoline station or a home heating oil company, then receives the product The last stage of fuel distribution is when a car receives gasoline and/or the home receives fuel oil for heating purposes.

14.6 ILLUSTRATIVE EXAMPLES

Six illustrative examples complement the material presented above regarding petroleum refining.

14.6.1 Illustrative Example 1

If a pumping system in the environmental control section of a refinery must have a reliability of 99.993%, how many pumps are required in a *parallel* system if each pump has a reliability of 94.2% due to pump leakage problems? Assume the reliabilities of all of the pumps are equal.

Solution: The required reliability for this parallel system is shown in Equation 14.7:

$$R_p = 1-(1-R_1)(1-R_2)...(1-R_i)...(1-R_n) \qquad (14.7)$$

where R_i is the fractional reliability of pump i. Since the reliabilities of all of the pumps are assumed equal,

$$R_1 = R_2 = \cdots = R_n = R = 0.942$$

then

$$R_p = 1-(1-R)^n = 0.99993 \qquad (14.8)$$

Solving yields:

$$1-0.99993 = (1-R)^n;\ 0.00007 = (1-0.942)^n;\ (0.00007)$$
$$= (0.058)^n;\ \ln(0.00007) = n\ln(0.058)$$

$$n = \ln(0.00007)/\ln(0.058) = -9.567/(-2.847) = 3.36$$

Therefore four pumps would be needed in parallel to ensure that the overall system reliability was 99.993%.

14.6.2 Illustrative Example 2

Successive failures of a chemical reactor's cooling system at a refinery are given by:

$$f(x) = 0.005\,e^{-0.005x},\ x > 0;\ f(x) = 0, \text{elsewhere} \qquad (14.9)$$

What is the probability that the time in months between successive cooling failures is greater than 2 but less than 5? Interest in this refinery's cooling system has surfaced recently because of the potential for a reactor explosion upon cooling system failure.

Solution: The probability that the time in months between successive cooling failures is greater than 2 but less than 5 is:

$$P(2 < X < 5) = \int_{2}^{5} 0.005 e^{-0.005x} dx \tag{14.10}$$

Solving yields the following:

$$P(2 < X < 5) = (0.005/0.005)\left[-\left(e^{-0.005(5)}\right) + \left(e^{-0.005(2)}\right)\right]$$
$$= (-0.9753 + 0.99) = 0.0145 = 1.45\%$$

14.6.3 ILLUSTRATIVE EXAMPLE 3

A compressor at a refinery has an average time to failure of approximately 450 weeks. Assuming the compressor failure distribution can be reasonably described by an exponential function, estimate the reliability that a compressor will survive 260 weeks.

Solution: For this calculation, the average time to failure, t_f, is 450 weeks. For an exponential model (refer to Chapter 5), a modification of Equation 5.37 can be written as:

$$f(t) = R = e^{-\lambda t} \tag{14.11}$$

where λ is the reciprocal of the average time to failure. Thus,

$$\lambda = \frac{1}{t_f} = \frac{1}{450} = 0.00222/\text{wk}$$

The reliability for this compressor lasting 260 weeks is then:

$$R = e^{-(0.00222/\text{wk})(260\,\text{wk})} = e^{-0.577} = 0.562 = 56.2\%$$

14.6.4 ILLUSTRATIVE EXAMPLE 4

A cargo ship containing crude oil arrives at a refinery port once a month. The oil may be contaminated with excessive amounts of sulfur (S), mercury (M), ash (A), cadmium (C), sodium (Na), and chlorine (Cl). If the probability of oil being contaminated with any of these is one-sixth, find the probability, P, of obtaining oil with S and M being exceeded exactly twice and the other four contaminants exactly once in the next eight shipments.

Solution: The multinomial distribution is applied to obtain this answer as follows:

$$P(2,2,1,1,1,1) = \frac{8!}{(2!)(2!)(1!)(1!)(1!)(1!)} \left(\frac{1}{6}\right)^2 \left(\frac{1}{6}\right)^2 \left(\frac{1}{6}\right)\left(\frac{1}{6}\right)\left(\frac{1}{6}\right)\left(\frac{1}{6}\right)$$

$$= \frac{40,320}{6,718,464} = 0.006 = 0.6\%$$

PROBLEMS

14.1 The fractional probability that a crude oil delivery to a refinery will contain sulfur above the regulatory requirement is 0.25. If there were six deliveries last month, find the probability that the sulfur content will be above the required limit exactly two times assuming this distribution can be described by a binomial distribution.

14.2 The parts per million concentration of a particular toxic compound in a refinery discharge stream is known to be normally distributed with mean $\mu = 400$ and standard deviation $\sigma = 8$. Calculate the probability that the toxic concentration, C, is between 392 and 416 ppm.

14.3 The probability that a fan in a refinery will not survive for more than 5 years is 0.09. How often should the fan be replaced? Assume that the time to failure is exponentially distributed and that the replacement time should be based on the fan's expected life.

14.4 Consider a valve at a chemical facility's water treatment plant whose time to failure, T, in hours, has a Weibull pdf (Section 5.3.1) with parameters $\alpha = 0.01$ and $\beta = 0.50$. This results in the following expression:

$$f(t) = (0.01)(0.5)t^{(0.5-1)}e^{(-0.01)t^{0.5}}$$

as the Weibull pdf of the failure rate of the component under consideration over its entire time domain. Estimate the probability that the valve will operate more than 8,100 days.

REFERENCES

Skipka, K., and Theodore, L. 2014. *Energy Resources: Availability, Management, and Environmental Impacts.* Boca Raton, FL: CRC Press, Taylor & Francis Group.

Theodore, L. 2012. *Chemical Reactor Analysis and Applications for the Practicing Engineer.* Hoboken, NJ: John Wiley & Sons.

Theodore, L. 2014. *Chemical Engineering,: The Essential Reference.* New York, NY: McGraw-Hill.

Theodore, L., and Ricci, F. 2011. *Mass Transfer Operations for the Practicing Engineer.* Hoboken, NJ: John Wiley & Sons.

US Energy Information Administration. 2021. *U.S. energy consumption in 2020 increased for renewables, fell for other fuels.* https://www.eia.gov/todayinenergy/detail.php?id=48236 (accessed November 23, 2021).

15 Energy and Power

15.1 INTRODUCTION

No discussion of energy would be complete reviewing the present major sources of energy available in the United States. The following eight key resources are listed below (Skipka and Theodore 2014):

1. Coal
2. Oil
3. Natural gas
4. Nuclear
5. Oil shale
6. Solar
7. Hydroelectric
8. Geothermal

It is likely that some additional discoveries will be made of new reserves in coming years, and new technologies will be developed that permit the recovery efficiency from already known resources to be increased. The supply of crude oil will probably extend to beyond the 21st century. Virtually no expectation exists among experts, however, that discoveries and inventions will extend the availability of cheap crude oil much beyond that period. In light of the reserves available and the dismal projections, it is apparent that alternative energy sources will be required to sustain the societies of the world in the future. The options are indeed few, however, when the massive energy requirements of the industrial world come to be appreciated. The various problems and potentials involved in such alternative sources as geothermal energy, solar energy, and nuclear energy are discussed later in this chapter.

This chapter presents information about the various sources of available energy and the production of power from these various energy supplies. The topics highlighted include fossil fuels, nuclear energy, solar energy, hydroelectric and geothermal energy, power generation, and air conditioning and refrigeration. Six illustrative examples and five problems are provided to demonstrate concepts presented throughout the chapter.

15.2 FOSSIL FUELS

Four subsections are included in the discussion of the various available fossil fuels addressing coal, oil, natural gas, and oil shale.

15.2.1 COAL

The United States' coal reserves are vast. They are estimated to range from 500 billion to more than one trillion tons. Such a bounty truly represents a major domestic energy supply for decades into the future, providing the technical and environmental problems associated with the use of this fuel can be overcome in a reasonable time and at an acceptable cost.

Access to coal reserves is based on either surface of underground mining. Approximately 30% of all coal is transported directly from a mine to the user. The remaining is washed to reduce the inorganic and ash content, producing approximately 100 million tons of waste annually. Most coal moves to power plants by rail, with a considerable amount of land devoted to railroad rights of way. A typical 1,000-MW coal-fired power plant requires approximately 100 carloads of coal every 24 hr. If power projections for the future were to be met by coal-fired plants alone, it would require the daily movement of approximately 100,000 railroad cars and the daily dumping of coal into billions of cubic feet of storage space. Coal at the power plant is burned to produce heat that is partially converted to electricity.

15.2.2 OIL

As described in Chapter 14, petroleum extraction involves drilling through overburden to the oil-bearing strata and removing the oil. Onshore oil production, except for accidental occurrences, does not present any difficult pollution problem. Nevertheless, nearly three barrels of brine must be disposed of for every barrel of oil produced. Environmental degradation resulting from offshore production, dramatized by the 2010 British Petroleum (BP) incident in the Gulf of Mexico, presents more difficult hazard and pollution problems, although much progress has been made in preventing and controlling oil pollution from spills and blowouts.

Approximately 42% of each barrel of oil is refined into gasoline to power this nation's vehicles. This figure approaches 54% when including the oil refined into diesel and jet fuels. Refined residual oil is usually transported directly to a power plant by barge or tanker. Transfer operations can result in oil spills; water contamination results if tankers discharge oil during bilge and tank cleaning operations. At the power plant, the burning of residual oil causes air pollution – primarily sulfur oxides and nitrogen oxides – and results in thermal discharges to water. Approximately 71% of the oil supply is used in transportation versus 22% used by industrial operations, while the remaining 7% goes to residential and commercial use and electricity power production.

15.2.3 NATURAL GAS

Natural gas extraction is in many ways similar to oil extraction. Indeed, both fuels are often taken from the same well. Gas extraction on land affects some acreage through the use of drilling rigs and associated equipment, and it also produces copious amounts of brine, posing a disposal problem. Pipelines, having extensive rights of way, then transport the gas to processing facilities where impurities are removed.

Energy and Power

Combustion of natural gas at power plants causes minor amounts of air pollution – mainly in the form of carbon monoxide and nitrogen oxides – and also results in thermal discharges to water. Natural gas is by far the least environmentally damaging of the fossil fuel alternatives. There is essentially no water pollution other than thermal discharges, and the amounts of solid wastes generated are not significant.

15.2.4 Oil Shale

Oil shale deposits are found in abundance in parts of Colorado, Utah, and Wyoming, where an estimated 600 billion barrels of oil could be extracted from thick oil shale seams to yield 20 or more gallons per ton of shale. Considerably more oil exists in less economic concentrations. In comparison, the nation's current proven reserves of liquid petroleum totaled only 47 billion barrels in 2019 (US Energy Information Administration 2021), although total resources are estimated at several hundred billion barrels.

Tapping oil shale requires the extraction of shale by underground or surface mining, more like mining coal than drilling for petroleum. This is followed by retorting or heating of the shale to produce crude oil. Alternatively, the shale could be heated in situ and then the oil can be withdrawn by drilling as with crude oil. All these techniques demand significantly more energy than that required for conventional liquid oil, producing more air and global warming pollution as a result.

15.3 NUCLEAR ENERGY

Nuclear power is still relatively young (relative to fossil fuels) but it has already produced some of the strongest environmental actions to date. Public controversy over thermal pollution, radioactivity releases, nuclear waste disposal, and nuclear accidents promises to grow in the coming years due to the projected expansion contemplated for the nuclear industry, the introduction of the fast breeder nuclear plants, and the search for carbon-free power to address climate change concerns.

At a light water reactor (LWR) power plant, fission energy is released in the form of heat and is transferred to a conventional steam cycle, which generates electricity. Because of coolant temperature limitations in the LWRs, their thermal efficiency is lower than modem fossil-fueled plants. This lower efficiency, as well as the absence of hot gaseous combustion products released through the stack, means that an LWR power plant discharges over 60% more heat to receiving waters than its fossil fuel counterparts. Extremely small amounts of radioactivity are routinely released to water bodies and to the atmosphere, but only enough to give an estimated annual worst-case ionizing radiation exposure in the range of 0.01–10.0% of the exposure received from natural background radiation.

The spent fuel, containing highly radioactive fission products, is stored at the reactor for several months while the radioactivity declines. It can then be transported to a reprocessing plant where the fuel is chemically treated to recover the remaining uranium and some plutonium that is produced during the fission process. Other fission products are also removed and concentrated.

To understand the factors which make nuclear chemical plant design different from conventional chemical plant design, two basic nuclear physics principles must be understood. The first is the nuclear chain reaction:

$$\text{Fission fuel} + 1\,\text{neutron} \rightarrow \text{fission fragments} + 2 - 3\,\text{neutrons} + \text{energy} \quad (15.1)$$

This simplified equation shows that more than one neutron is emitted for each one used in the fissioning of the fuel, e.g., U^{233}, U^{235}, or Pu^{239}. By proper design of a nuclear reactor or a chemical processing vessel containing fission fuel, the chain reaction can be controlled so that there is no net gain of neutrons available for fission with time. The second principle of prime importance to the design engineer is that of radioactivity. In simplified form:

$$\text{Fuel and fission fragments} \rightarrow \begin{Bmatrix} \text{negative beta particles}\,(\beta^-)\,\text{or electrons} \\ \text{positive beta particles}\,(\beta^+)\,\text{or positrons} \\ \text{gamma rays}\,(\gamma) \\ \text{alpha particles}\,(\alpha)\,\text{or helium nuclei} \\ \text{neutrons}\,(n) \end{Bmatrix} \quad (15.2)$$

The energy generated on the right-hand side of Equation 15.2 can be injurious to both man and material.

Summarizing, the two plant design factors of importance are: (1) protection of personnel and selection of materials of construction to avoid the injurious effects of radioactive energy and (2) controlling the inherent hazard of accidentally producing an uncontrolled nuclear chain reaction with the instantaneous release of vast quantities of energy and radioactivity.

15.4 SOLAR ENERGY

The Earth receives solar radiation from the Sun each day. Approximately 70% is absorbed by clouds, oceans, and land masses. This absorbed energy raises the temperature of the land surface, oceans, and atmosphere. Warm air containing evaporated water from the Earth's land and ocean rises due to buoyant forces, causing atmospheric circulation or *free convection* (Abulencia and Theodore 2009; Theodore 2011, 2014; Flynn, Akashige and Theodore 2019). When the air reaches a high altitude where the temperature is lower, water vapor condenses, providing rain. Radiant energy absorbed by the oceans and land masses maintains the surface at an approximate average temperature of 14°C. In relative terms, the total solar energy absorbed by Earth's atmosphere, oceans, and land masses is enormous. It has been reported that this energy over time is approximately twice the nonrenewable resources of coal, oil, natural gas, and mined uranium combined.

Solar energy can be converted to useful energy in different levels around the world. As one might suppose, geographical locations closer to the equator, in general, receive more solar energy than areas closer to the poles.

The Sun itself is a typical star of intermediate size and luminosity. Solar energy is radiant energy produced in the Sun as a result of nuclear fusion reactions. It is transmitted to the Earth through space in quanta of energy called *photons* which interact with the Earth's atmosphere and surface. The strength of solar radiation at the outer edge of the Earth's atmosphere when the Earth is taken to be at its average distance from the Sun is called the solar constant (1,353 W/m^2). Less than this intensity of energy is actually available at the Earth's surface, because of absorption and scattering of radiant energy as photons interact with the Earth's atmosphere.

The strength of the solar energy available at any point on the Earth depends, in a complicated but predictable way, on the day of the year, the time of day, and the latitude of the collection point. Furthermore, the amount of solar energy that can be collected depends on the orientation of the collecting object.

Approximately 30% of the solar energy reaching the outer edge of the atmosphere is consumed in the hydrologic cycle, which produces rainfall and the potential energy of water in mountain streams and rivers. The power produced by these flowing waters as they pass through modern turbines is called hydroelectric power that is discussed in the next section.

15.5 HYDROELECTRIC AND GEOTHERMAL ENERGY

Water power has been used to drive machinery mechanically for many years. Today, water power is used almost exclusively for the generation of electricity. Although it has historically been an important energy source, it currently meets only 7% of the total US needs. No major expansion of hydroelectric power in the United States is anticipated, partly because most of the readily available sites have been developed and partly because of growing concern for preservation of the remaining natural rivers. Due primarily to silting in the reservoirs behind the dams, hydroelectric plants have an expected lifetime of approximately 100–200 years. Although the impact is relatively small, hydroelectric power systems do cause environmental degradation in the form of the destruction of natural scenic views, fish and wildlife habitat, and deterioration of water quality. New technologies are also emerging to utilize water power, such as tidal power projects, water turbines in fast-moving currents, etc.

Geothermal steam or superheated water is produced when the Earth's heat energy is transferred to subsurface water from rocks in the Earth's crust. Recent explorations have revealed that the resource is larger and more extensive than had been supposed. There is evidence now that reservoirs of steam and hot water are actually widespread in the Earth's crust where the pressure and temperature are adequate, the steam output may be used in turbines for conversion to electricity. However, more often than not, the temperatures and pressures at which the steam emerges are well below those used in traditional electric generating plants, giving rise to inevitable inefficiencies in power production. In addition to electricity, geothermal steam or hot water can be applied to desalting seawater; to heating houses, greenhouses, and swimming pools; and to providing nonelectrical energy for refrigeration and air conditioning.

Geothermal energy sources are just now beginning to be exploited on a large scale, and they have the potential to generate significant amounts of electricity in some regions, especially if additional heat reservoirs are located and improved recovery techniques can be implemented.

15.6 POWER GENERATION

Electricity is a very important and useful form of power for the chemical process industry. Chemical process plants use electricity for driving pumps, compressors, agitators, and other mechanical equipment, for process instrumentation, and for lighting. Power can be either purchased from a public or private utility, produced at the plant site by steam-driven turbogenerators or natural gas-driven engines, or purchased from an adjacent industrial plant as a by-product. Important factors which should be considered in selecting a power supply include:

1. Proximity to existing utility power lines
2. Magnitude and type of power requirement
3. Demand for low-pressure steam for processing and heating
4. Availability of by-product fuel and heat
5. Competitive capital ventures

There are several factors which govern the design, installation, and maintenance of chemical process plant power systems. The major electrical design items include:

1. Power generation or purchased power substation and switching
2. Distribution systems – feeders, unit substations, transformers, switchgear, and overload protection
3. Power wiring for plant equipment – motors, heaters, furnaces, and welders
4. Lighting equipment – inside and outside buildings, yards, roadways, and protective lighting
5. Electrical process control systems
6. Communication equipment – intercommunication, public telephones
7. Safety equipment – fire alarms, burglar alarms, lightning and other static arresters
8. Environmental factors – excessive temperature, pressure, corrosion, and explosion hazards

Since the end of the 17th century, the use of steam has increased to such an extent that today steam furnishes the major portion of the total power developed in the chemical process industry. The most recent developments have been toward improvements in boiler construction, with the aim of producing higher-pressure steam, in central power stations. Such high pressures increase the overall efficiency in the production of electric power. The limiting factor is the failure of materials due to high temperature and pressure excursions.

There are two main types of boilers, the fire tube and the water tube (Santoleri, Reynolds and Theodore 2004). The *fire-tube boiler* is usually of small or medium capacity and is designed for the generation of steam at moderate pressure. In this type of boiler, the fire passes through the tubes. Fire-tube boilers have a low initial cost and a relatively large reservoir of hot water. This is of particular advantage in small chemical plants, where there may be a sudden demand on the steam plant. *Water-tube boilers* are used almost exclusively in stationary installations where service demands a large amount of evaporation at pressures above 150 psi. The water is in the tubes and can be converted to steam more quickly than with the fire-tube boiler ("quicker steaming"). High efficiencies are obtained by this type of boiler. Boiler feedwater should be conditioned before introduction into the boilers. Poor quality boiler water results in foaming, caustic embrittlement, corrosion, and scale formation, with a consequent loss of steam production and poor efficiency.

Steam is often regarded as being made solely for the generation of power. However, in the chemical process industries, particularly in those pertaining to the making of chemicals, so much heat is obtained by condensation of steam that the dual use of steam for power and for heat is of paramount importance. Engineering applications endeavor to coordinate and balance the two. In the chemical process industry, the object is often to expand the steam through the steam engine or turbine and then employ the exhaust or lower-pressure steam for heating purposes.

15.7 AIR CONDITIONING AND REFRIGERATION

15.7.1 Air Conditioning

The use of air conditioning in the chemical process industry has become more and more common in the past few years. Control of the temperature, humidity, and cleanliness of the air is very important in many chemical processes, particularly in artificial-fiber and paper manufacturing. Textile fibers are quite sensitive to changing conditions of the air. The comfort of workers is also important to efficient industrial organizations. This latter fact has led to the use of air conditioning in plants and offices where it is not essential for product quality.

Most air conditioning systems consist of a fan unit that forces air through a series of devices which act upon the air to clean it, increase or decrease its temperature, and increase or decrease its water vapor content. Air conditioning equipment can generally be classified into two broad types: central (also called field-erected) and unitary. At least a dozen types of central air conditioning and air distribution systems are commonly used in commercial and industrial applications. They usually have large cooling and heating equipment located in a central location from which many different spaces or zones are served.

Odors or pollutants arising from chemical processes in chemical plants must be controlled to maintain a safe working environment. Acceptable air quality can be maintained by localized exhaust of pollutants at their source, dilution with outdoor air that is free of such pollutants, or a combination of the two processes. Ventilation air requirements will vary because of the amounts of pollutant produced by different process activities. Outdoor ventilation air requires much more energy to condition than the recirculated air from the nearly constant temperature conditioned space. Occupancy sensors (which typically detect CO_2 levels) are often used to reduce space conditioning costs by regulating the amount of outdoor air when space occupancy may be highly variable.

Industrial air conditioning systems must often address harmful gases, vapors, dusts, or fumes that are released into the plant environment. These contaminants are best controlled by the exhaust systems located near the source before they can enter the working environment. Dilution ventilation may be acceptable where nontoxic contaminants come from widely dispersed points. Combinations of local exhaust and dilution ventilation may provide the least expensive installation. Dilution alone may not be appropriate for cases involving toxic materials or large volumes of contaminants, or where the employees must work near the contaminant source.

15.7.2 Refrigeration

All refrigeration processes involve work, specifically, the extraction of heat from a body of low temperature and the rejection of this heat to a body willing to accept it. Refrigeration generally refers to operations in the temperature range of 120–273 K, while cryogenics usually deals with temperatures below 120 K where gases, including methane oxygen, argon, nitrogen, hydrogen, and helium, can be liquefied.

One of the main cost considerations when dealing with refrigeration and cryogenics is the cost of building and powering the equipment. This is a costly element in the process, so it is important to efficiently transfer heat so that money is not wasted in lost heat in the cooling process. Cryogenics plays a major role in the chemical processing industry as it is important in the recovery of valuable feedstocks from natural gas streams, upgrading the heat content of fuel gas, purifying many process and waste streams, producing ethylene, as well as other chemical processes.

The development of refrigeration systems was rapid and continuous at the turn of the 20th century, leading to a history of steady growth. The purpose of refrigeration, in a general sense, is to make materials colder by extracting heat. As described earlier, heat moves in the direction of decreasing temperature (i.e., it is transferred from a region of high temperature to one of a lower temperature). When the opposite process needs to occur, it cannot do so by itself, and a refrigeration system (or its equivalent) is required.

The equipment necessary in refrigeration is dependent upon many factors, including the substances and fluids working in the system. One very important part of refrigeration is the choice of refrigerant being employed, which obviously depends on the system in which it will be used. The following criteria are usually considered in refrigerant selection:

1. Practical evaporation and condensation pressures
2. High critical and low freezing temperatures
3. Low liquid and vapor densities
4. Low liquid heat capacities
5. High latent heat of evaporation (enthalpy)
6. High vapor heat capacities

Energy and Power

Ideally, a refrigerant should also have a low viscosity and a high coefficient of performance (Theodore, Ricci and VanVliet 2009; Theodore 2014). Practically, a refrigerant should have:

1. A low cost
2. Chemical and physical inertness at operating conditions
3. No corrosiveness towards materials of construction
4. Low explosion hazard
5. Be non-poisonous and non-irritating

15.8 ILLUSTRATIVE EXAMPLES

Six illustrative examples complement the material presented above regarding energy and power in the chemical process industry.

15.8.1 ILLUSTRATIVE EXAMPLE 1

Let X denote the number of oil cargo ships that arrive adjacent to an oil-fired power plant on a given day. The pdf for X is given by:

$$f(x) = \frac{x}{55}; x = 1, 2, 3, 4, 5, 6, 7, 8, 9, 10 \tag{15.3}$$

Calculate the probability that at least three ships but less than six ships will arrive on a given day.

Solution: The probability that at least three ships but no more than six ships will arrive on a given day can be determined by:

$$P(3 \leq X \leq 6) = f(X=3) + f(X=4) + f(X=5) = \frac{3}{55} + \frac{4}{55} + \frac{5}{55} = \frac{12}{55} = 0.218 = 21.8\%$$

15.8.2 ILLUSTRATIVE EXAMPLE 2

Determine the reliability of the electrical system in a nuclear power plant shown in Figure 15.1 using the reliabilities indicated under the various components.

Solution: First identify the components connected in parallel: A and B are connected in parallel; D, E, and F are also connected in parallel. Then, compute the reliability of each subsystem of the components connected in parallel. The reliability of the parallel subsystem consisting of components A and B is:

$$R_p = 1 - (1 - 0.70)(1 - 0.70) = 0.91$$

The reliability of the parallel subsystem consisting of components D, E, and F is:

$$R_p = 1 - (1 - 0.60)(1 - 0.60)(1 - 0.60) = 0.936$$

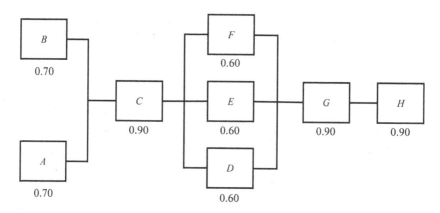

FIGURE 15.1 Design of a nuclear power plant electrical system.

Multiply the product of the reliabilities of the parallel subsystems by the product of the reliabilities of the components to which the parallel subsystems are connected in series:

$$R_s = (0.91)(0.9)(0.936)(0.9)(0.9) = 0.621$$

The reliability of the whole system is therefore 0.621% or 62.1%.

15.8.3 ILLUSTRATIVE EXAMPLE 3

A procuring agent for a coal-fired power plant is asked to sample a lot of 100 heat exchangers. The sample procedure calls for the inspection of 20 exchangers. If there are any bad exchangers in the sample population, the entire lot is generally rejected; otherwise, it is accepted. The probability of a bad exchanger is 4%. The chief engineer has asked the agent the following question. Suppose one bad exchanger is allowed in the sample population. What kind of protection would the plant have?

Solution: This application required the application of a binomial distribution (see Section 5.2.1). One needs to calculate the probability of accepting the lot if 0 or 1 defective exchanger is allowed in the sample population. Therefore:

$$P(X \leq 1) = P(X = 0) + P(X = 1)$$

$$= \frac{20!}{0!(20-0)!}(0.04)^0(0.96)^{20-0} + \frac{20!}{1!(20-1)!}(0.04)^1(0.96)^{20-1}$$

$$P(X \leq 1) = \frac{20!}{1(20)!}(1)(0.96)^{20} + (20)(0.04)(0.96)^{19} = (0.96)^{20} + (0.8)(0.96)^{19}$$

$$P(X \leq 1) = 0.442 + (0.8)(0.4604) = 0.442 + 0.368 = 0.81 = 81\%$$

Energy and Power

15.8.4 ILLUSTRATIVE EXAMPLE 4

Several multipurpose utilities have equal chance of employing fuel in any of the following six forms:

1. Coal
2. Oil
3. Natural gas
4. Nuclear
5. Solar
6. Geothermal

Seven utilities are involved. Find the probability, P, of two employing coal and oil and the others employing only one of the fuels available.

Solution: The multinomial distribution is applied to obtain this answer as follows:

$$P(2,2,1,1,1,1) = \frac{7!}{(2!)(2!)(1!)(1!)(1!)(1!)}\left(\frac{1}{6}\right)^2\left(\frac{1}{6}\right)^2\left(\frac{1}{6}\right)\left(\frac{1}{6}\right)\left(\frac{1}{6}\right)\left(\frac{1}{6}\right)$$

$$= \frac{5040}{6,718,464} = 0.00075 = 0.08\%$$

As expected, the probability is low.

15.8.5 ILLUSTRATIVE EXAMPLE 5

Consider a standby pumping redundancy system at a coal-fired power plant with one operating unit and two on standby, that is, a system that can survive two failures. If the failure rate is four units per year, what is the 6-month reliability of the system?

Solution: This problem can be solved using a Poisson probability distribution (Section 5.2.4). The required reliability is a 6-month reliability. The average number of failures in this 6-month period is μ and is the average failure rate/6-month time period. The given failure rate is four units per year or two per 6-month period. This value of μ is then substituted into the Poisson distribution to solve as follows:

$$f(x) = \frac{e^{-\mu}\mu^x}{x!}, x = 0,1,2,\ldots \tag{5.24}$$

The number of units in standby mode is $n = 2$. Therefore, the 6-month reliability is given by:

$$R = \sum_{x=0}^{2} \frac{e^{-2}(2)^x}{x!} = e^{-2} + 2e^{-2} + (4/2)e^{-2} = 0.135 + 0.271 + 0.271 = 0.677 = 67.7\%$$

15.8.6 ILLUSTRATIVE EXAMPLE 6

The time to "failure" in weeks of a newly designed power station is normally distributed with a mean and standard deviation of 250 and 5 weeks, respectively. Calculate the probability that the new system will fail during the 245th–260th week period.

Solution: Employing the data provided in the problem statement, and the equation used for the normal distribution (Equation 5.35), the standard normal variables Z are generated as follows:

$$P(T_1 < T < T_2) = P\left(\frac{T_1 - \mu}{\sigma} < \frac{T - \mu}{\sigma} < \frac{T_2 - \mu}{\sigma}\right) \quad (5.35)$$

$$Z_1 = \frac{T - \mu}{\sigma} = \frac{245 - 250}{5} = -1$$

$$Z_2 = \frac{T - \mu}{\sigma} = \frac{260 - 250}{5} = 2$$

Using values from Table 5.4, the probability of system failure during Weeks 245th–260th is:

$$P(-1.0 < Z < 2.0) = (0.5 - 0.159) + (0.5 - 0.023) = 0.341 + 0.477 = 0.818 = 81.8\%$$

PROBLEMS

15.1 Consider a component in a nuclear power plant whose entire time to failure, T, in hours, has a Weibull pdf (Equation 5.29) with parameters $\alpha = 0.01$ and $\beta = 0.50$. This gives:

$$f(t) = \alpha\beta t^{\beta-1} e^{-\alpha t^\beta} = (0.01)(0.50) t^{0.50-1} e^{-(0.01)t^{0.50}}; t > 0$$

as the Weibull pdf of the failure time of the component under consideration. Estimate the probability that the component will operate less than 10,000 hr.

15.2 The specification for the production of a toxic alloy for a battery employed at a utility plant calls for 23.2% copper. A sample of ten analyses of the product showed a mean copper content of 23.5% and a standard deviation of 0.24%. Can one conclude at the 0.01 significance level that the product meets the required specifications? Use a standard two-tailed t-test to compare the measured value to the required value.

15.3 Resolve Illustrative example 3 if no bad exchangers are allowed.

15.4 Refer to Illustrative example 6. Calculate the probability that the new system will fail during the 240th–260th week period.

15.5 Consider a critical air conditioning unit at a chemical process plant whose time to failure, T, in hours, has a Weibull pdf (Section 5.3.1) with parameters $\alpha = 0.02$ and $\beta = 0.55$. This results in the following expression:

$$f(t) = (0.02)(0.25) t^{(0.25-1)} e^{(-0.02)t^{0.25}}$$

as the Weibull pdf of the failure rate of the air conditioning unit under consideration over its entire time domain. Estimate the probability that the unit will operate more than 44,000 hr.

REFERENCES

Abulencia, P., and Theodore, L. 2009. *Fluid Flow for the Practicing Chemical Engineer*. Hoboken, NJ: John Wiley & Sons.

Flynn, A.M., Akashige, T., and Theodore, L. 2019. *Kern's Process Heat Transfer*, 2nd Edition. Salem, MA: Scrivener-Wiley.

Santoleri, J., Reynolds, J., and Theodore, L. 2004. *Introduction to Hazardous Waste Incineration*, 2nd Edition. Hoboken, NJ: John Wiley & Sons.

Skipka, K., and Theodore, L. 2014. *Energy Resources: Availability, Management, and Environmental Impacts*. Boca Raton, FL: CRC Press/Taylor & Francis Group.

Theodore, L. 2011. *Heat Transfer Operations for the Practicing Engineer*. Hoboken, NJ: John Wiley & Sons.

Theodore, L. 2014. *Chemical Engineering: The Essential Reference*. New York, NY: McGraw-Hill.

Theodore, L., Ricci, F., and VanVliet, T. 2009. *Thermodynamics for the Practicing Engineer*. Hoboken, NJ: John Wiley & Sons.

US Energy Information Administration. 2021. *Natural Gas*. https://www.eia.gov/naturalgas/crudeoilreserves/ (accessed December 25, 2021).

16 Pharmaceuticals

16.1 INTRODUCTION

A drug is a chemical used for treatment or prevention of disease. Drugs can also be defined as chemicals intended for use in the diagnosis, treatment, cure, or prevention of diseases in humans or animals. The study of the actions and disposition of drugs in the human body is called pharmacology. The use of drugs dates from prehistoric times; the first list of drugs with instructions for preparation, called a pharmacopeia, appeared in 1546 in Nuremberg, Germany. Drugs can be plant, mineral, animal, or synthetic in origin. Many early folk medicines were derived from plants, including aspirin, digitalis, ergot, opium, quinine, and reserpine. Minerals used as medicines include boric acid, Epsom salts, and iodine. Many hormones used to treat disease are currently obtained from animals. These include insulin and some vaccines. These drugs led to the development of the pharmaceutical industry.

In the past half century, emphasis has been placed on medicinal, chemical, biological, and pharmacological research to such an extent that this era is recognized, more than any other comparable period in history, for the new and effective drugs that have come from scientific laboratory investigators into the hands of physicians who prescribe them to improve the health of the public. The human life span has increased since 1900, from approximately 50 years to the present life expectancy of 76 years. In fact, one of the authors was 88 years old at the time of preparation of this book.

This chapter presents a brief summary of the pharmaceutical industry including its history, the creation of PhRMA, research and development, and some process descriptions and other operations and facilities of importance in pharmaceuticals production. Four illustrative examples and four problems are provided to highlight concepts presented throughout the chapter.

16.2 GENERAL COMMENTS (BARBOZA ET AL. 1977)

The pharmaceutical industry consists of manufacturing, packaging, and sales of chemicals used as medication for humans and animals including both "ethical" (i.e., prescription) and "proprietary" (i.e., nonprescription) drugs. Additionally, the industry manufactures related chemicals for nonpharmaceutical purposes, including preventive medicine and health-enhancing products, medicated and nonmedicated cosmetics, and food additives. Some facilities also manufacture medical apparatus, such as surgical and medical instruments and appliances, dental equipment and supplies, and various products used in diagnostic laboratories.

The predominant characteristic of the industry is diversity. Not only are a wide variety of products manufactured, but also a significant number of distinct chemical and manufacturing processes are used in producing these products. Consequently, no two manufacturing facilities are alike, thus making attempts to define the specific industry in terms of specific products and/or processes extremely difficult and complicated. The industry definition is further complicated by the manufacture of nonpharmaceutical products by some pharmaceutical facilities and pharmaceutical products, reagents, and intermediates by certain specialty synthetic chemical manufacturers whose operations would not ordinarily be classified as pharmaceutical. A more traditional definition of pharmaceutical manufacturing, which is limited to chemicals intended for use as medications (including both ethical and proprietary drugs) and as preventative medicine and health-enhancing agents, has been used. This approach allows classification of pharmaceutical manufacturing processes into a few more clearly defined categories.

Most pharmaceutical manufacturing operations tend to be relatively clean and closely controlled because of the importance of product purity and the presence of toxic and biologically active constituents. Controlling emissions is considered important to avoid inadvertent contamination of products and to prevent occupational and community exposure to potentially harmful substances. Using emission controls is also considered important in recovering expensive solvents used in the production and purification processes. For these reasons, the emissions from many pharmaceutical manufacturing facilities are well controlled, and serious environmental pollution from these facilities is relatively rare.

Nevertheless, problems that occur are usually associated with solvent extraction, fermentation operations, and general chemical synthesis operations. Other potential emission sources include product formulation and packaging operations, materials handling and storage, wastewater treatment, spent solvent treatment for recycling, and the regeneration of spent adsorbent carbon. Emission controls are available for most pharmaceutical operations, although applying controls is not always cost effective due to the high-volume, low-concentration airstreams that must be handled. Additionally, because manufacturing many pharmaceuticals involves batch operations, emission controls also must be operated in batch mode or operated continuously while handling intermittent, variable feed streams. These methods of operation can reduce the effectiveness of emission control systems, limit the availability of control system options, and increase control system operating costs.

The batch nature of pharmaceutical operations lends itself to decentralized emissions control, where control systems are applied to individual process units rather than to entire production lines. Consequently, regulatory actions tend to focus on controlling specific process units (e.g., reactors, distillation units, crystallizers, centrifuges, dryers, filters, and storage tanks) rather than on limiting emissions from an entire facility. Nevertheless, economical emission reduction for an entire facility is sometimes achieved by using collection manifolds and central emission control devices.

16.3 HISTORY

The use of drugs dates from prehistoric times. The use of these drugs to relieve pain and to ward off death is interwoven with some ancient superstitions that evil spirits cause disease. The American pharmaceutical industry had a modest beginning in the late 1700s. The synthetic organic chemicals ether and chloroform were not used for anesthesia until the 1840s. Three years after the end of the Civil War, the first integrated industrial synthetic organic manufacturing operation was established in the United States. The groundwork for modem pharmaceutical research was begun in 1881 with the establishment of a scientific division of Eli Lilly and Co. The shortage of important drugs, such as Novocain, caused by the entry of the United States into World War I, precipitated the expansion of the pharmaceutical industry into a successful effort to produce synthetic chemicals needed for both wartime and peacetime applications. Developments of insulin, liver extract, and the short-acting barbiturates were milestones of the next decade. Sulfa drugs and vitamins were rapidly added to many product lines during the 1930s. Blood plasma, new antimalarials, and the dramatic development of penicillin resulted from the demands of war. But the spectacular surge of new products, which included steroid hormones, tranquilizers, vaccines, and broad- and medium-spectrum antibiotics, came after World War II. Research has now been stepped up to find more efficient drugs for the unconquered host of viruses, and maladies including arthritis, cancer, heart conditions, high blood pressure, hepatitis, and mental and emotional diseases, among others (Shreve and Brink 1977).

16.4 PhRMA

PhRMA (phrma.org) was formed in 1958 to represent America's biopharmaceutical research companies and help promote smart public policy that supports medical research to address patient needs. Headquartered in Washington, DC, with offices in leading biopharmaceutical research communities, PhRMA advocates in the United States and around the world for policies that support the discovery and development of innovative medicines. PhRMA policy priorities include explaining the increased complexity and risk of the research and development (R&D) process, reinforcing the need for investment in R&D, ensuring broad access to and appropriate use of medicines, and emphasizing the importance of strong intellectual property incentives for new medicines. Today, PhRMA represents the country's leading innovative biopharmaceutical research companies which are devoted to discovering, developing, and manufacturing medicines that enable patients to live longer, healthier, and more productive lives. Since 2000, PhRMA member companies have invested more than a trillion dollars in the search for new treatments and cures, including an estimated $91 billion in 2020 alone.

More than 30 biopharmaceutical companies are members of PhRMA (phrma.org), and the top 10 companies in terms of 2020 revenues are listed in Table 16.1.

16.5 RESEARCH AND DEVELOPMENT

Research expenditures (over 20% of industry sales) are a primary factor in determining a pharmaceutical company's future share of industry sales and reflect the service rendered to the public. In the early 1960s an effort was made in Congress to place all aspects of the pharmaceutical industry under further governmental control. A proposal to limit the duration of patents did not receive widespread support. The discovery of new compounds followed by further R&D is one of the primary functions of the industry, and protection of these R&D expenditures through patents is critical to the long-term investment in new and improved pharmaceuticals.

TABLE 16.1
Top Ten Pharmaceutical Companies in Terms of Annual Revenue in 2020

Company
Johnson & Johnson
Roche
Novartis
Merck
AbbVie
Bristol-Meyers Squibb
Sanofi
Pfizer
GalxoSmithKline
AstraZeneca

The pharmaceutical production process starts with an extensive research stage, which can last several years. Following the discovery of a new drug that appears to have efficacy in treating or preventing illness, preclinical testing and clinical trials are conducted. Then a new drug application is submitted to the Federal Drug Administration (FDA) for approval. Market distribution of new drug in the United States can begin only after FDA approval. Normally it takes an average of 10 years to bring a new drug to market, from time of discovery to approval. Remarkably, the COVID-19 pandemic resulted in a significantly reduced time period for approval of vaccines for the virus to under 1 year. This will undoubtedly increase the development and marketing of essential drugs in the near future.

16.6 PROCESS DESCRIPTIONS

Pharmaceutical production (after R&D activities) can be broken into two major stages (Barboza et al. 1977):

1. Bulk pharmaceutical production
2. Pharmaceutical product formulation

Bulk pharmaceutical production involves the conversion of organic and natural substances into bulk pharmaceutical chemicals (BPCs) (i.e., active ingredients) through fermentation, extraction, and/or chemical synthesis. Pharmaceutical product formulation involves mixing, compounding, and/or formulating operations to convert manufactured BPC substances into final usable forms.

The manufacturing processes and operations that are most common in the pharmaceutical industry can be divided into the following four main manufacturing process categories: chemical synthesis, fermentation, extraction, and formulation. A rough breakdown of US pharmaceutical manufacturing facilities by these process categories are presented in Table 16.2.

TABLE 16.2
Percent of Total US Pharmaceutical Manufacturing Facilities Employing Various Manufacturing Process Categories

Manufacturing Process Category	Percent of US Facilities
Chemical Synthesis	10
Fermentation	1
Extraction	4
Formulation	60
Combination of Processes	25

Note that approximately 25% of pharmaceutical facilities utilize a combination of these process categories. The most common combination is chemical synthesis and formulation. These four main pharmaceutical manufacturing process categories are briefly described below.

16.6.1 Chemical Synthesis

Chemical synthesis processes use a variety of batch operations to produce high-purity, pharmaceutical-grade chemicals. Common production equipment includes reactors, dryers, distillation units, filters, centrifuges, and crystallizers. Unless a very-high-volume product is manufactured, the equipment is generally used to produce a number of different products over the course of a year and may be dedicated to a specific product for only a few weeks at a time.

Most pharmaceutical substances are synthesized utilizing batch processes (Theodore 2012). In a batch process, a particular substance or intermediate is manufactured for periods ranging from a few days to several months until sufficient material is manufactured to satisfy the projected demand. At the end of this batch production period, another pharmaceutical intermediate or substance is then made. The same equipment with potentially different configurations and the same operating personnel are often used to make a different intermediate or substance, utilizing different raw materials, executing different processes, and generating different waste streams. When the same equipment is used for manufacturing different intermediates and/or different bulk substances, the equipment must be thoroughly cleaned prior to its reuse.

16.6.2 Fermentation

Fermentation involves the combination of various species or strains of fungus or bacterium with selected raw materials and the reliance on biological processes to produce desired compounds. The principal pharmaceuticals produced by fermentation are antibiotics, vitamins, and steroids. The fermentation process takes place in the three steps:

1. *Inoculation and seed preparation.* Spores from a master stock are activated with water, nutrients, and heat. As the mass grows, it is transferred to a seed tank with a typical capacity of 350–7,500 L (100–2,000 gal), where further growth occurs.

2. *Fermentation.* Seeds from the inoculation and seed preparation step are combined with raw materials in a fermentation vessel, typically of 18,000–380,000 L (5,000–100,000 gal) in capacity. Air is sparged through the vessel to help sustain growth, and the action of the microorganisms on the raw materials results in the desired product. At the end of the fermentation step, filtration may be used to separate the microorganisms from the product. Fermentation typically requires a period of 12 hr to 1 week for completion depending on the product being manufactured.
3. *Product recovery.* Product recovery and purification are usually achieved through solvent extraction, direct precipitation, ion exchange, or adsorption.

16.6.3 Extraction

The pharmaceutical extraction process (Theodore and Ricci 2011; Theodore 2014) is unique in its low ratio of product weight to raw material weight (often less than 1%). Solvents are used to extract desired compounds from a variety of materials, including roots, leaves, animal glands, parasite fungi, and human blood. In extracting these compounds, solvents also are used to remove separate fats and oils. After extraction occurs several different types of separation devices (e.g., centrifuges, filters, evaporators, or dryers) can then be used to isolate the desired compounds. Numerous pharmaceuticals are produced by extraction, including insulin, morphine, digitalis derivatives, and allergy medications.

16.6.4 Formulation

Formulation processes convert bulk chemicals into refined products that include: tablets, capsules, liquids, and ointments (i.e., the final dose forms). Typical formulation operations include mixing, blending, granulating, drying, coating, polishing, tablet pressing, capsule filling, sorting, and packaging. Solvents are typically used during mixing and coating operations and to clean the equipment. Most of the dust generated by formulation operations is captured and sometimes returned to the process.

16.7 OTHER OPERATION AND FACILITY CONSIDERATIONS

A number of operations and facilities, although not considered pharmaceutical processes by definition, are components of many pharmaceutical manufacturing plants, especially some of the larger ones. The need for these operations varies from plant to plant. These other operations and facilities, which can contribute to releases from pharmaceutical manufacturing plants, include material storage, transfer, power and steam generation, waste disposal, as well as wastewater treatment. These various operations and facilities are briefly discussed below.

16.7.1 Storage and Transfer

Volatile compounds are often stored in tank farms, process storage tanks, and storage drums. These storage devices are susceptible to breathing and working losses, the magnitude of which depends on the type of compound stored, the size and design of the tanks, ambient temperature and diurnal temperature changes, and tank throughput. Emissions may also be associated with manual material transfer operations within the facility and with the transfer of liquids from tanker cars, trucks, and rail tank cars.

16.7.2 Power and Steam Generation

Pharmaceutical process plants require electricity as well as process heating. Many plants utilize industrial boilers, some with fairly large capacities to generate steam for space heating and process heating needs. In addition, plants may include standby boilers for emergency use as well as emergency generators for emergency electrical power. Some plants may also utilize cogeneration facilities for power and heating needs.

Boilers and process heaters represent an opportunity to burn a waste stream while simultaneously recovering the stream's heating value. Typical applications are combusting a pollutant-laden air stream as a source of combustion air. A fairly high-fuel-value vent gas stream is required for boilers or process heaters to function without added fuel (Santoleri, Reynolds, and Theodore 2004).

16.7.3 Waste Disposal

Pharmaceutical manufacturing generates wastes that may consist of both hazardous and nonhazardous materials. These wastes can include off-spec or obsolete raw materials or products, spent solvents, reaction residues, used filter media, still bottoms, used chemical reagents, dusts from filtration or air pollution control equipment, raw material packaging wastes, laboratory wastes, spills, as well as wastes generated during packaging of formulated products.

Filter cakes and spent raw materials (plants, roots, animal tissues, etc.) from fermentation and natural product extraction are two of the largest sources of residual wastes in the pharmaceutical industry. Other wastes include reaction residues and filtrates from chemical synthesis processes. These wastes may be stripped of any solvents remaining in them and then disposed of as either hazardous or nonhazardous wastes.

Facilities may perform shredding of wastes such as outdated or off-spec products. Mechanical shredders may be used to shred dry nonhazardous industrial solid waste material for destruction and volume reduction. Typically, solid wastes are shipped off-site for disposal or incineration. In some instances, facilities may utilize incinerators or waste heat boilers to dispose of solid wastes on-site. This allows recovery of potential heating value from some wastes.

The industry has been implementing practices to reduce waste generation and material losses. Typical practices include process optimization, production scheduling, materials tracking and inventory control, special material handling and storage procedures, preventive maintenance programs, and waste stream segregation.

16.7.4 Wastewater Treatment

Pharmaceutical manufacturers use water for process operations, as well as for other non-process purposes. However, the use and discharge practices and the characteristics of the wastewater will vary depending on the operations performed. In some cases, water may be formed as part of chemical reactions. Process water includes any water that, during manufacturing or processing, comes into direct contact with or results from the use of any raw material or production of an intermediate, finished product, by-product, or waste. Process wastewater includes water that was used or formed during the reaction, water used to clean process equipment and floors, and pump seal water. Non-process wastewater includes noncontact cooling water (e.g., used in heat exchangers), noncontact ancillary water (e.g., boiler blowdown, bottle washing), sanitary wastewater, and wastewater from other sources (e.g., stormwater runoff).

16.7.5 QA/QC

Additionally, in-process testing, as well as quality assurance/quality control (QA/QC) testing in on-site laboratories, is performed during drug product manufacturing. In-process testing may include simple pH measurements or checks on color, while QA/QC testing typically includes more sophisticated analyses such as gas or liquid chromatography. Upon completion of the manufacturing operation, batch production records are checked by competent and responsible personnel for actual yield against theoretical yield of a batch and to ensure that each step has been performed correctly and that product formulation standards have been met.

16.8 ILLUSTRATIVE EXAMPLES

Four illustrative examples complement the material presented above regarding the pharmaceutical industry.

16.8.1 Illustrative Example 1

Consider the case of a box of 500 pills from which a sample of 10 pills is to be drawn without replacement. If the box contains ten contaminated pills, what is the probability that the sample contains exactly two contaminated pills?

Solution: Let A denote the event that the first pill drawn is contaminated and B, the event that the second is contaminated. Then, the probability that the sample contains exactly two contaminated pills is $P(AB)$. By application of the multiplication theorem, one obtains:

$$P(AB) = P(A)P(B \mid A) = \left(\frac{10}{500}\right)\left(\frac{9}{499}\right) = (0.02)(0.018) = 0.0036 = 0.36\%$$

16.8.2 Illustrative Example 2

Suppose that an explosion at a pharmaceutical plant could have occurred as a result of one of three exclusive human causes: stupidity, carelessness, or laziness. It is estimated that such an explosion could occur with a probability 0.20 as a result of stupidity, 0.4 as a result of carelessness, and 0.75 as a result of laziness. It is also estimated that the prior probabilities of the three possible causes of the explosion are,

Pharmaceuticals

respectively, 0.50, 0.35, and 0.15. Using Bayes' theorem, determine the most likely cause of the explosion.

Solution: Let A_1, A_2, A_3 denote, respectively, the events in which stupidity, carelessness, and laziness are the problem. Let B denote the event of the explosion. Then,

$$P(A_1) = 0.50; P(B \mid A_1) = 0.20$$

$$P(A_2) = 0.35; P(B \mid A_2) = 0.40$$

$$P(A_3) = 0.15; P(B \mid A_3) = 0.75$$

Applying Bayes' theorem gives:

$$P(A_1 \mid B) = \frac{P(A_1)P(B \mid A_1)}{P(A_1)P(B \mid A_1) + P(A_2)P(B \mid A_2) + P(A_3)P(B \mid A_3)}$$

$$= \frac{(0.50)(0.20)}{(0.50)(0.20) + (0.35)(0.40) + (0.15)(0.75)} = \frac{0.10}{0.3525}$$

$$P(A_1 \mid B) = 0.284$$

$$P(A_2 \mid B) = \frac{P(A_2)P(B \mid A_2)}{P(A_1)P(B \mid A_1) + P(A_2)P(B \mid A_2) + P(A_3)P(B \mid A_3)}$$

$$= \frac{(0.35)(0.40)}{(0.3525)} = \frac{0.14}{0.3525} = 0.397$$

$$P(A_3 \mid B) = \frac{P(A_3)P(B \mid A_3)}{P(A_1)P(B \mid A_1) + P(A_2)P(B \mid A_2) + P(A_3)P(B \mid A_3)}$$

$$= \frac{(0.15)(0.75)}{(0.3525)} = \frac{0.1125}{0.3525} = 0.319$$

Therefore, carelessness is the most likely cause of the explosion!

16.8.3 Illustrative Example 3

Let X denote the number of new drugs developed by a pharmaceutical company. The probability density function of X is specified as:

$$f(x) = 0.25; x = 0$$

$$f(x) = 0.35; x = 1$$

$$f(x) = 0.24; x = 2$$

$$f(x) = 0.11; x = 3$$

$$f(x) = 0.04; x = 4$$

$$f(x) = 0.01; x = 5$$

a. What is the probability that the company will develop four or more new drugs in a year, given that two drugs have already been developed?
b. What is the probability that the company will develop one more new drug in a year, given that one drug has already been developed?

Solution: (a) This involves a conditional probability calculation (Equation 4.13). The describing equation and solution are:

$$P(x \geq 4 \mid x = 2) = \frac{P(x \geq 4 \text{ and } x = 2)}{P(x = 2)} = \frac{P(x \geq 4)}{0.24} = \frac{0.05}{0.24} = 0.208 = 20.8\%$$

(b) This also involves a conditional probability, with the describing equation and solution being:

$$P(x = 2 \mid x = 1) = \frac{P(x = 2 \text{ and } x = 1)}{P(x = 1)} = \frac{P(x = 2)}{0.35} = \frac{0.24}{0.35} = 0.686 = 69\%$$

16.8.4 ILLUSTRATIVE EXAMPLE 4

Determine the reliability of component G that is part of a process train shown in Figure 16.1 at a pharmaceutical manufacturing plant. The overall reliability of the system has been determined to be 0.42, and the reliability of other individual components is shown in the figure.

Solution: First identify the components connected in parallel: A and B are connected in parallel; D, E, and F are also connected in parallel. Then, compute the reliability

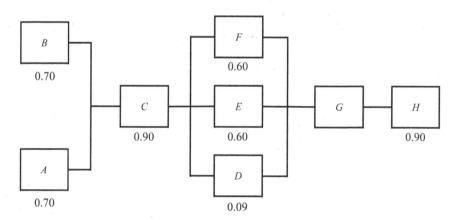

FIGURE 16.1 Diagram of a process train at a pharmaceutical manufacturing plant showing known component reliability values.

Pharmaceuticals

of each subsystem of the components connected in parallel. The reliability of the parallel subsystem consisting of components A and B is:

$$R_p = 1-(1-0.70)(1-0.70) = 0.91$$

The reliability of the parallel subsystem consisting of components D, E, and F is:

$$R_p = 1-(1-0.60)(1-0.60)(1-0.09) = 0.8544$$

Multiply the product of the reliabilities of the parallel subsystems by the product of the reliabilities of the components to which the parallel subsystems are connected in series:

$$R_s = (0.91)(0.9)(0.8544)(G)(0.9) = 0.42$$

Solving this equation for G yields:

$$G = 0.42/(0.63) = 0.667$$

PROBLEMS

16.1 Referring to Illustrative example 3, what is the probability that the company will develop three new drugs in a year given that they have already developed one new drug? What is this probability given that they have already developed two new drugs that year?

16.2 Five samples of a drug is drawn *with replacement* from a bin in which 5% are contaminated. Note that *with replacement* means that each sample is returned to the bin before the next sample is randomly drawn. What is the probability that exactly four contaminated samples are drawn from the bin? Assume a binomial distribution (Section 5.2.1), and associate *success* with drawing a contaminated sample.

16.2 Consider Problem 2. What is the probability that the number of contaminated samples drawn is *at most* two? What is the probability that the number of contaminated samples draw is *at least* two?

16.3 The downwind concentration of a hazardous chemical released from a pharmaceutical manufacturing plant can be described by the following equation:

$$C = 130\sqrt[x]{x} = 130(x)^{1/x}$$

where C is the compound concentration in $\mu g/m^3$, and x is downwind distance in km. Determine where the maximum concentration would occur, what the value of the maximum concentration is predicted to be and determine where downwind of the release would be a critical evacuation area where the concentration determined to be Immediately Dangerous to Life and Health (IDLH) of 150 $\mu g/m^3$ occurs. See Section 12.9.4 Illustrative example 4 for help with this problem.

REFERENCES

Barboza, M.J., Cheeboski, T.V., Crume, R.V., and Protzer, J.W. 1977. Pharmaceutical Industry. In *Chemical Process Industries*, 4th Edition. Eds.

Santoleri, J., Reynolds, J., and Theodore, L. 2004. *Introduction to Hazardous Waste Incineration*, 2nd Edition. Hoboken, NJ: John Wiley & Sons.

Shreve, N. and J. Brink, 725–732. New York, NY: McGraw-Hill.

Theodore, L. 2012. *Chemical Reactor Analysis and Applications for the Practicing Engineer*. Hoboken, NJ: John Wiley & Sons.

Theodore, L. 2014. *Chemical Engineering: The Essential Reference*. New York, NY: McGraw-Hill.

Theodore, L., and Ricci, F. 2011. *Mass Transfer Operations for the Practicing Engineer*. Hoboken, NJ: John Wiley & Sons.

17 Food Products Industry

17.1 INTRODUCTION

Food is defined as anything eaten to satisfy appetite and to meet physiological needs for growth, to maintain all body processes, and to supply energy to maintain body temperature and activity. Because foods differ markedly in the amount of the nutrients that they contain, they are classified on the basis of their composition and the source from which they are derived.

It is interesting to note that much of the technology used in other chemical process industries has been applied to food. Likewise, much of the technology developed for the food industry is applicable to other chemical process industries. The demand for more preprocessing and processing of food products for home use has arisen naturally because since World War II, many homemakers work away from home. The demand for uniform quality of food and high quality standards, even at consumption centers remote from production, has led to improved processing methods while affluence has led to a demand for greater variety.

This chapter presents a brief summary of the food products industry including its history, the role of the Food and Drug Administration, food processing and preservation techniques, refrigeration, and food additives important in the modern food products industry. Five illustrative examples and four problems are provided to highlight concepts presented throughout the chapter.

17.2 HISTORY

For about 99% of human history, human beings were hunters and gatherers of food, responding to their environment and restricted by it. The great variety of food items they used had the potential to provide essential nutrients, but the quantity varied greatly, thus limiting the number of people that could be supported on a given land area.

During the remaining 1% of human history, however, dynamic changes – the so-called cultural revolution – occurred. Humans began to recognize the advantages of organizing families into small social groups, and this resulted in cooperative efforts among individuals and the birth of the rudiments of government and social organization. Thus, humans began to gain a degree of control over their environment.

The first major result in terms of food supply was the agricultural revolution, occurred approximately 10,000 years ago. Over a period of several thousand years, many small social groups shifted from being hunters and gatherers to being producers of food. The global population at this time was approximately four million people, which was about the maximum that could readily be supported by a gathering and hunting way of life (Ponting 1991). The increasing difficulty in obtaining food is believed to be a major contributor to this sudden change. The farmer changed the landscape of the planet and was far more destructive than the hunter. While farming fostered the rise of cities and civilizations, it also led to practices that denuded the land of its nutrients and water holding capacity. Great civilizations flourished and then disappeared as once-fertile land was farmed into desert. The adoption of agriculture resulted in settled communities and a steadily rising population. The agriculturalist deliberately transformed nature in an attempt to simplify the planet's ecosystem.

The second major impact of cultural development was the scientific and industrial revolution that began about 400 years ago. In terms of human food supply and demand, this revolution had explosive effects that continue to this day. The application of scientific knowledge to food production resulted in spectacular opportunities to increase food output per land or animal unit. Basic medical advances brought improved health and greater life expectancy to people in many parts of the world and the total demand for food has continued to grow. Today, the materials produced on farms, which were formerly consumed or retailed there, are now seldom so handled. Production enterprises are now often single-product-oriented, buying much of their basic food from manufactured sources. At each step in the movement of the product to the consumer, consideration is given to factors influencing the quality of the product and its value to the ultimate user.

17.3 THE FOOD AND DRUG ADMINISTRATION

The Food and Drug Administration (FDA) is an agency of the US Department of Health and Human Services. Part of the Public Health Service, the FDA administers the Federal Food, Drug, and Cosmetic Act of 1938 and related laws to ensure that foods are pure and wholesome and produced under sanitary conditions; that drugs and therapeutic devices are safe and effective for their intended uses; that cosmetics are safe and made from appropriate ingredients; and that labels and packaging of products are truthful, informative, and not deceptive. The FDA also enforces the federal Hazardous Substances Act to ensure proper labeling and safety of chemical products, toys, and other articles used in the home. In 1969 the FDA became responsible for promoting sanitation in public eating places and interstate travel facilities and for federal/state programs to ensure safety of milk and shellfish.

In 1971 the FDA was given responsibility for enforcing the Radiation Control for Health and Safety Act of 1968. This law was designed to prevent unnecessary human exposure to radiation from electronic equipment ranging from television receivers to dental X-ray machines. In 1972 the agency was assigned to regulate biologic drugs, including vaccines, antitoxins, and serums.

The federal Food, Drug, and Cosmetic Act prohibits interstate traffic in adulterated or misbranded products. Defective products may be voluntarily destroyed or recalled from distribution by shippers or seized by US marshals on court orders obtained by the FDA. Persons responsible may be prosecuted in the federal district courts or enjoined from further violations. All court proceedings are brought by US district attorneys on evidence supplied by the FDA.

Inspection activities are centered at 19 district laboratories in major cities. The FDA inspectors periodically visit facilities and warehouses, and the chemists employed by the agency analyze the samples that inspectors collect. Facts so determined are the basis of regulatory decisions.

Specific products must be approved for safety prior to sale or use. Manufacturers submit samples of production batches of antibiotic drugs, insulin, or color additives to FDA laboratories for testing. The agency must certify their purity, potency, and safety before they may be shipped. New drugs and their labeling must also be approved for safety and effectiveness. Food additives must be generally recognized as safe or proven safe by scientific tests. Pesticide residues in food commodities must not exceed safe tolerances, which are set and enforced by the FDA. Such premarketing clearances are based on scientific data provided by manufacturers, subject to review and acceptance by FDA scientists.

The FDA maintains extensive educational programs in order to promote compliance by industry with its regulations and to enable consumers to benefit from its work. It is responsible for protecting the public health by ensuring the safety, efficacy, and security of human and veterinary drugs, biological products, and medical devices; and by ensuring the safety of our nation's food supply, cosmetics, and products that emit radiation. It also has responsibility for regulating the manufacturing, marketing, and distribution of tobacco products to protect the public health and to reduce tobacco use by minors. The FDA is responsible for advancing the public health by helping to speed innovations that make medical products more effective, safer, and more affordable and by helping the public obtain the accurate, science-based information they need to use medical products and foods to maintain and improve their health. Finally, the FDA plays a significant role in the Nation's counterterrorism capability; it fulfills this responsibility by ensuring the security of the food supply and by fostering development of medical products to respond to deliberate and naturally emerging public health threats.

17.4 FOOD PROCESSING AND PRESERVATION

Food processing and preservation involve processes used to protect food against microbes and other spoilage agents to permit its future consumption. The preserved food should retain palatable appearance, flavor, and texture, as well as its original nutritional value.

Many kinds of agents are potentially destructive to the agreeable or healthful characteristics of fresh foods. Microorganisms, such as bacteria and fungi, rapidly spoil food. Enzymes which are present in all raw food are catalytic substances that promote degradation and chemical changes especially affecting a food's texture and flavor. Atmospheric oxygen may react with food constituents, causing rancidity or color changes. Equally as harmful are infestations by insects and rodents, which account for tremendous losses in food stocks. No single method of food preservation affords protection against all hazards for an unlimited period of time. Canned food stored in Antarctica near the South Pole, for example, remained edible after 50 years of storage, but such long-term preservation cannot be duplicated in the hot climate of the Tropics. Besides canning and freezing, traditional methods of preservation include drying, salting, and smoking. Freeze-drying is a more recent method. Among recent experimental techniques are the use of antibiotics and exposure of food to nuclear radiation.

17.4.1 CANNING

The process of canning is sometimes called sterilization because the heat treatment of the food eliminates all microorganisms that can spoil the food and those that are harmful to humans, including directly pathogenic bacteria and those that produce lethal toxins. Most commercial canning operations are based on the principle that bacteria destruction increases tenfold for each 10°C (18°F) increase in temperature. Food exposed to high temperatures for only minutes or seconds retains more of its natural flavor. In the Flash 18 process, a continuous system, the food is flash-sterilized in a pressurized chamber to prevent the superheated food from boiling while it is placed in containers. Further sterilizing is not required.

17.4.2 FREEZING

Although prehistoric humans stored meat in ice caves, the food-freezing industry is more recent in origin than the canning industry. Freezing was used commercially for the first time in 1842, but large-scale food preservation by freezing began in the late 19th century with the advent of mechanical refrigeration.

Freezing preserves food by preventing microorganisms from multiplying. Because the process does not kill all types of bacteria, however, those that survive reanimate in thawing food and often grow more rapidly than before freezing. Enzymes in the frozen state remain active, although at a reduced rate. Vegetables are blanched or heated in preparation for freezing to ensure enzyme in-activity and thus to avoid degradation of flavor. Blanching has also been proposed for fish, in order to kill cold-adapted bacteria on their outer surface. In the freezing of meats various methods are used depending on the type of meat and the cut. Pork is frozen soon after butchering, but beef is hung in a cooler for several days to tenderize the meat before freezing.

Frozen foods have the advantage of resembling the fresh product more closely than the same food preserved by other techniques. Frozen foods also undergo some changes, however. Freezing causes the water in food to expand and can disrupt the cell structure by forming ice crystals.

Additional details on freezing are provided in the next section that addresses refrigeration.

17.4.3 DEHYDRATION

Vegetables, fruits, meat, fish, and some other foods that have moisture contents as high as 80% may be dried to one-fifth of the original weight and about one-half of the original volume. The disadvantages of this method of preservation include the time and labor involved in rehydrating the food before eating. Further, reconstituting the dried product may be difficult because it absorbs only about two-thirds of its original water content; this phenomenon tends to make the texture tough and chewy.

Drying was used by prehistoric humans to preserve many foods. Large quantities of fruits such as figs have been dried from ancient times to the present day. In the case of meat and fish, other preservation methods, such as smoking or salting, which

Food Products Industry   289

yielded a palatable product, were generally preferred. Commercial dehydration of vegetables was initiated in the United States during the American Civil War, but, as a result of the poor quality of the product, the industry declined sharply after the war. This cycle was repeated with subsequent wars, but after World War II the dehydration industry thrived. This industry is confined largely to the production of a few dried foods, however, such as milk, soup, eggs, yeast, and powdered coffee, which are particularly suited to the dehydration method.

Present-day dehydration techniques include the application of a stream of warm air to vegetables. Protein foods such as meat are of good quality only if freeze-dried. Liquid food is dehydrated usually by spraying it as fine droplets into a chamber of hot air, or occasionally by pouring it over a drum internally heated by steam where evaporation occurs.

17.4.4 MISCELLANEOUS METHODS

Other methods or a combination of methods may be used to preserve foods. Salting of fish and pork has long been practiced, using either dry salt or brine. Salt enters the tissue and, in effect binds the water, thus inhibiting bacteria that cause spoilage. Another widely used method is smoking, which frequently is applied to preserve fish, ham, and sausage. The smoke is obtained by burning hickory or a similar wood under low draft. In this case, some preservative action is provided by such bactericidal chemicals in the smoke as formaldehyde and creosote and by the dehydration that occurs in the smokehouse. Smoking usually is intended to flavor the product as well as to preserve it.

Sugar, a major ingredient of jams and jellies, is another preservative agent. For effective preservation the total sugar content should make up at least 65% of the weight of the final product. The sugar, which acts in much the same manner as salt, also inhibits bacterial contamination.

17.5 REFRIGERATION

Refrigeration and cryogenics have aroused considerable interest among those in engineering and science as was discussed earlier in Section 15.7.2. In addition to being employed for domestic purposes (when a small "portable" refrigerator is required), refrigeration and cryogenic units have been used for the storage of not only foods but also antibiotics and other medical supplies. Much larger cooling capacities than this are required in air conditioning equipment. Some of these units, both small and large, are especially useful in applications that require the accurate control of temperature. Most temperature-controlled enclosures are provided with a unit that can maintain a space below ambient temperature (or at precisely ambient temperature) as required. For example, the implementation of such devices led to the recognition that cooling units would be well suited to the refrigeration of electronic components and to applications in the field of instrumentation.

Refrigeration, in a commercial setting, not only refers to food preservation but also air conditioning. When food is kept at colder temperatures, the growth of bacteria and the accompanying spoiling of food is either reduced or prevented. People learned early on that certain foods had to be keep cold to maintain freshness and humans kept these foods in ice boxes where melting ice usually absorbed the heat from the foods. Household refrigerators became popular in the early 1900s and only the wealthy could afford them at the time. Freezers did not become a staple part of the refrigerator until after World War II when frozen food became popular.

Refrigeration equipment and ideal refrigerant characteristics were presented earlier in Section 15.7.2. Solid refrigerants are not impossible to use but liquid refrigerants are most often used in practice. Liquid refrigerants include non-ozone-depleting hydrofluorocarbons as well as non-halogenated hydrocarbon and non-hydrocarbon refrigerants. The most commonly used non-halogenated hydrocarbon refrigerants include:

1. Propane
2. Ethane
3. Propylene
4. Ethylene

Non-halogenated non-hydrocarbon liquid refrigerants include:

1. Nitrogen
2. Neon
3. Helium
4. Ammonia
5. Carbon dioxide

A basic refrigeration cycle (Theodore, Ricci and VanVliet 2009; Theodore 2014) is shown in Figure 17.1. The cycle begins when a refrigerant enters the compressor as a low-pressure gas (1). Once compressed, it leaves as a hot, high-pressure gas (2). Upon entering the condenser, the gas condenses to a liquid and releases heat to the outside environment, which may be air or water. The cool liquid then enters the expansion

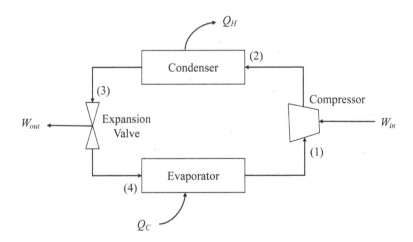

FIGURE 17.1 Basic components of a refrigeration cycle.

valve at a high pressure (3); the flow is restricted, and the pressure lowered. In the evaporator (4), heat from the source to be cooled is absorbed and the liquid becomes a gas. The refrigerant then repeats the process (i.e., the cycle continues).

In the refrigerator, the working fluid enters the evaporator in a wet condition and leaves dry and saturated (or slightly superheated). The heat absorbed, Q_C, by the evaporator can therefore be estimated by multiplying the change in the fluid's entropy (Theodore, Ricci and VanVliet 2009; Theodore 2014), ΔS, as it passes through the evaporator by the fluid's saturation temperature, T_S, at the evaporator pressure since the fluid's temperature will be constant while it is in a wet condition at constant pressure. Thus,

$$Q_C = T_S \Delta S \qquad (17.1)$$

where Q_C = heat absorbed by the evaporator, kJ/kg; T_S = fluid saturation temperature at the evaporator temperature, K; and ΔS = fluid entropy change, kJ/kg-K.

17.6 FOOD ADDITIVES

Food additives are natural and synthetic compounds added to food to supply nutrients, to enhance color, flavor, or texture, and to prevent or delay spoilage. Since ancient times table salt has been a preservative for fish, ham, and bacon, while sugar has been used to preserve jelly, fruit jams, and fruit preserves. At present more than 2,500 food additives of all kinds are in use. The principal preservative compounds used today are benzoic acid, C_6H_5COOH, sodium benzoate, C_6H_5COONa, and calcium propionate, $(C_2H_5COO)_2Ca$. Chlortetracycline and oxytetracycline are antibiotics used to delay spoilage of fresh fish and poultry. Nutrients – minerals and vitamins – are added to foods to increase their nutritional value or to restore vitamins or minerals lost during processing. Vitamin D is added to milk, vitamin C to orange drink, and vitamin A to margarine. Vitamin C is also used as an antioxidant for fresh fruit and frozen fish. Coloring agents, natural and synthetic flavorings, and flavor enhancers such as monosodium glutamate are added to make food more appealing. More than 750 synthetic flavorings are also available for use. Other additives such as agar and gelatin are used as texturizers or emulsifiers in foods such as ice cream and frozen desserts.

The use of food additives in the United States is regulated by the FDA. Coloring agents in particular are rigidly controlled, with only a few of them considered safe by the FDA. Not only were some colorings found to be toxic or carcinogenic to animals and possibly to humans, but they were also suspected of causing allergic reactions, including hyperactivity in children. The artificial sweetener calcium cyclamate was banned by the FDA as a possible carcinogen. Because of the growing awareness by consumers of the adverse effects of certain food additives on health, a trend in the production of more foods with fewer or no additives has become more popular since 2000.

17.7 ILLUSTRATIVE EXAMPLES

Five illustrative examples complement the material presented above regarding health and hazard risks in the food products industry.

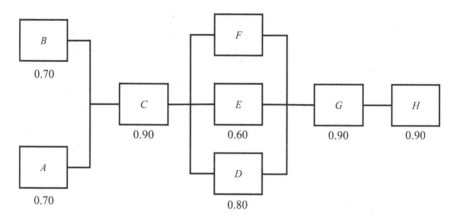

FIGURE 17.2 Diagram of a process train at a food processing plant showing known component reliability values.

17.7.1 Illustrative Example 1

Determine the reliability of component F that is part of a process train shown in Figure 17.2 at a food processing plant. The overall reliability of the system must be 0.82, and the reliability of other individual components is shown in the figure.

Solution: First identify the components connected in parallel: A and B are connected in parallel; D, E, and F are also connected in parallel. Then, compute the reliability of each subsystem of the components connected in parallel. The reliability of the parallel subsystem consisting of components A and B is:

$$R_p = 1-(1-0.70)(1-0.70) = 0.91$$

The reliability of the parallel subsystem consisting of components D, E, and F is:

$$R_p = 1-(1-0.80)(1-0.60)(1-F) = 1-(0.2)(0.4)(1-F) = 1-0.08(1-F)$$

Multiply the product of the reliabilities of the parallel subsystems by the product of the reliabilities of the components to which the parallel subsystems are connected in series:

$$R_s = (0.91)(0.9)(1-0.08(1-F))(0.9)(0.9) = 0.82$$

Solving this equation for F yields:

$$(0.663)(1-0.08+0.08F) = 0.82; (0.663-0.053+0.053F) = 0.82$$

$$(0.61+0.053F) = 0.82; F = 0.21/0.053 = 3.96$$

A component cannot be more than 100% reliable, so this system cannot be 82% reliable with the given reliability of all other components.

17.7.2 ILLUSTRATIVE EXAMPLE 2

A quality control engineer inspects a random sample of five food products drawn *without replacement* from each lot of 100 food products. A lot is accepted only if the sample contains no, that is, 0, off-spec products. What is the probability of accepting a lot containing 20 off-spec products?

Solution: This random variable X has a hypergeometric distribution whose pdf is specified as follows:

$$f(x) = \frac{\dfrac{a!}{x!(a-x)!} \dfrac{(n-a)!}{(r-x)!(n-a-r+x)!}}{\dfrac{n!}{r!(n-r)!}}, x = 0,1,\ldots,\min(a,r) \quad (5.23)$$

The term $f(x)$ is the probability of x successes in a random sample of r items drawn without replacement from a set of n items, a of which are classified as successes and $n - a$ as failures.

Associate success with drawing a defective sample, and failure with drawing a non-defective one. Each lot consists of 100 food products. Therefore, $n = 100$. For this problem, $a = 20$. Since a sample of five products is drawn without replacement from each lot, $r = 5$.

Substitute the values of N, n, and a in the pdf of the hypergeometric distribution, that is, Equation 5.23 becomes Equation 17.2:

$$f(x) = \frac{\dfrac{20!}{x!(20-x)!} \dfrac{(100-20)!}{(5-x)!(100-20-5+x)!}}{\dfrac{100!}{5!(100-5)!}}, x = 0,1,2,3,4,5 \quad (17.2)$$

Substituting the appropriate value of x (= 0 to accept the entire lot) to obtain the required probability, the probability of a lot containing 20 off-spec products if it passed QC is as follows:

$$P(X=0) = \frac{\dfrac{20!}{0!(20-0)!} \dfrac{(100-20)!}{(5-0)!(100-20-5+0)!}}{\dfrac{100!}{5!(100-5)!}} = \frac{\dfrac{20!}{(20)!} \dfrac{(80)!}{(5)!(75)!}}{\dfrac{100!}{5!(95)!}}$$

$$= \frac{\dfrac{(80)(79)(78)(77)(76)(75)!}{(5)!(75)!}}{\dfrac{(100)(99)(98)(97)(96)(95)!}{5!(95)!}}$$

$$P(X=0) = \frac{\dfrac{(80)(79)(78)(77)(76)}{(5)(4)(3)(2)(1)}}{\dfrac{(100)(99)(98)(97)(96)}{(5)(4)(3)(2)(1)}} = \frac{(80)(79)(78)(77)(76)}{(100)(99)(98)(97)(96)}$$

$$= \frac{2,884,801,920}{9,034,502,400} = 0.319 = 31.9\%$$

17.7.3 ILLUSTRATIVE EXAMPLE 3

Suppose 24 minor accidents occur at a food company's 40 international plants. Find the probability P that exactly three of these accidents occur at a given plant.

Solution: The number of accidents at one plant may be viewed as the number of successes. Since P is expected to be small, the Poisson approximation to the binomial distribution can be applied to the problem:

$$f(x) = \frac{e^{-\mu}\mu^x}{x!}, x = 0, 1, 2, \ldots \tag{5.24}$$

Here the average number of minor accidents per plant, $\mu = 24/40 = 0.60$. The probability of three accidents occurs at a given plant is:

$$P(X = 3) = \frac{e^{-0.6}(0.6)^3}{3!} = \frac{0.549(0.216)}{(3)(2)(1)} = \frac{0.1186}{6} = 0.0198 = 2\%$$

17.7.4 ILLUSTRATIVE EXAMPLE 4

The number of daily product tests conducted at a food processing plant, C, is normally distributed with a mean, μ, of 1,500 and standard deviation, σ, of 350. What is the probability that at least 2,000 tests will be conducted on a given day.

Solution: The term $(C - \mu)/\sigma$ is called a *standard normal variable* that is represented by Z (Section 5.3.2). The probability of a specific event for a normal distribution using the standard normal variable can be defined as:

$$P(\bar{C} \geq C_1) = P\left(\frac{\bar{C} - \mu}{\sigma} \geq \frac{C_1 - \mu}{\sigma}\right) \tag{17.2}$$

Because C is normally distributed with $\mu = 1,500$ and a standard deviation $\sigma = 350$, then $(C - 1,500)/350$ is a standard normal variable Z. Substituting given values into Equation 17.2 yields the following:

$$P(\bar{C} \geq 2,000) = P\left(Z \geq \frac{2,000 - 1,500}{350}\right) = P(Z \geq 1.43)$$

Referring to the values in the standard normal table, Table 5.4, the following probability can be calculated:

$$P(\bar{C} \geq 2,000) = (0.076) = 7.6\%$$

17.7.5 ILLUSTRATIVE EXAMPLE 5

Consider the problem of determining whether or not the occurrence of eight cases of food poisoning among 7,076 children in Nassau County, Long Island, is so unexpectedly large as to warrant further study by public health officials. An examination

Food Products Industry

of public health records in Nassau County (home of one of the authors) reveals that in the 2008–2013 period there were 286 diagnosed cases of food poisoning out of 1,152,685 children.

Solution: The historical probability that a child will be stricken with food poisoning in Nassau County is $P = 286/1{,}152{,}685 = 0.000248 = 0.0248\%$. The probability of eight or more cases of food poisoning occurring in a population of 7,076 children can be calculated using a Poisson distribution with $nP = 7{,}076\,(0.000248) = 1.75 = \mu$. Substituting for this problem results in the following:

$$P(X \le 7) = \sum_{x=0}^{7} \frac{e^{-1.75}(1.75)^x}{x!} = e^{-1.75}\left(\frac{(1.75)^0}{0!} + \frac{(1.75)^1}{1!} + \cdots + \frac{(1.75)^7}{7!}\right)$$

$$P(X \le 7) = 0.1738(1 + 1.75 + 1.53 + 0.893 + 0.39 + 0.137 + 0.04 + 0.01) = 0.99935$$

Consequently,

$$P(X \ge 8) = 1 - 0.99935 = 0.065 = 0.065\%$$

This probability is nearly three times the historical rate observed over a 5-year period, and the cause warrants further investigation by the County's public health officials.

PROBLEMS

17.1 Refer to Illustrative example 1. If component F has a high reliability of 90%, what is the maximum reliability of the current configuration?

17.2 Refer to Illustrative example 1. If all components can be selected to each have a 90% reliability, what then is the maximum reliability of the systems? How can the system shown in Figure 17.2 be modified to yield the required system reliability of 82%?

17.3 Referring to Illustrative example 2, what is the probability of accepting a lot with 20 off-spec products if the lot size is reduced to 50 and all other input values remain the same?

17.4 The regulatory specification on a toxin in a solid waste generated from a food processing plant calls for a level of 2.0 ppm or less. Earlier observations of the concentration, C, of the toxic waste constituent indicate a normal distribution with a mean of 0.40 ppm and a standard deviation of 0.08 ppm. Estimate the probability that the solid waste from the plant will exceed the regulatory limit.

REFERENCES

Ponting, C. 1991. *A Green History of the World*. London, England: Sinclair-Stevenson.

Theodore, L. 2014. *Chemical Engineering, The Essential Reference*. New York, NY: McGraw-Hill.

Theodore, L., Ricci, F., and VanVliet, T. 2009. *Thermodynamics for the Practicing Engineer*. Hoboken, NJ: John Wiley & Sons.

18 Nanotechnology

18.1 INTRODUCTION

The authors believe that nanotechnology is the second coming of the industrial revolution. It promises to make the nation that seizes the nanotechnology initiative the technology capital of the world. One of the main obstacles to achieving this goal will be to control, reduce, and ultimately eliminate any environmental and environmental-related problems associated with this technology. Unfortunately, the success or failure of this new industrial paradigm may well depend on the ability to effectively and efficiently address these environmental issues.

The environmental health and hazard risks associated with both nanomaterials and the applications of nanotechnology for industrial uses are not fully known. Some early studies indicate that nanoparticles *may* serve as environmental poisons that accumulate in organs. Although these risks *may* prove to be either minor, or negligible, or both, the engineer and scientist is duty bound to determine if there are in fact any health, safety, and environmental impacts associated with this new technology. Much of the material in this chapter is drawn from four sources: Theodore and Kunz (2005), Theodore (2006a), Stander and Theodore (2011), and Theodore and Theodore (2021).

This chapter summarizes a wide range of nanotechnology process topics by providing an introduction to the technology, associated environmental implications, health and hazard risk assessments, environmental regulations, and future trends in the industry. Five illustrative examples and five problems are provided to highlight concepts presented throughout the chapter.

18.2 NANOTECHNOLOGY

The dictionary (Dictionary.com 2022a) defines technology as "…the branch of knowledge that deals with the creation and use of technical means and their interrelation with life, society, and the environment, drawing upon such subjects as industrial arts, engineering, applied science, and pure science." Nano (Dictionary.com 2022b) receives less treatment: "Very small or at a microscopic level, … One billionth …" Therefore, in a very generic sense, nanotechnology is an applied science that is concerned with very tiny substances.

For the technical community, there are at least three definitions that have regularly appeared in the literature:

1. Molecular manufacturing at the atomic level (atom by atom in a stable pattern), i.e., bottom-up approach
2. Research at the 1–100 nanometer (nm) size range
3. Development and uses of nanoparticles in the 1–100-nm range, i.e., top-down approach

Terms (1) and (2) deal with futuristic activities that are beyond the scope of this text, and of little to no concern to the practicing engineer at this time. Applications and peripheral topics of (3) highlight the presentation to follow.

How does (3) above impact the practicing engineer? There are five major areas.

1. Developing new products
2. Improving existing products
3. Cost considerations
4. Health concerns
5. Hazard concerns

The first three areas have the potential to improve the quality of life. However, health and hazard concerns can adversely affect society, and these concerns need to be reduced and/or eliminated.

When familiar materials such as metals, metal oxides, ceramics and polymers, plus novel forms of carbon, are reduced to infinitesimally small particle sizes the resulting particles have some novel and special material properties, especially when compared to macroscopic particles of the same material. These properties can also vary with particle-size distribution (PSD), particle shape, particle density, process application, etc. These unique properties and property variations are certain to lead to a near infinite number of opportunities and applications of nanomaterials in the future.

18.2.1 NANOMATERIALS

Nanomaterials originated from entities that are now defined as *prime materials*. These prime materials essentially consist of (pure) elements and compounds. The elements and compounds that have been successfully produced and deployed as nanometer-sized particles include:

1. Metals such as iron, copper, gold, aluminum, nickel, and silver
2. Oxides of metals such as iron, titanium, zirconium, aluminum, and zinc
3. Silica sols, and fumed and colloidal silica
4. Clays such as talc, mica, smectite, asbestos, vermiculite, and montmorillonite
5. Carbon compounds, such as fullerenes, nanotubes, and carbon fibers

Each of these types of materials, along with the manufacturing methods used to render them into nanoscale particles, is discussed below. Some of the information presented was adapted from the literature (Boxall et al. 2008).

As noted earlier, when macro-sized metals are produced in the nanometer range, they exhibit properties not found in large particle sizes. This includes properties such as quantum effects, the ability to sinter at temperatures significantly below their standard melting points, increased catalytic activity due to higher surface area per unit mass, and more rapid chemical reaction rates. Additionally, when nanoscale particles of metals are incorporated into other larger structures, these larger structures often exhibit increased strength, hardness, and tensile strength when compared to structures formed from conventional micrometer-sized powders.

Nanotechnology

However, reducing some metals to nanoscale powder presents some problems; in particular, when reduced to sufficiently small particle sizes, many of the chemical properties of metals become more reactive and subject to oxidation – often explosively. Copper and silver powders are among the few metallic nanoparticles that are not explosive and thus can be handled in air. Many others must be stabilized with a passivation layer or handled in an inert, blanketed (neutral) environment. Moreover, in some applications, effort is required to minimize unwanted particle agglomeration.

Metals that have served as prime materials include the following:

1. Iron
2. Aluminum
3. Nickel
4. Silver
5. Gold
6. Copper

Mixed oxides used for nanoparticle production include the following:

1. Iron oxides (Fe_2O_3 and Fe_2O_4)
2. Silicon dioxide (silica; SiO_2)
3. Titanium dioxide (titania; TiO_2)
4. Aluminum oxide (alumina; Al_2O_3)
5. Zirconium dioxide (zirconia; ZrO_2) and
6. Zinc oxide (ZnO)

Additional details on both metals and metal oxides are available in the literature (Theodore and Kunz 2005).

18.2.2 Nanomaterial Production

Depending upon their application, nanomaterials may be manufactured so they are nanoscale in one dimension, such as for thin films, coatings, and quantum wells, which are used for applications such as computers and cell phones; in two dimensions, such as for nanowires and nanotubes, for applications such as transistors; and in three dimensions, such as nanoparticles, fullerenes, and dendrimers, for applications such as pollution prevention/remediation, sensors, water treatment and biomedical applications.

The ongoing challenge for the research community is to continue to devise, perfect, and scale up viable production methodologies that can cost effectively and reliably produce the desired nanoparticles with the desired particle size, PSD, purity, and uniformity in terms of both composition and structure.

In general, there are six widely used methods for producing nanoscaled particles, on the order of 1–100 nm in diameter, of various metal oxides (Fink et al. 2002; Wilson et al. 2004):

1. High-temperature processes such as plasma-arc and flame-hydrolysis methods (including flame ionization)
2. Chemical vapor deposition (CVD)

3. Electrodeposition
4. Sol-gel synthesis
5. Mechanical crushing via ball milling
6. Use of naturally occurring nanomaterials

Each of these production methods is described briefly below.

18.2.2.1 High-Temperature Processes

These high-temperature processes involve the use of a high-temperature plasma or flame ionization reactor. As an electrical potential difference is imposed across two electrodes in a gas, the gas, electrodes, or other materials ionize and vaporize if necessary and then condense as nanoparticles, either as separate structures or as surface deposits. An inert gas or vacuum is used when volatilizing electrodes.

During flame ionization, a material is sprayed into a flame to produce ions (Fink et al. 2002). Using flame hydrolysis, highly dispersed oxides can be produced via high-temperature hydrolysis of the corresponding chlorides. Flame hydrolysis produces extremely fine, mostly spherical particles with diameters in the range of 7–40 nm and high specific surface areas in the range of 50–400 m^2/g (Wilson et al. 2004).

In general, high-temperature flame processes for making nanoparticles are divided into two classifications – gas-to-particle or droplet-to-particle methods – depending on how the final particles are made.

In gas-to-particle processes, individual molecules of the product material are made by chemically reacting precursor gases or rapidly cooling a superheated vapor. Depending on the thermodynamics of the process, the molecules then assemble themselves into nanoparticles by colliding with one another or by repeatedly condensing and evaporating into molecular clusters. Gas-to-particle processes involve the use of flame, hot-wall, evaporation-condensation, plasma, laser, and sputtering-type reactors.

In droplet-to-particle processes, liquid atomization is used to suspend droplets of a solution or slurry in a gas at atmospheric pressure. Solvent is evaporated from the droplets, leaving behind solute crystals, which are then heated to change their morphology. Spray drying, pyrolysis, electrospray, and freeze-drying equipment are typically used in the droplet-to-particle production process.

18.2.2.2 Chemical Vapor Deposition (CVD)

In CVD, a starting material is vaporized and then condensed on a surface, usually under vacuum conditions. The deposit may be the original material, or a new and different species formed by chemical reaction.

18.2.2.3 Electrodeposition

Electrodeposition uses an approach where individual species are deposited from solution, with an aim to lay down a nanoscaled surface film in a precisely controlled manner.

18.2.2.4 Sol-Gel Synthesis

Sol-gel processing is a wet-chemical method that allows high-purity, high-homogeneity nanoscale materials to be synthesized at lower temperatures compared to competing high-temperature methods. A significant advantage that sol-gel science

Nanotechnology

affords over more conventional materials-processing routes is the mild conditions that the approach employs.

After mixing the reactants, organic or inorganic precursors undergo two chemical reactions: hydrolysis, and condensation or polymerization, typically with an acid or base as a catalyst, to form small solid particles or clusters in a liquid (either an organic or aqueous solvent). The resulting solid particles or clusters are so small (11,000 nm) that gravitational forces are negligible, and interactions are dominated by van der Waals, coulombic, and steric forces. These sols – colloidal suspensions of oxide particles – are stabilized by an electric double layer, or steric repulsion, or a combination of both. Over time, the colloidal particles link together by further condensation and a dimensional network occurs. As gelling proceeds, the viscosity of the solution increases dramatically.

The sol-gel can then be formed into three different shapes: thin film, fiber, and bulk. Thin (≈100 nm) uniform and crack-free films can readily be formed on various materials using lowering, dipping, spinning, or spray coating techniques.

18.2.2.5 Mechanical Crushing Via Ball Milling

Progressive particle-size reduction or pulverization using a conventional ball mill is one of the primary methods for preparing nanoscaled particles of various metal oxides. High-energy ball milling is in use today, but its use is considered by some to be limited because of the potential for contamination problems. The availability of tungsten carbide components and the use of an inert atmosphere and high-vacuum processes have helped operators reduce impurities to acceptable levels for many industrial applications (Fink et al. 2002). Other common drawbacks, however, include the highly polydisperse size distribution and partially amorphous state of nanoscaled powders prepared using the pulverization method.

18.2.2.6 Naturally Occurring Materials

Certain naturally occurring materials, such as zeolites, can be used as found or synthesized and modified by conventional chemistry to yield nanosized materials. A zeolite is a caged molecular structure containing large voids that can admit molecules of a certain size and deny access to other, larger molecules (Wilson et al. 2004). They find application as catalysts and adsorbents, among other uses.

18.3 CURRENT APPLICATIONS

An inventory of present-day applications of nanoparticles has been developed through the Wilson Center and Virginia Tech called the Nanotechnology Consumer Products Inventory (CPI) and can be accessed at https://www.nanotechproject.tech/cpi/. The current inventory includes some 1,833 chemical products that are organized into eight general product categories. Specific examples of items using nanomaterials in these eight product categories include:

1. Appliances – batteries; heating, cooling and air conditioning; large kitchen appliances; laundry and clothing care
2. Automotive – exterior, maintenance & accessories, watercraft, lubricants
3. Cross Cutting – coatings, bulk materials

4. Electronics and Computers – audio, cameras and film, computer hardware, displays, mobile devices and communications, televisions, video
5. Food and Beverages – cooking, food, storage, supplements
6. Goods for Children – supplies and cleaning agents, toys and games
7. Health and Fitness – clothing, cosmetics, filtration, personal care, sporting goods, sunscreen, supplements
8. Home and Garden – cleaning, construction materials, home furnishings, luggage, luxury items, paint, pets

Other applications of nanomaterials involving water and air purification systems can also be found on this Nanotechnology CPI. Other applications involving water purification and environmental remediation are certain to emerge in the future. For example, the protection of human health and ecosystems requires rapid, precise sensors capable of detecting pollutants at the molecular level. Major improvement in process control, compliance monitoring, and environmental decision-making could be achieved if more accurate, less costly, more sensitive contaminant monitoring techniques were available. Examples of research in sensors include the development of nanosensors for efficient and rapid in situ biochemical detection of pollutants and specific pathogens in water. Treatment options might include the removal of very fine particulates (under 300 nm). Substances of significant concern in groundwater, because of both their cancer and noncancer hazards, include heavy metals (e.g., mercury, lead, cadmium) and organic compounds (e.g., benzene, chlorinated solvents, creosote, toluene, polyfluorinated alkyl substances (PFAS)). Reducing releases to water, providing safe drinking water, and reducing quantities and exposure to hazardous wastes also are areas of interest for nanotechnology applications.

18.4 ENVIRONMENTAL IMPLICATIONS

Any technology can have various positive and negative effects on the environment and society. Nanotechnology is no exception, and its net benefits will be determined by the extent to which the technical community manages this technology. This is an area that has, unfortunately, been seized upon by a variety of environmental groups.

There are two thoughts regarding the environmental implications of nanotechnology: one is positive and the other is potentially negative. The positive features of this new technology are well documented. The other implication of nanotechnology has been dubbed by many in this diminutive field as, "potentially negative." The reason for this label is as simple as it is obvious. The technical community is dealing with a significant number of unforeseen effects that could have disturbingly disastrous impacts on society. Fortunately, it appears that the probability of such dire consequences actually occurring is near zero ... but *not* zero. This finite, but small probability is one of the key topics that is addressed here.

Air, water, and land (solid waste) concerns with emissions from nanotechnology operations in the future, as well as companion health and hazard risks, are discussed in brief below. All of these issues arose earlier with the Industrial Revolution, and then with the development/testing/use of the atomic bomb, the arrival of the Internet, Y2K, and so forth, and all were successfully (relatively speaking) resolved by the engineers and scientists of their period. Consistently large numbers of studies have reported associations between ultrafine particle exposure, i.e., PM2.5, and morbidity in elderly and compromised individuals (Gwinn and Vallyathan 2006). Therefore, there is a reason to suspect that nanoparticles with size and surface characteristics similar to ultrafine particles are likely to cause diseases – some with a long latency, although statements in the literature continue to refer only to *potential* health concerns (Papp et al. 2008).

Theodore (2006a) has speculated on the need for future nano regulations. Theodore (2006a) also noted that the ratio of pollutant nanoparticles (from conventional sources such as power plants) to engineered nanoparticles being released into the environment may be as high as a trillion to one (i.e., 10^{12}:1). If this is so, the environmental concerns for nanoparticles can almost certainly be dismissed.

It should also be noted that environmental concerns regarding nanomaterials are starting to be taken seriously around the globe. There are a variety of studies evaluating the health and environmental impacts of many applications of nanotechnology. Many believe it is in everyone's interest to ensure that any new compound is fully characterized, and the long-term implications are studied before it is commercialized. Class action suits in the United States against both tobacco companies and engineering companies, coupled with a new era of corporate responsibility, have ensured that most companies are well aware of this need. Now that potential risks that may have been overlooked are becoming widely known, these companies are more inclined to be proactive than they have been with risks in the past.

There has already been a considerable shift in both public and corporate attitudes toward the environment. Major scandals such as Enron and WorldCom have led not only to tighter corporate governance but also to calls for greater corporate responsibility. The end result of this shift will be to make companies increase their focus on the environment and look to leveraging nanotechnology as a way of not only improving efficiency and lowering costs but also doing this by reducing energy consumption and minimizing waste. A typical example would be in the use of nanoparticle catalysts that are not only more efficient, owing to more of the active catalysts being exposed (because of their large surface area to volume ratio), but also require less precious metal (thus reducing cost); it may also increase selectivity, i.e., produce more of the desired reaction product, rather than by-products.

Returning to the positive features of nanotechnology, it will be one of the key technologies used in the quest to improve the global environment in this century. While there will be some direct effects, much of the technology's influence on the environment will be through indirect applications of nanotechnology. Although any technology can always be put to both positive and negative uses, there are many areas in which positive aspects of nanotechnology look promising. These extend from pollution prevention and reduction through environmental remediation to sustainable development.

Though nanotechnology could have some significant effects on environmental technologies, environmental considerations have not historically been given anywhere near the priority in new developments that commercial considerations are given, and this imbalance, though swinging gradually more toward environmental considerations, still largely dominates. Many of the direct applications of nanotechnology relate to the removal of some element or compound from the environment, through for example, the use of nanofiltration, nanoporous sorbents (absorbents and adsorbents), catalysts in cleanup operation, and filtering, separating, and destroying environmental contaminants in processing waste products. Most effects, as with other technologies, are likely to be indirect.

18.5 HEALTH RISK ASSESSMENT

For some, the rapid progress of nanotechnology-related developments in recent years brings uncertainty. For instance, early studies on the transport and uptake of nanoscaled materials into living systems "suggest" that there may be harmful effects on living organisms. This has prompted many to call for further study to identify all of the potential environmental and health risks that might be associated with nanosized materials.

Studies raise questions about the potential health and environmental effects of nanoscaled materials, and while the initial toxicological data are preliminary, they underscore the need to learn more about how nanoscaled materials are absorbed, how they might damage living organisms, and what levels of exposure create unacceptable hazards.

At the time of the preparation of this chapter, the risks of nanotechnology were not definitively known, and it appears that they will not be known for some time. However, it should also be noted that health benefits, if any, are also not known. Furthermore, there are no specific nano-health-related regulations or rules at the US Environmental Protection Agency (EPA), Occupational Safety and Health Administration (OSHA), or other organizations, and it may be years before any definite regulations of nanomaterials are promulgated. However, the dark clouds on the horizon in this case are the environmental health impacts associated with these new and unknown operations, and the reality is that there is a serious lack of information on these impacts. Risk assessment studies in the future will be that path to both understanding and minimizing these effects.

As described earlier, perhaps the greatest danger from nanomaterials may be their escape and persistence in the environment, the food chain, and human and animal tissues. Although, the potential pollutants and the tools for dealing with them may be different, the methodology and protocols developed for conventional materials will probably be the same, bearing in mind that some instrumentation for nanomaterial characterization that is required for these studies may not yet be available (Theodore and Kunz 2005). Thus, environmental risk assessment remains "environmental risk assessment," using the same techniques described earlier regardless of the size of the alleged causative agent.

How is it possible to make decisions dealing with environmental risks from a new application, for example nanotechnology, in a complex society with competing interests and viewpoints, limited financial resources, and a lay public that is deeply concerned about the risks of cancer and other illness? Risk assessment constitutes a decision-making approach that can help the different parties involved and thus enable the larger society to work out its environmental problems rationally and with appropriate results. It also provides a framework for setting regulatory priorities and for making decisions that cut across different environmental areas. This kind of framework has become increasingly important in recent years for several reasons, one of which is the considerable progress made in environmental control. Nearly 40 years ago, it was not difficult to figure out where the first priorities should be. The worst pollution problems were all too obvious.

Nanotechnology

Health risk assessments provide an orderly, explicit way to deal with scientific issues in evaluating whether a health problem exists and what the magnitude of the problem may be. Typically, this evaluation involves large uncertainties because the available scientific data are limited, and the mechanisms for adverse health impacts or environmental damage are only imperfectly understood.

As discussed earlier in Chapters 2 and 3, most human or environmental health problems can be evaluated by dissecting the analysis into four parts: hazard identification, dose-response or toxicity assessment, exposure assessment, and risk characterization (see Figure 3.1). For many, the heart of a health risk assessment is toxicology. Toxicology is the science of poisons. It has also been defined as the study of chemical or physical agents that produce adverse responses in biological systems. Together with other scientific disciplines (such as epidemiology – the study of the cause and distribution of disease in human populations – and risk assessment), toxicology can be used to determine the relationship between an agent of interest and a group of people or a community (National Center for Environmental Research 2003).

Six primary factors affect human response to toxic substances or poisons. These are detailed below (Burke, Singh and Theodore 2000; Theodore and Kunz 2005; Theodore and Dupont 2012).

1. *The chemical itself.* Some chemicals produce immediate and dramatic biological effects, whereas other produce no observable effects or produce delayed effects.
2. *The type of contact.* Certain chemicals appear harmless after one type of contact (e.g., skin), but may have serious effects when contacted in another manner (e.g., lungs).
3. *The amount (dose) of a chemical.* The dose of a chemical exposure affects its impact on exposed individuals. A low dose may have no or beneficial effect, while a high dose may be toxic.
4. *Individual sensitivity.* Humans vary in their response to chemical substance exposure. Some types of responses that different persons may experience at a certain dose are serious illness, mild symptoms, or no noticeable effect. Different responses may also occur in the same person at different exposure levels.
5. *Interaction with other chemicals.* Toxic chemicals in combination can produce different biological responses than the responses observed when exposure is to one chemical alone.
6. *Duration of exposure.* Some chemicals produce symptoms only after one exposure (acute), some only after exposure over a long period of time (chronic), and some may produce effects from both kinds of exposures.

In a very real sense, the science of toxicology will be significantly impacted by nanotechnology. Unique properties cannot be described for particles in the nano-size range since properties vary with particle size. This also applies to toxicological properties. In effect, a particle of one size could be carcinogenic while a particle of the same material but of another size may not be carcinogenic. Alternatively, two different sized nanoparticles of the same substance could have different threshold limit values (TLVs).

This problem has yet to be resolved by toxicologists. Further toxicity assessment of these nanoparticles to weigh available evidence regarding the potential for particular materials to cause adverse effects in exposed individuals is still very much needed to provide, where possible, an estimate of the increased likelihood and/or severity of adverse effects of short-term and long-term exposures of nanomaterials.

18.6 HAZARD RISK ASSESSMENT

Although both health risk with hazard risk assessments employ a four-step method of analysis, the procedures are quite different, with each providing different results, information, and conclusions. As with health risk, there is a serious lack of information on the hazards and associated implications of these hazards for nanomaterials. The unknowns in this risk area are both larger in number and greater in potential consequences than in the health risk assessment. It is the authors' judgment that hazard risk has unfortunately received something less than the attention it deserves. However, hazard risk assessment (HZRA) details are available, and traditional approaches to HZRA, e.g., hazard and operability (HAZOP) studies, successfully applied in the past for other types of hazard risk are available in the literature (Theodore and Dupont 2012). Future work will almost definitely be based on this methodology.

Much has been written about Michael Crichton's powerful science-thriller novel entitled Prey. (The book was not only a best seller, but the movie rights were sold for $5 million.) In it, Crichton provides a frightening scenario in which swarms of nano-robots, equipped with special power generators and unique software, prey on living creatures. To compound the problem, the robots continue to reproduce without any known constraints. This scenario is an example of an accident and represents only one of a near infinite number of potential hazards that can arise in any nanotechnology application, particularly for bottom-up systems. Although the probability of the horror scene portrayed by Crichton, as well as other similar events, is extremely low, steps and procedures need to be put into place to reduce, control, and, it is hoped, to eliminate these events from actually happening.

The previous section defined both "chronic" and "acute" exposures. As indicated, when the two terms are applied to emissions, the former usually refers to ordinary, round-the-clock, everyday emissions while the latter term deals with short, out-of-the-norm, accidental emissions. Thus, acute problems normally refer to accidents and/or hazards. The Crichton scenario discussed above is an example of an acute problem, and one whose solution would be addressed/treated by a hazard risk assessment, rather than a health risk approach.

There are several steps in evaluating the risk of an accident (see Figure 3.2). These are detailed below if the system in question is a chemical plant. Note that this material can also be found in Part 1, Chapter 3, Section 3.3.

1. A brief description of the equipment and chemicals used in the plant is needed.
2. Any hazard in the system has to be identified. Hazards that many occur in a chemical plant include fire, toxic vapor release, slippage, corrosion, explosions, rupture of a pressurized vessel, and runaway reactions.

3. The event or series of events that will initiate an accident have to be identified. An event could be a failure to follow correct safety procedures, improperly repaired equipment, or failure of a safety mechanism for example.
4. The probability that the accident will occur has to be determined. For example, if a chemical plant has a given life span, what is the probability that the temperature in a reactor will exceed the specified temperature range over that lifetime? The probability can be ranked qualitatively from low to high. A low probability means that it is unlikely for the event to occur in the lifetime of the plant. A medium probability suggests that there is a possibility that the event will occur. A high probability means that the event will likely occur during the lifetime of the plant.
5. The severity of the consequences of the accident must be determined.
6. If the probability of the accident and the severity of its consequences are low, then the risk is usually deemed acceptable and the plant should be allowed to operate. If the probability of occurrence is too high or the damage to the surroundings is too great, then the risk is usually unacceptable and the system needs to be modified to minimize these effects.

The heart of the HZRA approach is enclosed in the dashed box shown in Figure 3.3 comprising Steps 3 through 6 above. The algorithm allows for re-evaluation of the process if the risk is deemed unacceptable (the process is repeated after system modification starting with Step 1 above). Once again, it is important to note that an accident generally results from a sequence of events. Each individual event, therefore, represents an opportunity to reduce the frequency, consequence, and/or risk associated with the accident culminating from the individual events.

18.7 ENVIRONMENTAL REGULATIONS

Many environmental concerns are addressed by existing health and safety legislation. Most countries require a health and safety assessment for any new chemical before it can be marketed. Further, the European Commission (EC) recently introduced the world's most stringent labeling system, EC 1907/2006, and established the European Chemicals Agency in Helsinki. This agency manages the Registration, Evaluation, Authorization and Restriction of Chemical (REACH) substances systems, which is a database of information provided by manufacturers and importers on the properties of their chemical substances. Detailed analysis of various US and EU laws and regulations are available in the literature (US EPA 2007; Breeggin et al. 2009; Fiorino 2010, 2011). Prior experience with materials such as PCBs and asbestos, and a variety of unintended effects of drugs such as thalidomide, means that both companies and governments have an incentive to keep a close watch on potential negative health and environmental effects of new material.

The principal US agencies concerned with environmental risks are the EPA and the OSHA. EPA's mission is to protect human health and the environment. One of its major goals is to ensure that all Americans are protected from significant risks to human health and the environment where they live, learn, and work, i.e., to closely monitor and control any potentially toxic or biohazardous conditions that could pose an unreasonable risk to human health or the environment. The mission of OSHA is to ensure safe and healthful working conditions for working men and women by setting and enforcing standards and by providing training, outreach, education, and assistance. Both agencies are, therefore, directly concerned with the environmental implications of nanotechnology.

It should be noted, however, that there are no nanoscale or nanoscale-related environmental regulations in the United States or the European Union at this time that require controls on process releases, production activities, or specific workspace safety measures. Completely new legislation and regulatory efforts may be necessary to protect the public and the environment from the potentially adverse effects of nanotechnology.

It is very difficult to predict what future regulations might come into play for nanomaterials. In the past, regulations have been both a moving target and confusing. What can be said is that there will be regulations, and there is a high probability that they will be contradictory and confusing. Past and current regulations provide a measure of what can be expected.

Commercial applications of nanotechnology are likely to be regulated under TSCA, which authorizes the EPA to review and establish limits on the manufacture, processing, distribution, use, and/or disposal of new materials that EPA determines to pose "an unreasonable risk of injury to human health or the environment." The term chemical is defined broadly by Toxic Substances Control Act (TSCA). Unless qualifying for an exemption under the law (a statutory exemption requiring no further approval by EPA), low-volume production, low environmental releases along with low volume, or plans for limited test marketing, a prospective manufacturer is subject to the full-blown procedure. This requires submittal of said notice, along with toxicity and other data to EPA at least 90 days before commencing production of the chemical substance.

Approval then involves recordkeeping, reporting, and other requirements under the statute. Requirements will differ, depending on whether EPA determines that a particular application constitutes a "significant new use" or a "new chemical substance." The EPA can impose limits on production, including an outright ban when it is deemed necessary for adequate protection against "an unreasonable risk of injury to health or the environment." The EPA may revisit a chemical's status under TSCA and change the degree or type of regulation when new health/environmental data warrant. If the experience with genetically engineered organisms is any indication, there will be a push for EPA to update regulations in the future to reflect changes, advances, and trends in nanotechnology (Theodore 2006b).

Workplace exposure to a chemical substance and the potential for pulmonary toxicity is subject to regulation by the OSHA, including the requirement that potential hazards be disclosed on a chemical Safety Data Sheet (SDS). (An interesting question arises as to whether carbon nanotubes, chemically carbon but with different properties because of their small size and structure, are indeed to be considered the same as or different from carbon black for SDS purposes.) Both governmental and private agencies can be expected to develop the requisite TLVs for workplace exposure. Also, the EPA may once again utilize TSCA to assert its own jurisdiction, appropriate or not, to minimize exposure in the workplace. Furthermore, the National Institute for

Occupational Safety and Health (NIOSH) provided initial workplace guidance for nanomaterial manufacturers and their employees in 2008 (NIOSH 2008) and has continued to update health and safety guidelines for manufacturers of nanomaterials with the most recent publications in 2018 (NIOSH 2018a, 2018b).

Another likely source of regulation would fall under the provisions of the Clean Air Act (CAA) for particulate matter less than 2.5 μm (PM2.5). Additionally, an installation manufacturing nanomaterials may ultimately become subject to the CAA's Section 112 governing hazardous air pollutants (HAP) as a "major source."

Wastes from a commercial-scale nanotechnology facility would be classified under the Resource Conservation and Recovery Act (RCRA), provided that it meets the criteria for RCRA waste. RCRA requirements could be triggered by a listed manufacturing process or the Act's specified hazardous waste characteristics. The type and extent of regulation would depend on how much hazardous waste is generated and whether the wastes generated are treated, stored, or disposed of on-site.

Fiorino (2010) published a review of the status of regulatory controls on the emerging nanotechnology industry, focusing on voluntary regulation and risk management self-imposed by the nanomaterials manufacturers, a model similar to that supported by the Pollution Prevention Act and EPA's Design for the Environment, Green Chemistry, Green Engineering and Safer Choice programs. EPA's nanomaterial-specific voluntary program, the Nanomaterial Manufacturing Stewardship Program (NMSP), was initiated in 2008 to obtain information on production, importation, and use; exposures; risk management practices; hazards; pollution prevention; and physical and chemical properties of nanomaterials from commercial manufacturers. EPA intended to use data collected through this program, to aid in determining how and whether certain nanoscale materials or categories of nanoscale materials present risks to human health and the environment. Although 29 companies provided data that described or identified 123 nanomaterials, only four companies were willing to participate in the portion of the program that encouraged them to sponsor the development of test data and provide that information to EPA. Based on the limited response to the program, EPA discontinued the NMSP in December 2009 and began developing regulatory approaches under Sections 5 and 8(a) of TSCA to collect nanomaterial data from manufacturers of industrial chemicals (US EPA 2011).

Fiorino's (2010) findings suggest that despite there being a community of technical, legal, management, and policy experts engaged in finding solutions to the challenges of nanotechnology oversight and that this group recognizes the unique characteristics of nanotechnology, there remains a need for coordination and development of a comprehensive voluntary industrial initiative to be carried out in parallel to ongoing regulatory efforts. It was suggested (Fiorino 2010) that this coordinated voluntary industrial initiative not be a substitute for regulatory action, but that it informs and prepares the groundwork for regulation, and complements existing and future regulatory controls to create an effective oversight system to minimize risk to human health and the environment of these highly innovative materials.

The reader is left to ponder the type, if any, of regulations required at this time for these nanomaterials, and the need to curb/eliminate liability concerns in the future.

18.8 FUTURE TRENDS

The unbridled promise of nanotechnology-based solutions has motivated academic, industrial, and government researchers throughout the world to investigate nanoscaled materials, devices, and systems with hope of commercial-scale production and implementation. Today, the private-sector companies that have become involved run the gamut from established global leaders throughout the chemical process industries, to countless small entrepreneurial start-up companies, many of which have been spun off from targeted research and development efforts at universities.

The governments of many industrialized nations are also keenly interested in nanotechnology. This stems in part from their desire to maintain technological superiority in an important evolving field and from the recognition that some applications of nanotechnology could have significant implications for national security.

In any event, future developmental efforts and advances will primarily be fueled by economic considerations. The greatest driving force behind any nano project is the promise of economic opportunities and cost savings over the long term. Hence, an understanding of the economics involved is quite important in making decisions at both the engineering and management levels.

From an environmental health and safety and pollution prevention perspective, this technology promises:

1. Use of less raw materials, some of which are nonrenewable
2. The generation of less waste/pollutants
3. Reduction of energy consumption
4. A safer environment with reduced risks

Other applications involving water purification/desalination are certain to emerge in the future. For example, the protection of human health and ecosystems requires rapid, precise sensors capable of detecting pollutants at the molecular level. Major improvement in process control, compliance monitoring, and environmental decision-making could be achieved if more accurate, less costly, more sensitive techniques were available. Examples of research in sensors include the development of nanosensors for efficient and rapid in situ biochemical detection of pollutants and specific pathogens in water. Treatment options might include the removal of very fine particulates (under 300 nm). Substances of significant concern in groundwater, because of both their cancer and noncancer hazards, include heavy metals (e.g., mercury, lead, cadmium) and organic compounds (e.g., benzene, chlorinated solvents, creosote, toluene). Reducing releases to water, providing safe drinking water, and reducing quantities and exposure to hazardous wastes are also areas of interest for nanotechnology applications.

Regarding regulations, both industry and government need to support reasonable policies and regulations. The traditional environmental precautionary approach needs to be supported by definitive data regarding health and environmental effects of these nanomaterials. For nanotechnology's most ardent supporters, the scope of this emerging field seems to be limited only by the imagination. However, considerable

Nanotechnology

technological and financial obstacles still need to be reconciled before nanotechnology's full promise can be realized (Murphy and Theodore 2006).

18.9 ILLUSTRATIVE EXAMPLES

Five illustrative examples complement the material presented above regarding health and hazard risks associated with the nanotechnology industry.

18.9.1 ILLUSTRATIVE EXAMPLE 1

Five samples of a nanomaterial are drawn *with replacement* from a process batch in which no more than 5% of the product can be contaminated. If three of the five samples drawn from the batch are contaminated what is the likelihood that the batch is more than 5% contaminated? Assume a binomial distribution (Section 5.2.1) and associate *success* with drawing a contaminated sample.

Solution: The probability of collecting three contaminated nanomaterials from a sample of five nanomaterials with the expected contamination rate being 5% can be determined with $n = 5$, $p = 0.05$, and $q = 0.95$ as:

$$f(X=3) = \frac{5!}{3!(5-3)!}(0.05)^3(0.95)^{5-3} = \frac{(5)(4)(3)(2)(1)}{(3)(2)(1)(2)(1)}(0.000125)(0.95)^2$$
$$= (0.00125)(0.9025) = 0.0011 = 0.11\%$$

This probability is very low and suggests that the contamination rate for the batch is much higher than 5%. Additional testing of the batch should be conducted to confirm the contamination rate, and if significantly higher than the acceptable contamination rate, further process analysis and optimization should be carried out to correct the production problem.

18.9.2 ILLUSTRATIVE EXAMPLE 2

The auto-ignition temperature (AIT) is the temperature at which a substance will catch fire in air without a source of ignition. In the hydrocarbon series of chemicals, the AIT falls as the boiling point rises. The AIT of ethylene is approximately 480°C, while for heavy fuel oil it is 250°C. Processing of these chemicals is often carried out at higher temperatures and any leaks that occur will ignite. Some concerns have arisen regarding the possibility of an explosion at a nanomaterials plant where the nanoparticles are present in a hydrocarbon mixture. Preliminary studies indicate that the difference between the actual AIT and a calculated value in a particulate-vapor mixture is a random variable X (°F) having the pdf:

$$f(x) = 1.2e^{-1.2x} \ @ \ x > 0; = 0 \text{ elsewhere} \tag{18.1}$$

Calculate the probability that the actual AIT for this hydrocarbon-nanoparticle mixture will be 2–6°F different than the calculated value.

Solution: Applying the definition of the cdf leads to:

$$P(2 < X < 6) = \int_2^6 f(x)\,dx = \int_2^6 1.2e^{-1.2x}\,dx$$

$$P(2 < X < 6) = -(1.2/1.2)\Big|_2^6 e^{-1.2x} = (-1)\left(e^{-7.2} - e^{-2.4}\right)$$
$$= (-1)(0.00075 - 0.091) = 0.090 = 9\%$$

18.9.3 Illustrative Example 3

The mean diameter of a sample of 200 nanoparticles produced by a machine is 0.502 μm and the standard deviation is 0.005 μm. The application of these particles is intended to allow a maximum tolerance in the diameter of 0.496–0.508 μm, otherwise the particles are considered unacceptable for use. Determine the percentage of unacceptable particles produced if the diameters are normally distributed.

Solution: Convert the data to the standard normal variable as follows:

$$P(T_1 < T < T_2) = P\left(\frac{T_1 - \mu}{\sigma} < Z < \frac{T_2 - \mu}{\sigma}\right) \tag{5.35}$$

Substituting yields:

$$P(0.496 < 0.502 < 0.508) = P\left(\frac{0.496 - 0.502}{0.005} < Z < \frac{0.508 - 0.502}{0.005}\right)$$
$$= P(-1.2 < Z < 1.2)$$

The probabilities of acceptable particles, *AP*, are therefore:

$$AP = \left(\text{Area under normal curve between } Z_1 = -1.2 \text{ and } Z_2 = 1.2\right)$$
$$= \left(\text{twice the area between } Z = 0 \text{ and } Z = 1.2\right)$$

Referring to the values in the standard normal table, Table 5.4, the following probability can be calculated:

$$AP = 2(0.5 - 0.115) = 2(0.385) = 0.77 = 77\%$$

Therefore, the percentage of unacceptable particles, *UP*, is:

$$UP = 1.0 - 0.77 = 0.23 = 23\%$$

This is a high percentage of unacceptable particles and the production process needs to be evaluated to improve the nanoparticle production process.

Nanotechnology

18.9.4 ILLUSTRATIVE EXAMPLE 4

A PSD measurement device at a nanoparticle manufacturing facility has a constant rate of failure of 0.005 per 1,000 hr. Find the probability that the device will operate for at least 25,000 hr.

Solution: This can be solved using a Weibull distribution (Section 5.3.1) and Equation 5.29:

$$f(t) = \alpha\beta t^{\beta-1} \exp\left(-\int_0^t \alpha\beta t^{\beta-1} dt\right) = \alpha\beta t^{\beta-1} \exp(-\alpha t^\beta); t > 0; \alpha > 0, \beta > 0 \quad (5.29)$$

For a constant failure rate, $\beta = 1$ and the failure rate is equal to α. Substituting the given failure rate into Equation 5.29 yields:

$$f(t) 0.005 \exp\left(-\int_0^t 0.005 dt\right) 0.005 e^{-0.005t}; t > 0$$

Because the failure rate is given in units of 1/(1,000 hr), the probability that the device will operate for at least 25,000 hr is given by:

$$P(T > 25) = \int_{25}^{\infty} -0.005 e^{-0.005t} dt = -e^{-\infty} + e^{-0.005(25)} = 0 + e^{-0.125} = 0.882 = 88.2\%$$

18.9.5 ILLUSTRATIVE EXAMPLE 5

Theodore Associates has been requested to conduct a risk assessment at a nanomaterial production plant that is concerned with the consequences of two incidents that occur at approximately the same location in the plant, and that are defined as follows.

1. An explosion resulting from the detonation of an unstable nanochemical
2. A continuous 240 g/s release of a resulting toxic chemical at an elevation of 125 m

Two weather conditions are envisioned, namely a northeast wind and a southwest wind (6.0 mph) with Stability Class B. Associated with these two wind directions are events IIA, and IIB, respectively, defined as follows:

IIA – Toxic cloud to the southwest
IIB – Toxic cloud to the northeast

Based on an extensive literature search, the probabilities and conditional probabilities of the occurrence of the defined events in any given year have been estimated by Dupont Consultants as follows:

$$P(I) = 10^{-6}$$

$$P(II) = 1/33,333$$

$$P(IIA \mid II) = 0.33$$

$$P(IIB \mid II) = 0.67$$

Note that P(IIA/II) represents the probability that Event IIA occurs *given* that Event II has occurred. The consequences of Events I, IIA, and IIB, in terms of number of people killed, are estimated as follows:

- I – All persons within 200 m of the explosion center are killed; all persons beyond this distance are unaffected.
- IIA – All persons in a pie-shaped segment 22. 5 degrees width (downwind of the source) are killed if the concentration of the toxic gas is above 0.33 µg/L; all persons outside of this area are unaffected.
- IIB – Same as IIA

Thirteen people are located within 200 m of the explosion center but not in the pie-shaped segment described above. Eight people are located within the pie-shaped segment southwest of the discharge center; five are 350 m downwind, three are 600 m away at the plant fence (boundary). Another six people are located 500 m away outside the pie-shaped segment but within the plant boundary. All individuals are at ground level.

Theodore Associates have been specifically requested to calculate the average annual individual risk (AAIR) based on the number of individuals potentially affected as well as the average risk based on all other individuals within the plant boundary, Hint: Perform atmospheric dispersion calculations at various distances from the emission source and combine these predicted concentrations with consequence information from the problem statement.

Solution: Draw a line diagram of the plant layout and insert all pertinent data and information (see Figure 18.1). An event tree for the accidents described above is presented in Figure 18.2.

First, calculate the probability of Event IIA occurring; also calculate the probability of Event IIB occurring as follows:

$$P(IIA) = P(IIA \mid II)P(II) = (0.33)(1/33,333) = 1/100,000 = 10^{-5} \quad (18.2)$$

$$P(IIB) = P(IIB \mid II)P(II) = (0.67)(1/33,333) = 2/100,000 = 2 \times 10^{-5} \quad (18.3)$$

Next, perform a dispersion calculation to determine the zones where the concentration of the chemical exceeds 0.33 µg/L. Assume a continuous emission for a point source. To maintain consistent units, convert wind speed from mph to m/s and concentration from µg/L to g/m³ as follows:

$$u = (6.0 \text{ mi/hr})(5280 \text{ ft/mi})(1 \text{ hr}/3600 \text{ s})(0.3048 \text{ m/ft}) = 2.68 \text{ m/s}$$

$$c = (0.33 \mu g/L)(1 g/10^6 \mu g)(10^3 L/m^3) = 3.3 \times 10^{-4} g/m^3$$

Nanotechnology

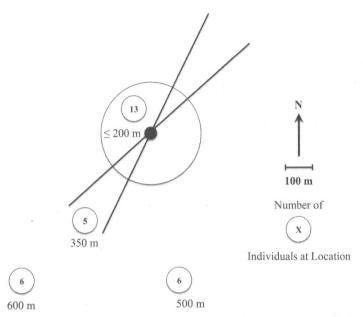

FIGURE 18.1 Source/receptor layout and pertinent data for Illustrative example 5.

These values are then used as input to Equation 18.4, the Pasquill-Gifford model for centerline, ground level concentrations of a continuous source of pollutant at an elevated emission height of H* (Turner 1970) as shown below:

$$c(x,0,0;125) = \frac{q}{\pi \sigma_y \sigma_z u} \left\{ \exp\left[-\frac{1}{2}\left(\frac{H^*}{\sigma_z}\right)^2\right]\right\}$$

$$= \left(\frac{240 \text{ g/s}}{\pi (\sigma_y)(\sigma_z)(2.68 \text{ m/s})}\right) \exp\left[-\frac{1}{2}\left(\frac{125 \text{ m}}{\sigma_z}\right)^2\right] \quad (18.4)$$

The downwind concentrations can be calculated based on the above equation. A linear interpolation indicates that the maximum ground level concentration (GLC) is approximately 1.01×10^{-3} g/m³ and is located at a downwind distance of about 800 m. In addition, the "critical" zone, where the concentration is above 3.3×10^{-4} g/m³, is located between 475 and 1,800 m. The concentration results for select downwind distances are provided in Table 18.1. It should be noted that the values of σ_y and σ_z in Table 18.1 are derived from dispersion coefficient values as a function of distance, x, from the source for Stability Class B from Turner (1970).

It should also be noted that only one "average" weather condition was considered in this example. However, one often selects the worst-case weather condition that has a reasonable probability of occurrence in the location of the site being evaluated. Employing this worst-case condition produces risk results on the conservative side. An analysis that includes a full spectrum of wind speeds, directions, and stability classes would obviously provide a more complete set of risk assessment calculations than is provided here.

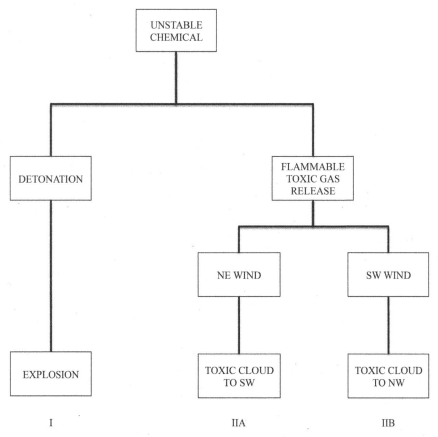

FIGURE 18.2 Event tree for accidents at a nanomaterial production facility for Illustrative example 5.

TABLE 18.1
Downwind Concentration Profile

x (m)	σ_y (m)	σ_z (m)	c (g/m³)
300	47	30	3.43×10^{-6}
400	60	41	1.11×10^{-4}
500	75	52	4.07×10^{-4}
550	80	60	6.78×10^{-4}
600	90	65	7.67×10^{-4}
700	105	77	9.44×10^{-4}
800	120	90	1.01×10^{-3}
900	150	110	9.06×10^{-4}
1,000	170	140	8.04×10^{-4}
1,500	250	240	4.15×10^{-4}
1,700	275	275	3.40×10^{-4}
2,000	300	380	2.37×10^{-4}

Nanotechnology

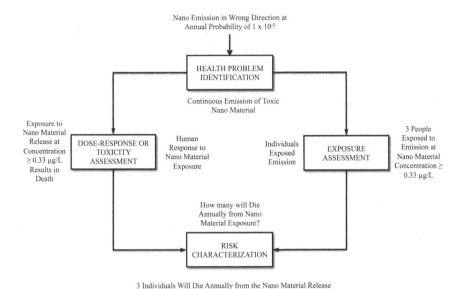

FIGURE 18.3 Plant health risk assessment for Illustrative example 5.

Determine which individuals within the pie-shaped segment downwind from the source will be killed if either accident (I or II) occurs. Referring to Figure 18.1, 13 individuals within the 200 m radius will die from Accident I. Three individuals located in the pie-shaped segment and 600 m southwest of the emission source will die from Accident II. The five individuals located 350 m southwest of the emission source are in the path of the dispersing plume but are all outside the critical zone. The six individuals located outside of the pie-shaped impact segment are within the plant boundary but are not potentially affected by either the explosion or the dispersing plume. The plant health risk and hazard risk assessment line diagrams are presented in Figures 18.3 and 18.4, respectively.

The total annual deaths, TAD, for the process if the accident occurs are therefore:

$$TAD = 13 + 3 = 16 \text{ deaths/yr}$$

The total annual risk, TAR, is obtained by multiplying the number of people in each impact zone by the probability of the event affecting that zone and summing the results. Thus:

$$TAR = (13)P(I) + (3)P(IIA) = (13)(10^{-6}) + (3)(10^{-5}) = 43 \times 10^{-6} = 4.3 \times 10^{-5}$$

The AAIR is obtained by dividing the above result by the number of people in the impact zone. The average annual risk, AAR, is calculated based only on the "potentially affected" people. Since 21 people are "potentially affected," i.e., are within the

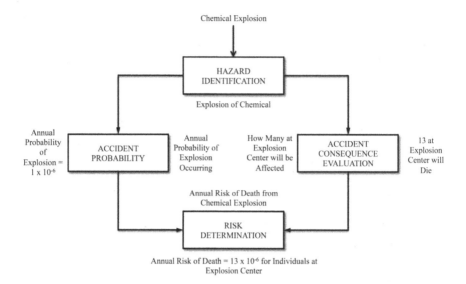

FIGURE 18.4 Plant hazard risk assessment for Illustrative example 5.

impact area of the explosion, or are in the path of the dispersing plume, the AAR is determined to be:

$$AAR = 4.3 \times 10^{-5}/21 = 2.05 \times 10^{-6}$$

The AAIR is based on all the individuals within the plant boundary. For this case study, this AAIR is now based on 27 rather than 21 individuals. Thus,

$$AAIR = 4.3 \times 10^{-5}/27 = 1.6 \times 10^{-6}$$

PROBLEMS

18.1 The time to failure for a heat exchanger in a nanochemical production facility is assumed to follow an exponential distribution with $\lambda = 0.15$/yr. What is the probability of a failure within the first 2 years?

18.2 The probability that a pressure gauge in a Canadian nanofacility will not survive more than 60 months = 0.95. How often should the gauge be replaced? Assume that the time to failure is exponentially distributed and that the expected replacement time should be based on the gauges expected operating life.

18.3 Referring to Illustrative example 3, determine what the standard deviation of the nanoparticle distribution must be to ensure less than 5% unacceptable particles are generated in this nanoparticle production process.

18.4 Refer to Illustrative example 4. What would the failure rate of the PSD measurement device have to be to ensure that there is a 95% probability that the device will last for at least 25,000 hr?

18.5 EPA's *Nanotechnology White Paper* (US EPA 2007) and *Nanotechnology for Site Remediation Fact Sheet* (US EPA 2008) have a range of case studies highlighting potential pollution prevention opportunities for nanomaterials. Go to these industrial sector resources and using the various pollution prevention opportunity descriptions and case studies presented in them and other resources, answer the following open-ended questions.
- a. Discuss additional source reduction techniques that may be applied to this industrial sector.
- b. Discuss additional recycling/reuse options that may be applied to this industrial sector.
- c. Discuss how ultimate disposal options may be either enhanced or eliminated for this industrial sector.
- d. Discuss how energy conservation measures may be applied to this industrial sector.
- e. Discuss how health, safety, and accident prevention measures may be applied to this industrial sector.

REFERENCES

Boxall, A., Chaudhry, Q., Jones, A., Jefferson, B., and Watts., C.D. 2008. *Current and Future Predicted Environmental Exposure to Engineered Nanoparticles*. Sand Hutton, UK: Central Science Laboratory.

Breeggin, L., Falkner, R., Jasoers, N., Pendergrass, J., and Porter, R. 2009. *Securing the Profits of Nanotechnologies: Towards Transatlantic Regulatory Cooperation*. London: Chatham House.

Burke, G., Singh, B., and Theodore, L. 2000. *Handbook of Environmental Management and Technology*, 2nd Edition. Hoboken, NJ: John Wiley & Sons.

Dictionary.com. 2022a. "technology," in The Free On-line Dictionary. https://www.dictionary.com/browse/technology (accessed January 3, 2022).

Dictionary.com. 2022b. "nano," in The Free On-line Dictionary. Source: The American Heritage® Science Dictionary, ©2011. Published by Houghton Mifflin Harcourt Publishing Company. https://www.dictionary.com/browse/nano (accessed January 3, 2022).

Fink, U., Davenport, R.E., Bell, S.L., and Ishikawa, Y. 2002. *Nanoscale Chemicals and Materials—An Overview on Technology, Products and Applications*. Menlo Park, CA: Specialty Chemicals Update Program, SRI Consulting.

Fiorino, D.J. 2010. *Voluntary Initiatives, Regulation, and Nanotechnology Oversight: Charting a Path*. Pen 19. Washington, DC: Project on Emerging Nanotechnologies, Woodrow Wilson International Center for Scholars (http://www.nanotechproject.org/process/assets/files/8347/pen-19.pdf).

Gwinn, M.R., and Vallyathan, V. 2006. Nanoparticles: Health Effects – Pros and Cons. *Environmental Health Perspectives* 114(12): 1818–1825.

Murphy, M., and Theodore, L. 2006. Environmental impacts of nanotechnology: Consumer issues. In *Consumers Union Symposium on Nanotechnology in Consumer Products*, Yonkers, NY (March).

National Center for Environmental Research 2003. *Nanotechnology and the Environment: Applications and Implications*, STAR Progress Review Workshop, Office of Research and Development. Washington, DC: National Center for Environmental Research.

NIOSH 2008. *Safe Nanotechnology in the Workplace: An Introduction for Employers, Managers, and Safety and Health Professionals.* Publication No. 2008-112. Cincinnati, OH: US Department of Health and Human Services, Centers for Disease Control and Prevention, National Institute for Occupational Safety and Health, DHHS (NIOSH). https://www.cdc.gov/niosh/docs/2008-112/pdfs/2008-112.pdf?id=10.26616/NIOSHPUB2008112

NIOSH 2018a. *Protecting Workers during the Handling of Nanomaterials.* Publication No. 2018-121. Cincinnati, OH: US Department of Health and Human Services, Centers for Disease Control and Prevention, National Institute for Occupational Safety and Health, DHHS (NIOSH). DOI: https://doi.org/10.26616/NIOSHPUB2018121.

NIOSH 2018b. *Protecting Workers during Intermediate and Downstream Processing of Nanomaterials.* Publication No. 2018-122. Cincinnati, OH: US Department of Health and Human Services, Centers for Disease Control and Prevention, National Institute for Occupational Safety and Health, DHHS (NIOSH). DOI: https://doi.org/10.26616/NIOSHPUB2018122.

Papp, T., Schiffmann, D., Weiss, D.G., Castranova, V., Vallyathan, V., and Rahman, Q. 2008. Human Health Implications of Nanomaterial Exposure. *Nanotoxicology* 2(1): 9–27.

Stander, L., and Theodore, L. 2011. Environmental Implications of Nanotechnology – An Update. *International Journal of Environmental Research and Public Health* 8(2): 470–479.

Theodore, L. 2006a. *Nanotechnology: Basic Calculations of Engineers and Scientists.* Hoboken, NJ: John Wiley and Sons.

Theodore, L. 2006b. *Nanotechnology: Environmental Implications and Solutions.* Presentation to US EPA Nanotechnology and OSWER: Opportunities and Challenges, Washington, DC, July 12–13. https://archive.epa.gov/oswer/nanotechnology/web/pdf/nd-waste-management-presentation-slides.pdf.

Theodore, L., and Dupont, R.R. 2012. *Environmental Health Risk and Hazard Risk Assessment: Principles and Calculations.* Boca Raton, FL: CRC Press/Taylor & Francis Group.

Theodore, L., and Kunz, R. 2005. *Nanotechnology: Environmental Implications and Solutions.* Hoboken, NJ: John Wiley and Sons.

Theodore, M.K., and Theodore, L. 2021. *Major Environmental Issues Facing the 21st Century*, 2nd Edition. Boca Raton, FL: CRC Press/Taylor & Francis Group.

Turner, D. 1970. *Workbook of Atmospheric Dispersion Estimates*, rev., AP-26. Research Triangle Park, NC: US EPA, Office of Air Programs.

US EPA 2007. *Nanotechnology White Paper.* EPA 100/B-07/001. Washington, DC: Science Policy Council, Office of the Science Advisor, US Environmental Protection Agency (http://nepis.epa.gov/Exe/ZyPDF.cgi/60000EHU.PDF?Dockey=60000EHU.PDF).

US EPA 2008. *Nanotechnology for Site Remediation Fact Sheet.* EPA 542-F-08-009. Washington, DC: Office of Solid Waste and Emergency Response, US Environmental Protection Agency (ttp://nepis.epa.gov/Exe/ZyPDF.cgi/P1001JIB.PDF?Dockey=P1001JIB.PDF).

US EPA 2011. *EPA Needs to Manage Nanomaterial Risks More Effectively.* Report No. 12-P-0162. Washington, DC: Office of the Inspector General, US Environmental Protection Agency (http://www.epa.gov/sites/production/files/2015-10/documents/20121229-12-p-0162.pdf).

Wilson, M., Kannangara, K., Smith, G., Simmons, M., and Raguse, B. 2004. *Nanotechnology Basic Science and Emerging Technologies.* Boca Raton, FL: Chapman & Hall/CRC Press.

19 Military and Terrorism

19.1 INTRODUCTION

The military is defined as the forces of a nation, assembled, drilled, disciplined, and equipped for offense and defense in maneuvers in warfare. The term may refer either to the entire body of military personnel in a nation or to a specific unit under a military commander. The composition of the military often reflects the attitudes toward war of the civilizations and societies they represent. In ancient Greece, for example, men up to the age of 60 were expected to serve in the Greek army. More importance was attached to military than to civil office. In ancient Rome, the citizen soldier army of the Republic changed to a professional force as social conditions changed and the Republic gave way to the Empire.

In prehistoric and early historic times, the military consisted of groups engaged sporadically in combat for the purpose of defending land desired for hunting or pasture. For example, the Greek city-states maintained bodies of militia capable of being united into one great force. The superior organization and strict discipline of these citizen-soldiers helped achieve the great victories won at such battles as Marathon and Plataea during the Persian Wars of the 5th-century BC.

Terrorism in general terms is the actual or threatened use of violence for political goals, directed not only against the victims themselves but also against larger, related groups often transcending national boundaries.

This chapter summarizes a wide range of topics associated with the military and terrorist activities that have a potential health and environmental impact. These topics include a discussion of the US military, explosives, terrorism consideration, and current risks and priorities for risk reduction associated with this sector. Five illustrative examples and five problems are provided to highlight concepts presented throughout the chapter.

19.2 THE US MILITARY

Colonial militias were the first American forces to do battle with the British. Their lack of reliability, however, caused Congress, at the urging of George Washington, to create the Continental Army on June 14, 1775. With the raising of ten companies, the army achieved a maximum strength of 35,000 men in November 1778. On June 2, 1784, however, Congress abolished the army on the basis that "standing armies in time of peace are inconsistent with the principles of republican government ..." In 1789 the War Department was established to oversee and administer military forces. After the US Constitution was ratified in 1789, Congress again authorized a small standing army to guard US frontiers, and in 1802 it established the United States Military Academy at West Point to train regular army officers. State militias, however, provided the main manpower resource during the American Civil War when military conscription was first adopted and then abolished. Until Wood War I, and the establishment of the draft, federalized state troops and volunteers provided the manpower needed in times of crisis. The first peacetime conscription was instituted in 1940, continued throughout the Korean and Vietnam wars, and ended on January 27, 1973. The Military Selective Service Act expired June 30, 1973. Since that time military power in the United States has been dependent on an all-volunteer force.

Technical demands made of individual soldiers will likely require in the future a small, elite corps of well-trained volunteers to operate under conditions of peace or limited war. This corps would be reinforced by large numbers of conscripted soldiers performing traditional functions during large-scale operations. Women are also now assigned to service and combat-support and combat branches.

19.3 EXPLOSIVES

Many practicing engineers and scientists tend to view the subject of explosives from a military standpoint. Although explosives have contributed much to the destruction of humans and their property, they have also allowed many great engineering feats to be performed, which would have been physically or economically impossible without their use. Engineering projects such as the Hoover Dam and the St. Bernard Tunnel would have taken a near infinite amount of time to accomplish by hand labor alone. It should also be remembered that explosives are among the most powerful servants of man. Mining of all kinds depends on blasting, as does the clearing of stumps and large boulders from land for agriculture production.

Explosives are chemical compounds or mixtures that undergo rapid burning or decomposition with the generation of large amounts of gas and heat and the subsequent production of sudden pressure effects (Theodore 2014). The primary use of explosives in peacetime is for blasting and quarrying, but explosives are also used in fireworks and signaling equipment. Explosives are used as propellants for projectiles and rockets and as bursting charges for demolition purposes and for projectiles, bombs, and mines.

The first explosive known was gunpowder, also called black powder. It was likely discovered in the 13th century and was the only explosive known for over 500 years. Nitrocellulose and nitroglycerin, both discovered in 1846, were the first modem explosives. Since then nitrates, nitrocompounds, fulminates, and azides have been the chief explosive compounds used alone or in mixtures with fuels or other agents.

Explosives are grouped into two main classes, low explosives, which bum at rates of cm/sec (in/sec), and high explosives, which undergo detonation at rates from 914 to 9,140 m/sec (1,000–10,000 yd/sec). Explosives vary in other important characteristics that influence their use in specific applications. Among these characteristics are the ease with which they can be detonated and their stability to conditions of heat, cold, and humidity. The shattering effect of an explosive depends upon the velocity of detonation. Some of the newer high explosives are extremely effective for military demolition and certain types of blasting. On the other hand, for quarrying and mining, when it is desirable to dislodge large pieces of rock or ore, explosives with a lower detonation velocity must be employed. Explosives used as propellants in rifles and cannon should burn still more slowly, as they are required to deliver a steadily increasing push to the projectile in the barrel of a gun rather than a sudden shock which, if strong enough, might break the gun. Special types of explosives that are sensitive to heat or shock are used to initiate the detonation of less sensitive high explosives. High explosives are often mixed with inert materials to reduce sensitivity to environmental conditions, as in the case of dynamite.

Two types of explosives are in general use for the propulsion of projectiles in firearms and rockets, and both are commonly called by the generic name of smokeless powder. The term is properly applied to the low explosive, gelatinized nitrocellulose. The other type of smokeless powder, which consists of a mixture of nitrocellulose with a high explosive such as nitroglycerin, is known correctly as double-base powder or compound powder. A common double-based explosive is cordite, which contains 30–40% nitroglycerin and a small quantity of petroleum jelly as a stabilizer. The term smokeless powder applied to either type of explosive, however, is misleading, because neither is free from smoke when exploded and neither takes the form of a true powder.

The rate of burning of either type of smokeless powder is controlled by the shaping of the powder grains. Because the powder grains burn from the surface inward, it is possible to produce grains that burn progressively more slowly, at an even rate, or progressively more quickly depending on the shape and dimensions of the grains. For example, spherical grains have progressively smaller surface areas as they burn and therefore burn progressively more slowly. Such degressive burning powders are used in such short-barreled small arms as pistols.

A great number of explosives undergo detonation. Some of these, such as trinitrotoluene (TNT), have a high resistance to shock or friction and can be handled, stored, and used with comparative safety. Others, such as nitroglycerin, are so sensitive that they are almost invariably mixed with an inert desensitizer for practical use. To obtain desirable characteristics, explosives of different characteristics are often mixed.

During World War I, TNT was the high explosive most generally employed, but before and during World War II, a number of new, extremely efficient high explosives were developed. Among the most important are cyclonite and pentaerythritol tetranitrate.

Cyclonite, also called RDX, is used in detonators. A mixture with TNT and wax is called Composition B and is used in bombs. A similar mixture, containing aluminum and called torpex, has an underwater effect about 50% greater than that of TNT. A plastic composition containing cyclonite and an explosive plasticizer is used for demolition charges.

Pentaerythritol tetranitrate, also called PETN, has characteristics similar to those of cyclonite and is mixed with TNT to form the explosive pentolite. It also forms the core of the explosive primacord fuses used for detonating demolition charges and the booster charges used in blasting. Two types of high explosives introduced since 1955 have largely replaced dynamite. A mixture of ammonium nitrate and fuel oil is very cheap and has explosive strength 25% greater than that of TNT.

For detonating charges of comparatively insensitive high explosives, compounds are used that will themselves detonate under a moderate mechanical shock or heat with sufficient force to explode the main charge. For many years, mercury fulminate, $Hg(ONC)_2$, was the compound chiefly employed for this purpose, either alone or mixed with other substances such as potassium chlorate. Its manufacture, however, is hazardous, and it cannot be stored at high temperatures without decomposition. In addition, mercury may be difficult to obtain in time of war. As a result, the fulminate has been replaced almost entirely in commercial and military detonators by lead azide, PbN_6, diazodinitrophenol, and mannitol hexanitrate. These initiators are used in conjunction with a charge of cyclonite or PETN, which have largely replaced the tetryl (trinitrophenylmethylnitramine) used previously. These sensitive explosives have high detonation pressure and explosive strength values. They are also frequently used as booster charges between the detonator and the major charge of high explosive in large shells and bombs. A blasting cap or exploder is a small charge of a detonator designed to be embedded in dynamite and ignited either by a burning fuse or a spark.

In coal mining the use of ordinary high explosives is hazardous because of the danger of igniting gases or suspended coal dust that may be present underground. For blasting under such conditions, several special types of safety explosives have been developed that minimize the danger of fires or explosions by producing flames that last for a very short time and are relatively cool. The types of safety explosives approved for work in coal mines are chiefly mixtures of ammonium nitrate with other ingredients such as sodium nitrate, nitroglycerin, nitrocellulose, nitrostarch, carbonaceous material, sodium chloride, and calcium carbonate. Another type of blasting charge for use in mining has grown in favor, because it produces no flame whatsoever. This charge is a cylinder of liquid carbon dioxide that can be converted into gas almost instantaneously by an internal chemical heating element. One end of the cylinder contains a breakable seal through which the gas can expand. The carbon dioxide charge is not a true explosive and absorbs heat rather than evolving it. It has the additional advantage that the force of the explosion can be directed at the base of the bore hole, thus lessening the shattering of the coal.

19.4 TERRORISM

Using the Global Terrorism Database (https://www.start.umd.edu/gtd), which now includes more than 200,000 terrorism incidents from 1970 to 2021, the Wilson Center (Veilleux and Dinar 2018) developed a method to codify water-related terrorism across the globe and found 675 water-related incidents in 71 countries, conducted by 124 known terrorist organizations, resulting in approximately 3,400 dead or wounded people. The most common target of water-related terrorism was found to be water infrastructure: the pipes, dams, weirs, levees, and treatment plants associated with water storage, treatment, and delivery. Terrorists target infrastructure to inconvenience government authorities, influence populations, and cripple corporations. While these terrorist incidents are not distributed evenly across the world, i.e., more than 200 events have occurred in South Asia (primarily Pakistan, Afghanistan, and India), these events are not uncommon in South America (>60 in Columbia, >40 in Peru) and have been attempted in the United States in the recent past (Veilleux and Dinar 2018). In 2016, the Justice Department revealed that Iran's Islamic Revolutionary Guards had hacked into the control system of a small dam in Rye Brook, NY, north of New York City (Thompson 2016). Although the attack was not successful because during the attack the sluice gate was offline for maintenance, this event did underscore the vulnerability of critical infrastructure systems even in developed nations.

In 2003, President Bush issued Homeland Security Presidential Directive 7 (HPSD-7), which affirmed US Environmental Protection Agency (EPA) as the lead federal agency for coordinating the protection of the nation's critical infrastructure. Under this directive, EPA is responsible for developing and providing tools and training on improving security to roughly 52,000 community water systems and 16,000 municipal wastewater treatment facilities.

EPA established a Water Security Division within the Office of Ground Water and Drinking Water in 2003. This Division works with drinking water and wastewater utilities, states, tribes, and other stakeholders to improve the security of these utilities and improve their ability to respond to security threats and breaches. Among its responsibilities and activities, the Water Security Division provides security and antiterrorism-related technical assistance and training to the water sector.

Security-related activities undertaken by EPA have fallen into five general categories, including: (1) establishing an information center for drinking water alerts or incidents, (2) developing vulnerability assessment tools, (3) identifying actions to minimize vulnerabilities, (4) revising emergency operations plans, and (5) supporting research on biological and chemical contaminants considered to be potential weapons of mass destruction.

19.4.1 INTERNATIONAL TERRORISM

International terrorism has been recurrent during periods of political and social upheaval. In the 19th century, anarchists in rural parts of Italy and Spain used terrorism, as did their counterparts in France. The Russian revolutionary movement before World War I had a strong terrorist element. In the 20th century, such groups as the Internal Macedonian Revolutionary Organization, the Croatian Ustashi, and the Irish Republican Army often carried their terrorist activities beyond the boundaries of individual countries. They were sometimes supported by established governments, such as those of Bulgaria and of Italy under the fascist leader Benito Mussolini.

The wave of international terrorism that developed after the mid-1960s differed from earlier events in its broader ramifications and greater impact. Several elements combined to make international terrorism easier and more effective: technological advances, resulting in both greater destructiveness and smaller size of weapons; the means available to terrorists for quick movement and rapid communication; and the more extensive worldwide connections of the groups' chosen victims.

The origins of the terrorist wave that began in the 1960s can be traced to the unresolved Middle East conflict between the Arab nations and Israel. Some Jewish radicals, such as the Stem Gang and the Irgun Zvai Leumi, resorted to terrorism during their struggle for an independent Israel in the late 1940s. Their Arab adversaries in the 1960s and 1970s chose to use that weapon much more systematically. The expulsion of Palestinian guerrillas from Jordan in September 1970 was commemorated by the creation of an extremist terrorist arm called the Black September.

The spread of terrorism beyond the Middle East in the 1960s was most conspicuous in the three industrial nations where transition from authoritarianism to democracy after World War II had been the most rapid and traumatic: West Germany, Japan, and Italy. Inspired by vague Communist ideologies and typically supported by fashionably leftist sympathizers in the affluent middle classes, the terrorists aimed to bring about the collapse of the state by provoking its violent, self-destructive reaction.

In West Germany, the so-called Red Army Faction, better known as the Baader-Meinhoff Gang, robbed numerous banks and raided US military installations. Its most spectacular exploits were the 1977 kidnapping and murder of a prominent industrialist, Hans-Martin Schleyer, and the subsequent hijacking by Arab sympathizers of a Lufthansa airliner to Mogadishu, Somalia. As did the Japanese Red Army terrorist group, the members of the West German gang frequently cooperated with Palestinian terrorists, notably in the murder of Israeli athletes at the Olympic Games in Munich in 1972. By the late 1970s, most activists of the Red Army Faction were either imprisoned or dead.

Unlike terrorist movements in West industrial nations and in the Middle East, those in Latin America owed more to long-standing local traditions of political violence than to new problems common to advanced industrial societies. In Latin America, the main new development was the rise in so-called urban guerrilla movements as terrorist activities shifted from the countryside into Latin America's sprawling cities.

Direct support by a few governments, particularly those of Libya, Southern Yemen, and Algeria, and indirect support by others (including Russia, East Germany, and Czechoslovakia) accounted for much of the rise of international terrorism in the 1970s. Its continuation, which might include blackmail by nuclear explosives acquired by terrorists, presents a clear danger for the future. Technological innovations have made countermeasures by the authorities more effective, but the best hope in the struggle against terrorism remains the stabilization of international order and the internal consolidation of the affected societies.

19.4.2 THE NEED FOR EMERGENCY RESPONSE PLANNING

In response to the terrorist events of September 11, 2001, across the United States, the 107th Congress passed the Public Health Security and Bioterrorism Preparedness and Response Act of 2002 (P.L. 107-188, the Bioterrorism Act of 2002) to address a wide range of security issues. For example, Title IV of the Bioterrorism Act amended the Safe Drinking Water Act (SDWA) to address threats to drinking water security. One key provision of the Act requires that all Public Water Systems (PWSs) serving populations of more than 3,300 persons conduct assessments of their vulnerabilities to terrorist attack or other intentional acts and to defend against adversarial actions that might substantially disrupt the ability of a system to provide a safe and reliable supply of drinking water. In addition to the vulnerabilities assessment (VA), the act requires a PWS to certify and submit a copy of the VA to the EPA Administrator, prepare or revise an emergency response plan (ERP) based on the results of the VA, and certify that an ERP has been completed or updated within 6 months of completing the assessment. The Law also requires the EPA to conduct research on preventing and responding to terrorist or other attacks

Emergencies have occurred in the past and will continue to occur in the future. The frequency of intense storms and natural disasters is increasing, and unsuccessful acts of terrorism have occurred in the United States in the past. Both natural and man-made disasters are likely to continue to occur in the future, and planning for these types of emergencies is a must. A few of the many commonsense reasons to plan ahead for emergencies are as follows (Krikorian 1982; WEF 2013):

1. Emergencies will happen; it is only a question of when.
2. When emergencies do occur, minimization of loss and the protection of people, property, and the environment can be achieved through the proper implementation of an appropriate emergency response plan.
3. Minimizing the losses caused by an emergency requires planned procedures, understood responsibilities, designated authority, accepted accountability, and trained, experienced people. A fully implemented plan can do this.
4. If an emergency occurs, it may be too late to plan. Lack of preplanning can turn an emergency into a disaster.

A particularly timely reason to plan ahead is to ease the "chemophobia" or fear of chemicals which is so prevalent in society today. So much of the recent attention to emergency planning and newly promulgated laws are a reaction to the tragedy at Bhopal. Either a total lack of information or misinformation is the probable cause of "chemophobia." Fire is hazardous, and yet it is used regularly at home. Most adults have understood the hazard associated with fire since the time of the caveman. By the same token hazardous chemicals, necessary and useful in today's technological society, are not something to be afraid of. Chemicals need to be carefully used and their hazards understood by the general public. An emergency plan that is well designed, understood by the individuals responsible for action, and understood by the public can ease the concern over emergencies and reduce chemophobia. People will react during an emergency; how they react can be somewhat controlled through education. The likely behavior during an emergency when ignorance is pervasive is panic.

An emergency plan can minimize loss by helping to ensure the proper response in an emergency. "Accidents become crises when subsequent events and the actions of people and organizations with a state in the outcome combine in unpredictable ways to threaten the social structures involved" (Beranek et al. 1987). The wrong response can turn an accident into a disaster as easily as no response. One example is a chemical fire that is doused with water, causing the fire to emit toxic fumes; this fire would be better left to burn itself out. Another example is the evacuation of people from a building into the path of a toxic vapor cloud; they might well be safer staying indoors with closed windows to let the vapor cloud pass. Still another example is the members of a rescue team becoming victims because they were not wearing proper breathing protection. The proper response to an emergency requires an understanding of the hazards. A plan can provide the right people with the information needed to respond properly during an emergency.

Other than these mentioned commonsense reasons to plan, there are legal reasons. Recognizing the need for better preparation to deal with chemical emergencies, Congress enacted the Superfund Amendments and Reauthorization Act (SARA) of 1986. One part of SARA is a free-standing act called Title III, the Emergency Planning and Community Right-to-Know Act of 1986. This act requires federal, state, and local governments to work together with industry in developing emergency plans and "community right-to-know" reporting on hazardous chemicals. These requirements build on the EPA's Chemical Emergency Preparedness Program and numerous state and local programs that are aimed at helping communities deal with potential chemical emergencies. As indicated, the Bioterrorism Act of 2002 and the recent America Water Infrastructure Act (AWIA) of 2018 both extend legal requirements for water utilities to complete risk assessments and emergency response plans.

Most larger industries have long had emergency plan designed for on-site personnel. The protection of people, property, and thus, profits has made emergency plans and prevention methods common in industry. On-site emergency plans are often a required by insurance companies. Expansion of these existing industry plans to include all significant hazards and people in adjacent communities is a way to minimize the effort required for additional emergency planning.

19.4.3 ANTI-TERRORISM EFFORTS

Characterization of a water system should include the mission and objectives of the facility, including the highest priority services provided by the utility and critical facilities, customers, processes, and assets to achieve these objectives. Priority services provided by a facility may include water provision to the public, the government, the military, industrial or critical care facilities, retail operations, and firefighting. In identifying critical facilities and assets, the following should be considered: critical customers, dependence on other infrastructure (e.g., electricity, transportation, other water utilities), contractual obligations, single points of failure (e.g., critical aqueducts, transmission systems, aquifers etc.), chemical hazards and other aspects of the utility's operations, or availability of other utility capabilities that may increase or decrease the criticality of specific facilities, processes, and assets (US EPA 2002).

Specific consequences to be identified should be of significant concern in that they could substantially disrupt the ability of the system to provide a safe and reliable supply of drinking water or otherwise present significant public health concerns to the surrounding community. Characteristics of an event that should be considered when identifying "significant" consequences include:

1. Magnitude of service disruption
2. Economic impact (such as replacement and installation costs for damaged critical assets or loss of revenue due to service outage)
3. Number of illnesses or deaths resulting from an event
4. Impact on public confidence in the water supply
5. Chronic problems arising from specific events

Consider the operation of critical facilities, assets, and/or processes and assess what an adversary could do to disrupt these operations. Such acts may include physical damage to or destruction of critical assets, contamination of water, intentional release of stored chemicals, and interruption of electricity or other infrastructure interdependencies. Critical assets that must be evaluated include pipes and other constructed conveyances; physical barriers; water collection, pretreatment and treatment facilities; storage and distribution facilities; electronic, computer, or other automated systems that are used by a PWS; the use, storage, or handling of various chemicals; and the operation and maintenance processes and procedures for all of these systems.

It should also be noted that there are a wide range of hazard identification and risk assessment methodologies that have been used for many years in the chemical and process industry that may be helpful in developing qualitative risk assessments in the water sector. These techniques include fault and event trees, hazard and operability studies (HAZOPs), and qualitative risk assessments as described by Theodore and Dupont (2012).

Depending on countermeasures already in place, some critical assets may already be sufficiently protected. Evaluation of existing countermeasures will aid in identification of the areas of greatest concern and help to focus priorities for further risk reduction in water facilities. A facility should identify the capabilities it currently employs for detection, delay, and response to security incidents. Current intrusion detection systems, water quality monitoring, operational alarms, guard post orders, and employee security awareness programs should be identified. Delay mechanisms, including locks and key control, fencing, structural integrity of critical assets, and vehicle access checkpoints should be cataloged along with existing policies and procedures for evaluation and response to intrusion and system malfunction alarms, adverse water quality indicators, and cyber system intrusions. Above all, it is important to determine the performance characteristics of all these systems and policies as poorly operated and maintained security technologies provide little or no protection.

19.5 CURRENT RISKS AND PRIORITIZATION FOR RISK REDUCTION

Information gathered on threat, critical assets, water utility operations, consequences, and existing countermeasures should be analyzed to determine the current level of risk. The utility should then determine whether current risks are acceptable or risk reduction measures should be pursued. Recommended actions should measurably reduce risks by reducing vulnerabilities or consequences through improved deterrence, delay, detection, and/or response capabilities or by improving operational policies or procedures. Selection of specific risk reduction actions should be completed prior to considering the cost of the recommended action(s). Utilities should carefully consider both short- and long-term solutions. An analysis of the cost of short- and long-term risk reduction actions may impact which actions the utility chooses to reach its security goals.

Utilities may also want to consider security improvements in light of other planned or needed improvements. Security and general infrastructure may provide significant multiple benefits. For example, improved treatment processes or system redundancies can both reduce vulnerabilities and enhance day-to-day operations.

Generally, strategies for reducing vulnerabilities fall into three broad categories: sound business practices, system upgrades, and security upgrades. Sound business practices affect policies, procedures, and training to improve the overall security-related culture at a facility. System upgrades include changes in operations, equipment, processes, or infrastructure itself that make the system fundamentally safer.

Security upgrades improve capabilities for detection, delay, or response to an accident or emergency event.

An action plan defines the specific actions that should be taken to respond to events where high priority vulnerabilities have been compromised. Even if a VA did not identify any vulnerabilities, contingency planning should be considered for the possibility of at least the following high consequence events:

1. Contamination of the drinking water
2. Structural damage/physical attack
3. Supervisory Control and Data Acquisition (SCADA) system, computer, or cyber attack
4. Intentional hazardous chemical release (e.g., release of chlorine or ammonia from storage facilities, etc.)

Example action plans for the four intentional events listed here are provided by the US EPA (2004).

19.6 ILLUSTRATIVE EXAMPLES

Five illustrative examples complement the material presented above regarding health and hazard risks associated with the military and terrorism.

19.6.1 Illustrative Example 1

Consider the Picatinny arsenal parallel system shown in Figure 19.1. Determine the reliability of this system employing the information provided in the figure.

Solution: Because this is a parallel system, employing the information from Figure 19.1 yields the following:

$$R_p = 1-(1-R_A)(1-R_B) = 1-(1-0.92)(1-0.95) = 1-(0.08)(0.05) = 1-0.004 = 0.996$$
$$R_p = 99.6\%$$

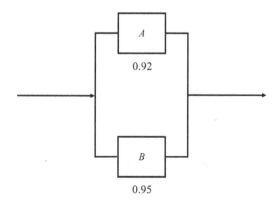

FIGURE 19.1 Parallel system for Illustrative example 1.

Military and Terrorism

19.6.2 ILLUSTRATIVE EXAMPLE 2

A Central Intelligence Agency (CIA) agent's ability to distinguish a terrorist from a nonterrorist is tested independently on ten different occasions. What is the probability of eight correct identifications if the agent is only guessing?

Solution: For guessing purposes, the probability of making a correct terrorist identification may be reasonably assumed to be 0.5. Let X denote the number of correct identifications. If the agent is only guessing, X has a binomial distribution with $n = 10$ and $p = 0.5$. The probability of at least eight correct identifications is:

$$f(x) = \frac{n!}{x!(n-x)!} p^x q^{n-x}; x = 0, 1, \ldots, n \qquad (5.3)$$

$$P(X \geq 8) = P(X = 8) + P(X = 9) + P(X = 10)$$

$$P(X = 8) = \frac{10!}{8!(2)!} 0.5^8 0.5^2 = \frac{10(9)(8!)}{8!(2)!} 0.5^8 0.5^2$$
$$= \frac{90}{2}(0.0039)(0.25) 45(0.000976) = 0.044$$

$$P(X = 9) = \frac{10!}{9!(1)!} 0.5^9 0.5^1 = \frac{10(9!)}{9!(1)!} 0.5^9 0.5^1$$
$$= 10(0.00195)(0.5) 10(0.000976) = 0.0098$$

$$P(X = 10) = \frac{10!}{10!(0)!} 0.5^{10} 0.5^0 = \frac{10!}{10!(0)!} 0.5^9 0.5^1$$
$$= 1(0.00195)(0.5) = 1(0.000976) = 0.00098$$

Therefore:

$$P(X \geq 8) = 0.044 + 0.0098 + 0.00098 = 0.055 = 5.5\%$$

19.6.3 ILLUSTRATIVE EXAMPLE 3

A drone hits its target with a probability $p = 30\%$. Find the number of drones that should be deployed so that there is at least a 95% probability of hitting a target.

Solution: The probability of missing the target is $q = 1 - p = 0.7$. Hence, the probability that n drones miss the target is $(0.7)^n$. Thus, one seeks the smallest n for which

$$1 - (0.7)^n > 0.95$$

or, equivalently,

$$(0.7)^n < 0.05$$

This can be solved by trial and error or using the Solver function in Excel. With the Solver function, $n = 8.399$ to yield a probability of failure of <0.05. This indicates that at least nine drones are required to ensure at least 95% probability of hitting a select target if all drones have the same probability of success.

19.6.4 Illustrative Example 4

A small explosive product has a weight normally distributed with a mean of 16 lb. It may be regarded as appreciably under the mean weight and an environmental hazard if it is less than 15.75 lb. If the risk of this is to be less than 0.01 (1%), what is the maximum allowable value of the standard deviation?

Solution: This is a one-sided normal distribution test. For this case,

$$P(T_1 < T < T_2) = P\left(\frac{T_1 - \mu}{\sigma} < \frac{T - \mu}{\sigma} < \frac{T_2 - \mu}{\sigma}\right) \tag{5.35}$$

$$P(X \leq 15.75) = 0.01$$

Normalizing:

$$P(X \leq 15.75) = P\left(\frac{T - \mu}{\sigma} \leq \frac{15.75 - 16}{\sigma}\right) = P\left(Z \leq \frac{-0.25}{\sigma}\right) = 0.01$$

From Table 5.4, the Z value for which the probability = 0.01 for this one tailed normal distribution is -2.31. Solving for σ yields:

$$Z \leq \frac{-0.25}{\sigma} = -2.31; \sigma = \frac{-0.25}{-2.31} = 0.108$$

Therefore, the maximum value of the standard deviation for the weight distribution of the explosive product = 0.108 lb.

19.6.5 Illustrative Example 5

Defective bombs pose a major environmental risk. Ten percent of the bombs produced at a certain arsenal turn out to be defective. Find the probability that in a sample of ten bombs chosen at random exactly two will be defective.

Solution: Note that this may be solved using either the binomial distribution or the Poisson distribution. Apply the Poisson distribution for this application. First note that for a Poisson distribution (Section 5.2.4):

$$\mu = np = (10)(0.1) = 1$$

Then the Poisson distribution is described as:

$$f(x) = \frac{e^{-\mu}\mu^x}{x!}, x = 0, 1, 2, \ldots \tag{5.24}$$

Then:

$$P(X=2) = \frac{e^{-1}1^2}{2!} = \frac{(0.368)1}{(2)(1)} = 0.184 = 18.4\%$$

PROBLEMS

19.1 Referring to Illustrative example 2, what is the probability that the agent can determine seven correct identifications if the agent is only guessing?

19.2 Referring to Illustrative example 5, repeat these calculations using the binomial distribution.

19.3 If the probability that nine soldiers will suffer a bad reaction from handling a poisonous gas is 0.001, determine the probability that out of 2,000 soldiers exactly three will suffer a bad reaction. Also calculate that more than two will experience a bad reaction. Assume a Poisson distribution for the solution to this problem.

19.4 If the probability of a control valve employed at a military facility being defective is 10%, what is the probability of obtaining less than four defects in a sample size of four? Assume a Poisson distribution for the solution to this problem.

19.5 A random sample of 64 observations of rocket travel distance is normally distributed with a mean $\bar{X} = 101$ miles and a standard deviation of 4 miles. Find $P(X > 103$ miles).

REFERENCES

Beranek, W., McCullough, J.P., Pine, S.H., and Soulen, R.L. 1987. Getting Involved in Community-Right-To Know. *Chemical & Engineering News* 65(43): 62–65.

Krikorian, M. 1982. *Disaster and Emergency Planning*. Loganville, GA: Institute Press.

Theodore, L. 2014. *Chemical Engineering: The Essential Reference*. New York, NY: McGraw-Hill.

Theodore, L., and Dupont, R.R. 2012. *Environmental Health Risk and Hazard Risk Assessment: Principles and Calculations*. Boca Raton, FL: CRC Press/Taylor & Francis Group.

Thompson, M. 2016. Iranian Cyber Attack on New York Dam Shows Future of War. *Time. Com*. http://time.com/4270728/iran-cyber-attack-dam-fbi/ (accessed January 6, 2022).

US Environmental Protection Agency 2002. *Vulnerability Assessment Factsheet*. Washington, DC: Office of Water. https://nepis.epa.gov/Exe/ZyPDF.cgi/P1004AYS.PDF?Dockey=Pl004AYS.PDF.

US Environmental Protection Agency 2004. *Emergency Response Plan Guidance for Small and Medium Community Water Systems to Comply with the Public Health Security and Bioterrorism Preparedness and Response Act of 2002*. Washington, DC: Office of Water. https://www.epa.gov/sites/production/files/2015-04/documents/2004_04_27_watersecurity_pubs_small_medium_erp_guidance040704.pdf.

Veilleux, J., and Dinar, S. 2018. New Global Analysis Finds Water-Related Terrorism Is on the Rise. *NewSecurityBeat, the Blog of Environmental Change and Security Program*. Washington, DC: Wilson Center. https://www.newsecuritybeat.org/2018/05/global-analysis-finds-water-related-terrorism-rise/ (accessed January 6, 2022).

Water Environment Federation (WEF). 2013. *Emergency Planning, Response, and Recovery*. WEF Special Publication. Alexandria, VA: Water Environment Federation. 283 pp.

20 Weather and Climate

20.1 INTRODUCTION

Three terms need to be defined before proceeding to the heart of this chapter: weather, climate, and meteorology. These terms can be defined as follows:

1. *Weather* is a general term describing the condition of the atmosphere at a particular time and place with regard to temperature, pressure, moisture, cloudiness, etc.
2. *Climate* is defined by the long-term effect of the Sun's radiation on the rotating Earth's varied surface and atmosphere. It can be understood most easily in terms of annual or seasonal averages of temperature and precipitation.
3. *Meteorology* is the scientific study of the Earth's atmosphere. It includes the study of day-to-day variations of weather conditions (synoptic meteorology); the study of electrical, optical, and other physical properties of the atmosphere (physical meteorology); the study of average and extreme weather conditions over long periods of time (climatology); the variation of meteorological elements close to the ground over a small area (micrometeorology); and studies of many other phenomena associated with the Earth's atmosphere.

This chapter summarizes a wide range of topics related to weather and climate that impact health and hazard risk assessment. Topics include a brief history of weather observations, information on climate and climate variability, engineering consideration of meteorological factors affecting pollutant transport and fate, a brief history of the National Weather Service, weather observations, and weather forecasting. Four illustrative examples and four problems are provided to highlight concepts presented throughout the chapter.

20.2 HISTORY

The scholars of ancient Greece were greatly interested in the atmosphere. As early as 400 BC Aristotle wrote a treatise called *Meteorologica*, dealing with the "study of things lifted up;' about one-third of the treatise is devoted to atmospheric phenomena, and it is from this work that the modern term meteorology is derived. Throughout history much of the progress in the discovery of the laws of physics and chemistry was stimulated by curiosity about atmospheric phenomena.

Weather forecasting has challenged the human mind from the earliest times, and much of the worldly wisdom that people have displayed has been identified with weather lore and weather almanacs. Little progress was made in scientific forecasting, however, until the 19th century, where developments in the fields of thermodynamics (Theodore, Ricci and VanVliet 2009) and hydrodynamics (Abulencia and Theodore 2009; Theodore 2014) provided the theoretical basis for meteorology.

Exact measurements of the atmosphere are also of great importance in meteorology, and the advance of the science has been furthered by the invention of suitable instruments for observation and by the organization of networks of observing stations to gather weather data. Weather records for individual localities were made as early as the 14th century, but not until the 17th century were any systematic observations made over extended areas. Slowness of communications also hampered the development of weather forecasting, and it was not until the invention of the telegraph in the middle of the 19th century that weather data from an entire country could be transmitted to a central point and correlated for the development of a forecast.

One of the most significant milestones in the development of the modern science of meteorology occurred in the World War I period, when a number of Norwegian meteorologists, led by Wilhelm Bjerknes, made intensive studies of the nature of fronts and discovered that interacting air masses generate the cyclones that create the typical storms of the northern hemisphere. Later meteorological work was aided by the invention of apparatus such as the rawinsonde which made possible the investigation of atmospheric conditions at extremely high altitudes. Immediately following the World War I period, a British mathematician, Lewis Fry Richardson, made the first significant attempt to obtain numerical solutions of atmospheric equations for the prediction of meteorological elements. Although his efforts were not successful at the time, they helped pave the way for the explosive progress in numerical weather prediction that is taking place today.

20.3 CLIMATE

Land and sea areas, being so variable, react in many different ways to the atmosphere, which is constantly circulating in a state of dynamic activity. Weather is measured by thermometers, rain gauges, barometers, and other instruments, but the study of climate relies on statistics. Today such statistics are handled efficiently by computers. A simple, long-term summary of weather changes, however, is still not a true picture of climate. To obtain this requires the analysis of daily, monthly, and yearly patterns. Investigation of climate changes over geologic time is the province of paleoclimatology, which requires the tools and methods of geological research.

The word *climate* comes from the Greek *klima*, referring to the inclination of the Sun. Besides the effects of solar radiation and its variations, however, climate is also influenced by the complex structure and composition of the atmosphere and by the ways in which it and the ocean transport heat. Thus, for any given area on Earth, not only the latitude (the Sun's inclination) must be considered but also the altitude, terrain, distance from the ocean, relation to mountain systems and lakes, and other such influences. Another consideration is scale: a macroclimate refers to a broad region, a mesoclimate to a small district, and a microclimate to a minute area. A microclimate, for example, can be specified that is good for growing plants underneath large shade trees.

Climate has profound effects on vegetation and animal life, including humans. It plays statistically significant roles in many physiological processes, from conception and growth to health and disease. Humans, in turn, can affect climate through the alteration of the Earth's surface and the introduction of pollutants and chemicals such as carbon dioxide into the atmosphere.

Climates are also described by agreed-upon codes or by descriptive terms that are somewhat loosely defined but nevertheless are somewhat useful. On a global scale, climate can be spoken of in terms of zones, or belts, that can be traced between the equator and the pole in each hemisphere. To understand them, the circulation of the upper atmosphere, or *stratosphere*, must be considered, as well as that of the lower atmosphere, or *troposphere*, where weather takes place. Upper atmospheric phenomena were little understood until the advent of such advanced technology as rocketry, high-altitude aircraft, and satellites.

Ideally, hot air can be thought of as rising by convection along the equator and sinking near the poles. Thus, the equatorial belt tends to be a region of low pressure and calms, interrupted by thunderstorms associated with towering cumulus clouds. Because of the calms, this belt is known as the doldrums. It shifts somewhat north of the equator in the northern summer and south in the southern summer. By contrast, air sinks in the polar regions. This leads to high atmospheric pressure, and the dry, icy winds tend to radiate outward from the poles.

Complicating this simplistic picture is the Earth's rotation, which deflects the northerly and southerly components of the atmosphere's circulation. Thus, the tropical and polar winds both tend to be easterlies, and two intermediate belts develop in each hemisphere. Around latitude 30° N and S is a zone of high pressure, where the upper air sinks and divides, sending air streams toward the equator. Steady northeast trade winds blow in the northern hemisphere, and southeast trade winds in the southern hemisphere. These high-pressure areas lead to arid areas on the continents but to moist air over the oceans, because of evaporation. If these trade winds meet an island or mainland coast, moist air is pushed up into cooler elevations, and heavy rainfall usually occurs.

20.4 METEOROLOGICAL FACTORS

The atmosphere has been labeled the dumping ground for air pollution. Industrial society can be thankful that the atmosphere cleanses itself (up to a point) by natural phenomena. Atmospheric dilution occurs when the wind moves because of wind circulation or atmospheric turbulence caused by local Sun intensity. Interestingly, pollutants are removed from the atmosphere by precipitation and by other reactions (both physical and chemical) as well as by gravitational fallout.

The atmosphere is the medium in which air pollution is carried away from its source and diffuses. Meteorological factors have a considerable influence over the frequency, length of time, and concentrations of air pollutants to which the general public may be exposed. The variables that affect the severity of an air pollution problem at a given time and location are wind speed and direction, insolation (amount of sunlight), lapse rate (temperature variation with height), mixing depth, and precipitation. Unceasing change is the predominant characteristic of the atmosphere; for example, temperatures and winds vary widely with latitude, season, and surrounding topography (Gilpin 1963; Hesketh 1972).

Atmospheric dispersion depends primarily on horizontal and vertical transport. Horizontal transport depends on the turbulent structure of a wind field. As the wind velocity increases, the degree of dispersion increases with a corresponding decrease in the ground-level concentration of a contaminant at a receptor site. This is a result of the emissions being mixed into a larger volume of air. The dilute effluents may, depending on the wind direction, be carried out into essentially unoccupied terrain away from any receptors. Under different atmospheric conditions, the wind may funnel the diluted effluent down a river valley or between mountain ranges. If an inversion (temperature increases with height) is present aloft that would prevent vertical transport, the pollutant concentration will continually build up.

One can define atmospheric turbulence as those vertical and horizontal convection currents or eddies that mix process gaseous discharges (plumes) with the surrounding air. Several generalizations can be made regarding the effect of atmospheric turbulence on plume dispersion. Turbulence increases with increasing wind speed and causes a corresponding increase in horizontal dispersion. Mechanical turbulence is caused by changes in wind speed and wind shear at different altitudes. Either of these conditions can lead to significant changes in the concentration of a pollutant within a plume at different elevations.

Topography can also have a considerable influence on the horizontal transport and thus pollutant dispersion. The degree of horizontal mixing can be influenced by sea and land breezes. It can also be influenced by man-made and natural terrain features such as mountains, valleys, or even a small ridge or a row of hills. Low spots in the terrain or natural bowls can act as sites where pollutants tend to settle and accumulate because of the lack of horizontal movement in the land depression(s). Other topographical features that can affect horizontal transport are city canyons and isolated buildings. City canyons occur when the buildings on both sides of a street are fairly close together and are relatively tall. Such situations can cause funneling of pollutants from one location to another. Isolated buildings or the presence of a high-rise building in a relatively low area can cause redirection of dispersion patterns and route emissions into an area in which many receptors live.

Dispersion of air contaminants is strongly dependent on the local meteorology of the atmosphere into which the pollutants are emitted. Most mathematical formulation for the design of pollutant dispersal is associated with open-ground terrain free of obstructions.

20.4.1 PLUME RISE

A plume of hot gases emitted vertically has both a momentum and a buoyancy. As the plume moves away from the stack, it quickly loses its vertical momentum (owing to drag by and entrainment of the surrounding air). As the vertical momentum declines, the plume bends over in the mean wind direction. However, quite often the effect of buoyancy is still significant, and the plume continues to rise for a long time after bending over. The buoyancy term is due to the less-than-atmospheric density of the stack gases and may be temperature or composition induced. In either case, as the plume spreads out in the air (at the same time mixing with the surrounding air), it becomes diluted by the ambient air.

Modeling the rise of the plume of gases emitted from a stack into a horizontal wind is a complex mathematical problem. Plume rise depends not only on such stack gas parameters as temperature, molecular weight, and exit velocity but also on such atmospheric parameters as wind speed, ambient temperature, and stability conditions adjacent to the stack (Cooper and Alley 1986).

The behavior of plumes emitted from any stack depends on localized air stability. Effluents from tall stacks are often injected at an effective height of several hundred feet to several thousand feet above the ground because of the added effects of buoyancy and velocity on the plume rise. Other factors affecting the plume behavior are the diurnal variations in the atmospheric stability and the long-term variations that occur with changing seasons.

20.4.2 EFFECTIVE STACK HEIGHT

Reliance on atmospheric dispersion as a means of reducing ground-level concentrations is not foolproof. Inversions can occur with a rapid increase in ground-level pollutant concentrations. One solution to such situations is the tall stack concept. The goal is quite simple: inject the effluent above any normally expected inversion layer. This approach is used for exceptionally difficult or expensive treatment situations because tall stacks are quite expensive. To be effective, they must reach above the inversion layer to avoid local plume fallout. The stack itself does not have to penetrate the inversion layer if the emissions have adequate buoyancy and velocity to carry the plume through the inversion. In such cases, the effective stack height will be considerably greater than the actual stack height. The effective stack height (equivalent to the effective height of emission) is usually considered the sum of the actual stack height, the plume rise due to velocity (momentum) of the issuing gases and the buoyancy rise which is a function of the temperature of the gases being emitted relative to the atmospheric conditions.

The effective stack height depends on a number of factors. These factors include the gas flow rate, the temperature of the effluent at the top of the stack, and the diameter of the stack opening. The meteorological factors influencing plume rise are wind speed, air temperature, shear of the wind speed with height, and the atmospheric stability. No theory on plume rise presently takes into account all these variables, and it appears that the number of equations for calculating plume rise varies inversely with one's understanding of the process involved. Even if such a theory was available, measurements of all parameters would seldom be available. Most of the equations that have been formulated for computing the effective height of an emission stack are semiempirical in nature. When considering any of these plume rise equations, it is important to evaluate each in terms of assumptions made and the circumstances existing at the time the particular correlation was formulated. Depending on the circumstances, some equations may definitely be more applicable than others.

The effective height of an emission rarely corresponds to the physical height of the source or the stack. If the plume is caught in the turbulent wake of the stack or of buildings in the vicinity of the source or stack, the effluent will be mixed rapidly downward toward the ground. If the plume is emitted free of these turbulent zones, a number of stack emission factors and meteorological factors will influence the rise of the plume. The influence of mechanical turbulence around a building or stack can significantly alter the effective stack height. This is especially true with high winds when the beneficial effect of the high stack gas velocity is at a minimum and the plume is emitted nearly horizontally. Details regarding a host of plume rise models and calculation procedures are available in literature (Cooper and Alley 1986; Theodore 2008).

20.5 ATMOSPHERIC DISPERSION MODELING

The initial use of dispersion modeling occurred in military applications during World War I. Both sides of the conflict made extensive use of poison gases as a weapon of war. The British organized the Chemical Defense Research Establishment at Parton Downs during the war. Research at this institute dominated the field of dispersion modeling for more than 30 years through the end of World War II.

With the advent of the potential use of nuclear energy to generate electrical power, the United States Atomic Energy Commission invested heavily in understanding the nature of atmospheric transport and diffusion processes. Since about 1950 the United States has dominated researching the field. The US Army and Air Force have also studied atmospheric processes to understand the potential effects of chemical and biological weapons.

The Pasquiil-Gifford model has been the basis of many models developed and accepted today (Gifford 1961; Pasquill 1961; Cota 1984). This model has served as an atmospheric dispersion formula from which the path downwind of emissions can be estimated after obtaining the effective stack height. There are many other dispersion equations (models) presently available, most of them semiempirical in nature. Calculation details regarding the use of this Pasquiil-Gifford model are available in the literature (Turner 1970; Cooper and Alley 1986; Theodore 2008).

The problem of having several models is that various different predictions can be obtained. To establish some reference, a standard was sought by the government. The *Guideline on Air Quality Models* (US EPA 1978, 1986) is used by the Environment Protection Agency (EPA), by the states, and by private industry in reviewing and preparing prevention of significant deterioration (PSD) permits and in state implementation plans (SIP) revisions. The guidelines are a means by which consistency is maintained in air quality analyses. In October 1986, EPA proposed to include four different changes to this guideline: (1) addition of a specific version of the rough terrain diffusion model (RTDM) as a screening model, (2) modification of the downwash algorithm in the industrial source complex (ISC) model, (3) addition of the offshore and coastal dispersion (OCD) model to EPA's list of preferred models, and (4) addition of the AeroVironment Air Pollution Model for Complex Terrain Applications (AVACTA) II model as an alternative model in the guideline. In industry today, the ISC models are the preferred models for permitting and therefore are used in many applications involving normal or "after the fact" releases, depending on which regulatory agency must be answered to.

Current status of atmospheric modeling for regulatory purposes, modeling guidance and support, a list of acceptable air quality models, modeling applications and tools, and training and additional air modeling resources can be found at EPA's *Support Center for Regulatory Atmospheric Modeling (SCRAM)* (https://www.epa.gov/scram).

Weather and Climate

In order to prepare for and prevent the worst, screening models (e.g., AERSCREEN, etc.; https://www.epa.gov/scram/air-quality-dispersion-modeling-screening-models) are applied in order to simulate the worse-case scenario. One difference between the screening models and the refined models mentioned earlier is that certain variables are set that are estimated to be values to give the worst conditions. In order to use these screening models, the parameters of the model must be fully grasped. In short, these models are necessary to predict the behavior of the atmospheric dispersions. These predictions may not necessarily be correct; in fact, they are rarely completely accurate. In order to choose the most effective model for the behavior of an emission, the source and the models have to be well understood.

20.6 THE NATIONAL WEATHER SERVICE

The National Weather Service is a government agency engaged in studying, reporting, and predicting the weather, including temperature, moisture, barometric pressure, and wind speed and direction, throughout the United States and its territories. Established in 1870 under the direction of the Signal Corps of the US Army, the Weather Bureau was transferred to the Department of Agriculture in 1891 and to the Department of Commerce in 1940. In 1965 it was made a branch of the Environmental Science Services Administration within the Department of Commerce. Under the Reorganization Plan of 1970, the Weather Bureau became a part of the new National Oceanic and Atmospheric Administration in the Department of Commerce and was officially renamed the National Weather Service.

Information on the weather is gathered from more than 180,000 weather stations across the country including nearly 2,200 interagency Remote Automatic Weather Stations, and 10,600 Cooperative Weather Stations, 5,000 of which make up the "climate" network strategically located throughout the United States. Observations are recorded at the stations at varying intervals ranging from 1 to 6 hr. Observations over the oceans are made by volunteer cooperative observers and transmitted from more than 2,000 ships. In addition, automatic meteorological observing systems are used to obtain data at locations too remote for regular use by human observers.

Most of the supplementary stations send reports every hour, enabling the Weather Service to furnish information necessary for the operation of airlines and to prepare weather maps, i.e., simplified maps or charts showing the principal weather conditions at a given hour over an extended area. For more long-range studies, the Weather Service receives statistics from volunteer weather observers who keep daily records of air temperature, rainfall, and in some cases river stages and tides. The activities of these field stations and individuals are coordinated by six regional offices, four supervising the contiguous lower 48 states, and one each for Alaska and Hawaii.

20.7 WEATHER OBSERVATIONS

Improved observations of high-level winds during and following World War II provided the basis for new theories of weather forecasting and revealed the necessity for changes in older concepts of general atmospheric circulation. During this period major contributions to meteorological science were made by the Swedish born meteorologist Carl-Gustav Rossby and his collaborators in the United States. The so-called jet stream, a fast-moving river of air circling the globe high in the atmosphere, was discovered. In 1950, using electronic computers, it became possible to apply the fundamental theories of hydrodynamics and thermodynamics to the problem of weather forecasting, and today computers are employed regularly to provide weather predictions for industry, agriculture, and the general public. Observations made at ground level are more numerous than those made at upper levels. They include the measurement of air pressure, temperature, humidity, wind direction and speed, the amount and height of cloud cover, visibility, and precipitation.

For the measurement of air pressure, the mercury barometer is the accepted standard. Aneroid barometers, although less accurate, are also useful, particularly on board ships and when used in the recording form called a barograph to show the trend of pressure change over a period of time. All barometer readings used in meteorological work are corrected for variations resulting from temperature and the height of the station, so that pressures from different stations may be directly compared

For the observation of temperature many different types of thermometers are employed. For most purposes an ordinary thermometer covering an appropriate range is satisfactory. It is important to place the thermometer in such a way as to minimize the effects of sunshine during the day and of heat loss by radiation at night, thus yielding values of the air temperature representative of the general area.

The instrument most often used at weather observatories is the hygrometer. A special type of this instrument, known as the psychrometer, consists of two thermometers to measure the dry-bulb and wet-bulb temperatures. A more recent device to measure humidity is based on the fact that certain substances undergo changes in electrical resistance with changes in humidity. Instruments making use of this principle commonly are used in rawinsondes, high-level atmospheric sounding devices.

The most common instrument for measuring the direction of the wind is the ordinary weather vane, which keeps pointing into the wind and which is connected to an indicating dial or to a series of electronic switches that light small bulbs at the observers' stations to show the wind direction. Wind speed is measured by means of a cup anemometer, an instrument consisting of three or four cups mounted about a vertical axis. The anemometer spins faster as the speed of the wind increases, and some form of device for counting its revolutions is used to gauge the wind speed.

Precipitation is measured by a rain gauge or a snow gauge. The rain gauge consists of an upright cylinder, open at the top to catch the rain and is calibrated either in inches or in millimeters, so that the total depth of the precipitation may be measured. The snow gauge also is a cylinder, which is thrust into the snow to collect a core of snow. This core is melted and measured in terms of the equivalent depth of water.

20.8 WEATHER FORECASTING

National Weather Service field stations provide weather forecasts, warnings, and advice to the general public, as well as to specialists in agriculture, business, commerce, and industry. Hourly weather observations are updated every hour between 10 and 15 minutes after the hour. The 7-day public forecast is normally prepared twice a day, at about 4 am and 4 pm. However, the forecast is updated between the

Weather and Climate

regularly scheduled times as needed, often in the late morning (9–11 am) and evening (8–10 pm). Known as the "Voice of the National Weather Service," the NOAA Weather Radio network has more than 750 transmitters, covering nearly 90% of the 50 states, along with the adjacent coastal waters, Puerto Rico, the US Virgin Islands, and the US Pacific Territories. NOAA Weather Radio broadcasts continuous weather information directly from National Weather Service offices across the country. The broadcasts include warnings, watches, forecasts, current weather observations, and other hazard information, 24 hr a day. Working with the Federal Communications Commission's Emergency Alert System, NOAA Weather Radio is an "all hazards" radio network, making it the single source for the most comprehensive weather and emergency information available to the public. It broadcasts warning and post-event information for all types of hazards – both natural (such as tornadoes, earthquakes, and tsunamis) and technological (such as chemical releases or oil spills). NOAA Weather Radio is also used to broadcast AMBER alerts for missing children.

The Weather Service also provides certain specialized weather forecast information. These include aviation weather forecasts to assist aircraft operating in the United States and over transoceanic routes. Tracking storms at sea and on the Great Lakes is the province of the marine meteorological service. A closely related function is performed by observation centers in the Tropics, which, as part of the hurricane and storm-warning service, issue advice and warnings from June through October, the critical hurricane season.

During severely dry seasons, rangers, campers, and now more frequently, homeowners and communities in forested regions are also guided by a special service forecasting fire weather. Another important activity is the river and flood forecasting system; daily observation of river levels and rainfall at more than 9,900 stations results in extremely accurate predictions and a high degree of protection. Finally, the importance of agricultural production to the nation's economy is attested to by the existence of weather services to farmers and ranchers for frost warnings and spraying advice, and the climatological and climate forecasting services, which distribute statistical summaries of climatologic data and weekly reports on crop progress.

In addition to its regular and special services to the public, the Weather Service conducts research projects. In its meteorological investigations, primarily concerned with forecasting techniques and storm behavior, the agency works closely with various government research laboratories. It utilizes in its forecasting service the findings from studies of mathematical modeling of the general circulation of the atmosphere, advances in radar meteorology, high-speed computer methods, and earth-orbiting satellites.

The accuracy of weather forecasts is relative, and published percentages have little meaning without a detailed description of the ground rules that have been followed in judging the validity of a prediction. The use of complex numerical models has resulted in considerable improvement of forecast accuracy when compared to predictions previously made by subjective methods, especially for periods of more than 1 day. At the present time, a 7-day forecast can accurately predict the weather about 80% of the time and a 5-day forecast can accurately predict the weather approximately 90% of the time. However, a 10-day or longer forecast is only right about half the time. Since data cannot be collected from the future, models have to use estimates and assumptions to predict future weather. The atmosphere is changing all the time, so those estimates are less reliable the further the forecasts are into the future.

20.9 ILLUSTRATIVE EXAMPLES

Four illustrative examples complement the material presented above regarding health and hazard risks associated with the weather.

20.9.1 ILLUSTRATIVE EXAMPLE 1

Let X denote the annual number of hurricanes in a certain region. The pdf of X is specified as:

$$f(x) = 0.25; x = 0$$

$$f(x) = 0.35; x = 1$$

$$f(x) = 0.24; x = 2$$

$$f(x) = 0.11; x = 3$$

$$f(x) = 0.04; x = 4$$

$$f(x) = 0.01; x = 5$$

What is the probability of at least four (4) hurricanes in any year?

Solution: For at least four hurricanes in any year, one may write:

$$P(x \geq 4) = \sum_{4}^{\infty} f(x) = \sum_{4}^{5} f(x) = f(4) + f(5) = 0.04 + 0.01 = 0.05 = 5\%$$

20.9.2 ILLUSTRATIVE EXAMPLE 2

The Enhanced Fujita Scale or EF Scale, which became operational on February 1, 2007, is used to assign a tornado a "rating" based on estimated wind speeds and related damage. When tornado-related damage is surveyed, it is compared to a list of Damage Indicators (DIs) and Degrees of Damage (DoD) which help estimate better the range of wind speeds the tornado likely produced. From that, a rating (from EF0, wind gusts 65 to 85 mph to EF5, wind gusts > 200 mph) is assigned. The probability of tornado intensity in a given location is estimated to be described by a random variable X having the following pdf:

$$f(x) = 1.5 e^{-3x}$$

Calculate the probability of a tornado that will result in severe damage with an EF rating greater than or equal to 3 ($P(X \geq 3)$).

Solution: Applying the definition of the cumulative distribution function (cdf) leads to:

$$P(X \geq 3) = \int_3^\infty f(x)dx = \int_3^5 f(x)dx = \int_3^5 1.5e^{-3x}dx$$

$$P(Z \geq 3) = -1.5(e^{-15} - e^{-9}) = -1.5(3x10^{-7} - 1.23x10^{-4}) = 0.0001227 = 0.0123\%$$

20.9.3 ILLUSTRATIVE EXAMPLE 3

The frequency of Categories 4 and 5 hurricanes globally has been estimated to be 18/yr since 1990. If this frequency is normally distributed with a variance of 4/yr, calculate the probability that a Category 4 or 5 hurricane will occur more than 23 times in a given year.

Solution: For this application

$$\sigma = \sqrt{4} = 2$$

This is a one-tailed calculation so that:

$$P(X > 23) = P\left(\frac{X-18}{2} > \frac{23-18}{2}\right) = P(Z > 2.5)$$

Referring to the values in the standard normal table, Table 5.4, the following probability can be calculated:

$$P(Z > 2.5) = 0.006 = 0.6\%$$

There is a very low chance of 23 or more Category 4 or 5 hurricanes happening globally in a year and is somewhat encouraging as these storms have sustained winds of >130 mph and cause catastrophic damage to the areas where they occur.

20.9.4 ILLUSTRATIVE EXAMPLE 4

The number and cost of weather-related disasters are increasing over time due to a combination of increased exposure (i.e., values at risk of possible loss), vulnerability (i.e., how much damage does the intensity [wind speed, flood depth] at a location cause), and that climate change is increasing the frequency of some types of extremes that lead to billion-dollar disasters. Go to NOAA's National Centers for Environmental Information (NCEI) US Billion-Dollar Weather and Climate Disasters website at https://www.ncdc.noaa.gov/billions/ and comment on the changes in disaster frequency over time and how the Year 2021 compared to the recent historical weather-related record.

Solution: NOAA's US Billion-Dollar Weather and Climate Disasters website contains a vast amount of data regarding significant weather-related disasters that have affected the United States since 1980. Over the total period of record (1980–2021) the average of costly extreme weather events was 7.4 events/yr (CPI-adjusted); the annual average for the most recent 5 years (2017–2021) was 17.2 events (CPI-adjusted). The total costs for these events in the last 5 years ($742.1 billion) are more than one-third of the disaster cost total of the last 42-year period (1980–2021), which exceeds $2.155 trillion (inflation-adjusted to 2021 dollars). This reflects the most recent 5-year cost average of nearly $148.4 billion/year – a new record.

In 2021, there were 20 weather/climate disaster events with losses exceeding $1 billion affecting the United States. These events included drought (1 event), floods (2 events), severe storms (11 events), tropical cyclones (4 events), wildfires (1 event), and severe winter storms (1 event). The total cost from these events was $145.0 billion and is the third most costly year on record, behind 2017 and 2005. Overall, these events resulted in the deaths of 688 people and had significant economic effects on the areas impacted. These data indicate that the frequency of the loss of life and personal property damage due to extreme weather events is expected to increase in the future. This will undoubtedly impact the siting, design, and operation of new chemical process industry plants, and the upgrading and retrofitting of existing facilities to protect against the potential damage these weather-related events may cause.

PROBLEMS

20.1 The probability is 0.80 that an airplane crash due to weather conditions is diagnosed correctly. Suppose, in addition, that the probability is 0.30 that an airplane crash not due to weather is incorrectly attributed to adverse weather conditions. If 35% of all airplane crashes are actually due to adverse weather, what is the probability that an airplane crash was due to weather conditions given that it has been so diagnosed?

20.2 The annual number of storms that strike the Florida coast is normally distributed with a mean equal to 15 and a standard deviation equal to 3.5. What is the probability that fewer than seven (7) storms will hit the Florida coast in a given year?

20.3 The normalized wind speeds at a certain hurricane test site during the past 10 years are summarized in Table 20.1. If the wind speeds are assumed to follow a log-normal distribution, predict the level that would be exceeded once in 20 years.

20.4 The average annual wind speed for a city, Y, has a log-normal distribution. If Y has a mean of 4.5 mph and a variance of 2.72 mph, find the probability that the wind speed will exceed 85 mph characteristic of an EF1 tornado.

TABLE 20.1
Wind Speed Data for Problem 3

Year	1	2	3	4	5	6	7	8	9	10
Wind Speed (fpm)	23	38	17	210	62	142	43	29	71	31

REFERENCES

Abulencia, P., and Theodore, L. 2009. *Fluid Flow for the Practicing Chemical Engineer.* Hoboken, NJ: John Wiley & Sons.

Cooper, C.D., and Alley, F.C. 1986. *Air Pollution Control: A Design Approach.* Prospect Heights, IL: Waveland Press, 493–515, 519–552.

Cota, H.M. 1984. Basic Computer Program for the Gaussian Equation for a Point Source. *Journal of the Air Pollution Control Association* 34(3): 253.

Gifford, F.A. 1961. Uses of Routine Meteorological Observations for Estimating Atmospheric Dispersion. *Nuclear Safety* 2(4): 47–51.

Gilpin, A. 1963. *Control of Air Pollution.* New York, NY: Butterworth, 326–333.

Hesketh, H.B. 1972. *Understanding and Controlling Air Pollution.* Ann Arbor, MI: Ann Arbor Science Publishers, 33–70.

Pasquill, F. 1961. The Estimation of the Dispersion of Windborne Material. *Meterology Magazine* 90(33): 33–49.

Theodore, L. 2008. *Air Pollution Control Equipment Calculations.* Hoboken, NJ: John Wiley & Sons.

Theodore, L. 2014. *Chemical Engineering: The Essential Reference.* New York, NY: McGraw-Hill.

Theodore, L., Ricci, F., and VanVliet, T. 2009. *Thermodynamics for the Practicing Engineer.* Hoboken, NJ: John Wiley & Sons.

Turner, D. 1970. *Workbook of Atmospheric Dispersion Estimates*, rev., AP-26. Research Triangle Park, NC: US EPA, Office of Air Programs.

US Environmental Protection Agency 1978. *Guideline on Air Quality Models.* EPA-450/2-78-027. Research Triangle Park, NC: US EPA, Office of Air Programs.

US Environmental Protection Agency 1986. *Industrial Source Complex (ISC) Dispersion Model User's Guide*, 2nd Edition. Volumes 1 and 2. EPA-450/4-86-005a and EPA-450/4-86-005b. Research Triangle Park, NC: US EPA, Office of Air Programs.

21 Architecture and Urban Planning

21.1 INTRODUCTION

Architecture is the practice of building design and its resulting products; customary usage refers only to those designs and structures that are culturally significant. Architecture is to building as literature is to the printed word. More prosaically, one could say today that architecture must satisfy its' intended uses, must be technically sound, and must convey esthetic meaning. But the best buildings are often so well constructed that they outlast their original use. They then survive not only as beautiful objects, but as documents of the history of cultures, achievements in architecture that testify to the nature of the society that produced them. These achievements are never wholly the work of individuals. Architecture is a social art.

Architectural form is inevitably influenced by the technologies applied, but building technology is conservative and knowledge about it is cumulative. Precast concrete, for instance, has not rendered brick obsolete. Although design and construction have become highly sophisticated and are often computer directed, this complex apparatus rests on preindustrial traditions inherited from millennia during which most structures were lived in by the people who erected them. The technical demands on building remain the elemental ones to exclude enemies, to circumvent gravity, and to avoid discomforts caused by an excess of heat or cold or by the intrusion of rain, wind, or vermin. This is no trivial assignment even with the best of modem technology.

Urban planning involves community planning and development programs that address issues such as elimination of slums, increase of middle-income housing, preservation of existing housing, and improvement of public services. Many urban development action grants are made to severely distressed areas in order to stimulate private investment in a community. These actions can include low-income housing plans, mortgage insurance, rent-supplement programs, rehabilitation of property, housing loan programs for the elderly, plans to facilitate independent living for the disabled, etc.

As environmental concerns present some of the most pressing issues to the world, both professional and academic architects have begun to address how planning and built form affect the environment. The term *built environment* has come to mean the result of human activities that impact the environment. It essentially includes everything that is constructed or built, i.e., all types of buildings, chemical plants, roads, railways, parks, farms, gardens, bridges, etc. Thus, the built environment includes everything that can be described as a structure or "green" space. Generally, the built environment is organized into six interrelated components:

1. Products
2. Interiors

3. Structures
4. Landscapes
5. Cities
6. Regions

While architecture may appear to be one of the many contributors to the current environmental state, in reality, the energy consumption and pollution affiliated with the materials, the construction, and the use of buildings contributes to most major environmental impacts. In fact, architectural planning, design, and building significantly contribute to the depletion of nonrenewable energy sources and exposure to carcinogens and other hazardous materials. Where one chooses to build, which construction materials are selected, how a comfortable temperature is maintained, or what type of transportation is needed to reach it – each issue, decided by both architect and user, significantly impacts the overall environment. Sadly, despite these opportunities to shape a healthier future, an analysis of American planning and building describes a continuing assault on the environment.

There are signs of hope that this industrial sector is moving toward lower environmental impacts, however, as most architects have committed to build green. New buildings will incorporate a range of green elements including: radiant ceiling panels that heat and cool, saving energy and improving occupant comfort; cogeneration plants that utilize waste heat; green roofs that are irrigated exclusively with rainwater and mitigate the heat island effect; use of materials that are rapidly renewable and regionally manufactured, etc. Additionally, buildings are being designed to maximize daytime lighting and air circulation. For example, a bird nest's design has been employed that is efficient, withstanding wind loads and wind shear while simultaneously enabling light and air to move through it. Throughout the building process, construction and demolition waste is recycled. Measurement and verification plans are also being employed to track utility usage for sustainability purposes.

This chapter summarizes a wide range of topics related to architecture and urban planning that impact health and hazard risk assessment. Topics include a brief history of the sector; considerations for changing the approach to building design, construction, and operation; and considerations for improving urban planning aspects of siting, design, and materials use. Four illustrative examples and four problems are also provided to highlight concepts presented throughout the chapter.

21.2 HISTORY

A systematic review of US architectural expansion reveals a strict adherence to a grid. While facilitating the organization of a new country, the gridding of land parcels and urban plans made few allowances for existing conditions. In fact, "slapped down anywhere," the grid imposed a man-made order on nature. From the New England town, to the first cities of Philadelphia, New York, and Washington, DC, the grid etched an order atop the country with little acknowledgment of or regard for the natural landscape. Instead, in the case of the earliest urban examples, the grid contained nature in the form of the village or town "green." Unfortunately, the green did not retain a reserve of natural landscape. Rather, as nature controlled, it set the precedent for the simulation and subjugation of nature. Today, many housing developments raze forests only to turf and replant the area with something else. The simulation of landscapes, rather than preserving or using the existing landscapes, increases net land usage, energy consumption, and pollution. The retention of untouched and undeveloped land protects more than just trees. Each area – forest, wetland, prairie, coastal plain – sustains a complete ecosystem of plant and animal life. In examining the clearing of a forest, not only are the trees lost, but also the birds that used to live in and off of them, the plants that needed the trees' shade to survive, the animals that ate those understory plants, and so on. These losses reflect the chain reaction of ecological destruction caused by land development. With this in mind, the reports of multiple species eradication loom that much larger.

Early examples of architecture, in both indigenous and colonial cultures, exhibit tremendous adaptation to both site and climate. But as buildings evolved from dwellings necessary for survival to conveyors of status and wealth, architectural planning and forms increasingly ignored the existing environment. A study of contemporary architecture, particularly housing developments, shows the mass production of styles transplanted everywhere. These styles originally became categorized because they evolved from an architectural response to climatic conditions. The stick style's steep roofs and projected eaves respond to climatic conditions while its diagonal "stick work" suggestively reflects the structural frame. But when these architectural elements appear on the surface of an airtight, concrete box in a development in Dallas, they cease to have any real function. In order to convey sociocultural meaning, the architect/developer and homeowner lose the opportunity to have a building that responds to and respects the local natural environment.

The quintessential American architecture – the suburban house complemented by a lawn, paved driveway, and two-car garage – evolved from a long history of antiurban development celebrating a frontier sense of independence and isolation. Unfortunately, this evolution of American housing, combined with the mass production and purchasing of the car, led to the present-day condition of major suburban and extra urban growth. Necessitating car use for practically every activity outside the home, the suburban house's auto-reliance causes massive fossil fuel consumption, road building, and parking paving. The extensive development of the American suburb has spawned other enclaved architectural forms: the mall, the retail park, the industrial park, the business park, and the leisure complex. All create a greater dependence on the car and disturb more land. The ecological repercussions are enormous. Considering net hours spent in today's home – families are smaller, more households have both partners working, more people live alone – the increase in square area of living space per person exemplifies society's tendency toward excessive expenditures of money, energy, and other resources. These wastes extend to the land. Each house typically occupies a cleared lot of land, destroying an enormous portion of existing ecological environments. Because an individual normally does not need or use that much land, current efforts encourage a reduction of that private land while increasing community land in the form of public green spaces like parks and undeveloped zones.

The desire for more (land, space, money, things, and so on) seems human, but in fact identifies the most important environmental concern. Conservation represents the most significant means of addressing environmental problems. Whether it be car use, private green space, or total built square footage, *less is environmentally more*. Beginning with less built space starts a whole chain of environmental reductions in energy and materials consumption.

21.3 CURRENT DEBATE ON THE NEED FOR SUSTAINABLE ARCHITECTURE

Reviewing actions of current political, governmental, and legislative bodies reflects the desire and urgency for change. Green parties, groups, and leaders with environmental agendas aid in public awareness and implementing change. For example, both the former Vice President of the United States Al Gore and the recent Presidents of the American Institute of Architects have raised many concerns to the national level. Within the government, the Environmental Protection Agency (EPA) has researched and implemented change in a broad range of issues from hazardous materials found in the built environment, like asbestos, lead, radon, and mercury (found in paints) to energy sources and consumption. Particularly significant and innovative are the new city ordinances, like that of Austin, which encouraged energy conservation through financial incentives. The Green Builder Program, sponsored by the Environmental and Conservation Services Department of Austin, Texas, uses a rating system encouraging environmentally sensitive building practices and products in new homes. Large organizations, like the North Carolina Recycling Association and the National Audubon Society, have publicized environmental concerns and new practices through the design of their buildings. Both aim to conserve natural resources and to be as energy efficient and nontoxic as possible.

As architects have struggled to come to terms with the environmental implications of their buildings, the term *sustainability* has become the catchword. While sustainability will not answer all environmental concerns, it provides a program to address current practices. With the present rates of fossil fuel consumption and non-renewal resource depletion, the Earth's systems may not be able to support life in the not too distant future. This risk of extinction necessitates examination and change. As Solow (1991) stated "... it is an obligation to conduct ourselves so that we leave to the future the option or the capacity to be as well off as we are ... Sustainability is an injunction not to satisfy ourselves by impoverishing our successors." How can the species sustain itself, i.e., secure a viable environment for future generations? Through analyzing the environmental impact of architectural siting, design, construction, and material use, a greater understanding of feasible alternatives can suggest ways to alter practices in order to do the least possible damage to the environment. The following discussion suggests environmentally conscious practices specifically related to architecture.

Recycling another building normally offers the most significant environmental savings. Particularly in places where there are unused and vacant buildings, to build more of the same represents one of the greatest environmental wastes. The initial energy spent in construction – through preparing the site, manufacturing the building materials, transporting them to the location, and then assembling them in construction – normally exceeds years of operational costs. Therefore, barring the least efficient structures, reuse – even with renovation – is the most sustainable choice. But if this is not possible, there are many ways in which the traditional building process can be improved with sustainability in mind.

21.4 SITING

As described above, the sprawl of development has threatened or destroyed many ecosystems. Therefore, one of the first site concerns is to avoid clearing previously untouched land. Once the land or place has been chosen, the existing landscape, the topography, wind movements, and context should be analyzed. First, one should use the given resources. Retaining existing trees and other plant life does the least ecological damage, while additionally saving later expenditures for artificial landscaping. Next, topography and wind movements can be used to naturally assist in creating a more comfortable environment. Careful siting, in relation to the given landscape, reduces the building's heating and cooling loads. For example, tree groupings can provide wind barriers in the winter, while others can direct winds into the building during the summer. Along the same lines, deciduous trees offer a building summer shading, while still allowing for passive solar heating in the winter. Additional concern for solar orientation can provide the building with natural lighting. With a balance in relation to heating/cooling gains and losses, fenestration uses daylight to produce a more comfortable, healthy, and energy-efficient space. The building's context must also be examined as a potential source of environmental hazards and opportunities. The surrounding buildings and structures can significantly influence siting. Like the natural elements mentioned above, built forms create shade and redirect wind. They also effect site hydrology. Buildings and nonporous surfaces (like asphalt) change how water moves through and drains from a site. In general, a site should be well drained with adequate flooding and erosion control for proper building maintenance and a healthy living environment. The best siting will not disturb the normal patterns of water flow and drainage. But if this is unavoidable, the effect of redirected water should be analyzed to avoid upsetting existing ecologies and conditions.

Another important site consideration is potential pollution sources. For the most part, industry and transportation create the greatest amounts of air, noise, and water pollution: roadways, cars, airports, oil refineries, power plants, and so on. Siting analysis of these potential sources should either suggest the use of another site or a way to avoid exposure to the hazards.

21.5 DESIGN CONSIDERATIONS

Environmentally conscious design offers perhaps the greatest opportunity for ecological improvement. For the most part, the current "green" trends concentrate on materials, products, and energy systems. This emphasis allows architects to ignore the environmental implications of their buildings. Responsible behavior requires more than a substitution of traditional building materials with recycled or nontoxic products. An ethical response to the environmental concerns necessitates change at the core, i.e., in the architectural theory of design. Environmental concerns must be completely incorporated into architectural thinking. Then, as an integral aspect of the architectural process, sustainability can shape design decisions and form buildings.

Several key concerns shape an environmentally conscious design' strategy: minimizing the building's effect on the existing ecosystem, minimizing the use of new resources, increasing the energy efficiency of the building in its form and operation, and creating a healthy environment for the users.

Minimizing the building's effect on the existing environment has been discussed above, specifically in the siting analysis section. Additionally, those aspects of the designed landscape can complement and enhance the viability of the existing ecosystems. The use of the traditional turf lawn represents a seriously destructive design practice. It removes the existing, natural environment at the risk of plant and animal biodiversity. Also, lawn maintenance requires irrigation, fertilization, and mowing, which increases water and fossil fuel consumption. Mowing and the use of pesticides both contribute to pollution. Instead, to complement the existing landscape, drought resistant native plantings enhance an outdoor environment to the benefit of resource management and ecosystems. Native plants thrive with a minimum of watering, chemicals (pesticides and fertilizers), and cutting. They also aid in maintaining or restoring an ecosystem's biodiversity. In areas that must be cleared for parking and walkways, the substitution of pervious paving materials (gravel, crushed stone, open paving blocks, and pervious paving blocks) minimizes runoff and increases infiltration and groundwater recharge.

Minimizing the use of resources, particularly new resources, can be achieved in several ways. Again, to recycle that modernist line, less is more and smaller is better. Beginning with the preliminary design, the interior space should be kept to a minimum. This reduces land use, building materials, and operational energy expenditures.

Increasing energy efficiency through reduced operational expenditures can be achieved in several ways. Passive systems, such as solar heating, daylighting, and natural cooling (berms, shade, and ventilation), produce a more energy-efficient building with minimal expenditures. As suggested above in the siting considerations, a building should be designed to work with the surrounding climate and natural energy sources. A building that responds to and takes advantage of what is naturally given results in a more sustainable design. A multitude of opportunities exist. To begin, what will create a comfortable environment? Orientation, built forms (like shading devices), and window and door locations can reduce heating and cooling loads while simultaneously enhancing living conditions. Also, a more systematic evaluation of heating and cooling loads reduces the energy operating expenses of a building. High levels of insulation, high performance windows, and a tight construction (but not at the expense of indoor air quality) create a more energy-efficient building.

Considering that people generally spend about 90% of their time indoors, the quality of indoor air crucially impacts well-being and comfort. Indoor air pollution comes from many different sources, both indoor and out. One of the more serious threats to indoor air quality and health is radon. Radon rises from subsurface uranium deposits through and into buildings. Posing a tremendous threat, radon is the nation's second leading cause of lung cancer. Other outdoor pollutants – like pesticides and car exhausts – threaten many buildings' indoor air quality. All three pollutants – radon, pesticides, and car exhaust – can be significantly reduced by good planning and design. First, an adequate ventilation system prevents accumulation within the building. While the building should open up for natural ventilation, an airtight construction will avoid many problems. For example, radon usually enters a building through cracks in the foundation. In the case of pesticides, the building's envelope works doubly. Careful detailing of the building, particularly in its corners and where it meets the ground, prevents many pests from entering. As a result, toxic interior pesticides and fumigants become unnecessary. Additionally, an airtight construction through detailing prevents many pollutants, particularly exterior pesticides and car exhaust, from entering the interior. Another preventive measure, the removal or avoidance of the pollution source, improves indoor air quality. Detaching a garage or parking structure from inhabited spaces eliminates direct exhaust infiltration into the building. In the cases where the pollution sources cannot be removed, ventilation intakes should be situated to avoid contaminants: other building's exhausts, car pollution, and pesticides.

Indoor air pollutants, like outdoor pollutants, pose more serious problems when buildings have inadequate, poorly maintained, or improperly located ventilation systems. An adequate ventilation system lessens the harmful effects of pollutants like lead, formaldehyde, carcinogenic wood finishes, smoke, and biological contaminants (bacteria, molds, mildew, and viruses). Especially in the case of lead dust and biological contaminants, keeping interiors clean and dust-free improves indoor air quality. In addition to increased ventilation and maintenance, source removal eliminates many problems. Smoking, a major indoor air pollution source, should be prohibited in interiors. Exposure to other pollutants, like lead, mercury, and volatile organic compounds (VOCs), can be more easily avoided through the greater availability of nontoxic building materials and finishes.

In smaller scale residential projects where there is little threat of on-site or near-site pollution sources, natural ventilation may suffice. But with larger scale projects, or those that are exposed to other sources of pollution (traffic, the exhausts of other buildings, the outgassing of building materials) conditions necessitate mechanical ventilation systems. Particularly in buildings like offices, with a large number of users, successful mechanical ventilation becomes crucial to maintaining indoor air quality. Without proper ventilation or systems maintenance, problems like outgassing or sick building syndrome can significantly affect the health and productivity of the building's users.

The building's design should incorporate recycling into the program so that it is easy and available. For example, a kitchen or an office can be designed to include recycling containers or cabinets for glass, aluminum, plastic, and paper. Composting systems for green waste and food waste can be specified and located. Also, saving water can serve as a recycling opportunity. The recycled water from clothes washers, baths, showers, and nonkitchen sinks can be redirected for irrigation use. Another way to save water is through harvesting rainwater.

Finally, the greatest recycling opportunity exists in the building itself. It should be designed with reuse in mind. A building should be adaptable with no or minimal renovation. As mentioned above, this may help avoid the enormous energy and material expenditures required by a new building's construction.

21.6 MATERIALS CONSIDERATIONS

Informed material selections greatly enhance the resource and energy savings created by an ecologically aware design. The use of each building material impacts both the global and local environments through its extraction, manufacture, and use. For example, the lumber and mining industries have devastated ecological systems. Therefore, the selection of a material should be made only after an analysis of its removal or extraction impacts. This type of thinking has led to some changes in the lumber industry. For example, to minimize the use of old-growth timber, sustainably produced lumber or recycled plastic lumber products have been introduced. While this represents an improvement, the greatest ecological savings occur through using less.

In addition to the raw material itself, the material's processing should be considered. Thinking in terms of total environmental costs has led to the analysis of materials' embodied energy. Manufacturing a building material often requires large expenditures of water, fossil fuels for energy and transportation, and human labor. Many environmental experts suggest choosing low embodied energy materials, i.e., materials that need less energy and resources and generates less pollution in its production. Normally, this type of material is closer to its natural state. For example, natural stone has a low embodied energy, while plastics, steel, and aluminum have high embodied energies. Additionally, when available, using materials found on or near the site normally reduces energy expenditures. For example, using local stone in place of brick eliminates not only the manufacturing energies and pollution associated with the manufactured material but also higher transportation costs and transportation-related pollution.

Regarding materials with low embodied energy and from local sources, it is crucial to also consider the net energy used during the useful life of the building. For example, a certain type of insulation may be completely synthetic, requires a large amount of energy to manufacture and must be transported from elsewhere. But the energy savings resulting from its use may exceed the preliminary energy expenditures for its production and transport.

Many building materials outgas, i.e., release harmful, airborne materials that pose a risk to the building's indoor air quality. The VOCs most often found in floor finishes, paints, stains, adhesives, synthetic wallpapers, plywood, and chip-boards should be avoided to maintain a healthy environment. Many new VOC-free product alternatives are now available.

As always, recycling represents an important opportunity for waste reduction. When possible, salvaged building materials should be used. On the other end of the building process, using building materials that can eventually be recycled will also eliminate further resource expenditures. Along similar lines, products and materials need to last; durability increases net energy savings and represents an important pollution prevention option.

21.7 ILLUSTRATIVE EXAMPLES

Four illustrative examples complement the material presented above regarding health and hazard risk-related topics associated the architecture and urban planning.

Architecture and Urban Planning

21.7.1 ILLUSTRATIVE EXAMPLE 1

Describe urban planning in layman's terms and note the risk factors that arise in the planning process.

Solution: Urban planning is the design and regulation of the uses of space that focus on the physical form, economic functions, and social impacts of the urban environment and on the location of different activities within it. Risk factors that arise in the urban planning process have their roots in the disconnects between the planners and community inhabitants. These disconnects can include the following:

1. When planning results in urban sprawl and inefficient use of land and materials rather than preserving community identity and design for the human scale.
2. When planning is dominated by roads and does not support public transportation and is unfriendly for walking and cycling.
3. When planning results in isolation, lack of familiarity with neighbors (which contributes to crime) and many trips out of the neighborhood.
4. When planning results in single use rather than mixed use developments.
5. When planning results in affordability and accessibility issues leading to community segregation and isolation.

21.7.2 ILLUSTRATIVE EXAMPLE 2

If the probability of a high-rise building (>30 stories) in New York City will suffer structural damage in a given year is 0.001, calculate the probability that out of 2,000 high-rise buildings exactly three will suffer structural damage in a year. Solve this problem utilizing a binomial distribution.

Solution: If a binomial distribution (Equation 5.3) applies, one may note that:

$$f(x) = \frac{n!}{x!(n-x)!} p^x q^{n-x}; x = 0, 1, \ldots, n \quad (5.3)$$

$$n = 2,000; p = 0.001; q = 0.999; x = 3.$$

Substituting and solving yields:

$$f(x) = \frac{2000!}{x!(2000-x)!}(0.001)^x (0.999)^{2000-x}; x = 0, 1, 2, 3\ldots, 2000$$

$$f(X = 3) = \frac{2000!}{3!(1997)!}(0.001)^3 (0.999)^{2000-3} = \frac{(2000)(1999)(1998)}{(3)(2)(1)}(1 \times 10^{-9})(0.1356)$$

$$= (1.33 \times 10^9)(1 \times 10^{-9})(0.1356) = 0.1805 = 18\%!$$

21.7.3 ILLUSTRATIVE EXAMPLE 3

Answer the previous problem assuming that a Poisson distribution described the likelihood of building structural failure.

Solution: If a Poisson distribution (Equation 5.24) applies, one may note that:

$$f(x) = \frac{e^{-\mu}\mu^x}{x!}, x = 0, 1, 2, \ldots \quad (5.24)$$

$\mu = (2,000)(0.001) = 2$ building structural failures/yr; $x = 3$.

Substituting and solving yield:

$$P(X = 3) = P(3) = \frac{e^{-2}(2)^3}{3!} = \frac{0.1353(8)}{(3)(2)(1)} = \frac{1.0824}{6} = 0.1804 = 18\%$$

As expected the results are essentially the same and are high due to the large number of high-rise buildings in the city.

21.7.4 ILLUSTRATIVE EXAMPLE 4

The mean height of buildings in a small midwestern town (containing a total of 50 multistory buildings) is 75 ft with a standard deviation of 8 ft. Assuming that the heights of the multistory buildings are normally distributed, estimate how many buildings are between 60 and 80 ft high.

Solution: Note that the value of the standard normal variable for this problem is:

$$Z = \left(\frac{X - 75}{8}\right)$$

The standard normal variables for the range of building heights are as follows:

$$Z_1 = \left(\frac{60 - 75}{8}\right) = -1.875$$

$$Z_2 = \left(\frac{80 - 75}{8}\right) = 0.625$$

Referring to the values in the standard normal table, Table 5.4, the following probability can be calculated:

$$P(-1.875 < Z < 0.625) = (0.5 - 0.0305) + (0.5 - 0.266) = 0.4695 + 0.234 = 0.7035$$

Based on a total of 50 multistory buildings in the town, the total number of buildings between 60 and 80 ft high is:

$$\text{Number of buildings} = 50(0.7035) = 35 \text{ buildings}$$

PROBLEMS

21.1 The probability that a building's boiler will not survive for more than 144 months is 0.99. What is the expected life of this boiler? Use the exponential distribution for the solution to this problem.

21.2 Refer to Illustrative example 2. Calculate the probability that out of the 2,000 high-rise buildings no more than two buildings will suffer structural damage in a year. Solve using the binomial distribution.

21.3 Refer to Problem 2. Solve the problem using the Poisson distribution.

21.4 Refer to Illustrative example 4. Estimate the number of buildings in this small midwestern town that would have a height greater than 90 ft.

REFERENCE

Solow, R.M. 1991. *Sustainability: An Economist's Perspective.* Eighteenth J. Seward Johnson Lecture in Marine Policy, Marine Policy Center. Woods Hole, MA: Woods Hole Oceanographic Institution. June 14.

22 Environmental Considerations

22.1 INTRODUCTION

Since the beginning of time, humans have been coping with balancing requirements to survive and flourish with the environmental impacts of these resource needs. In effect, in trying to provide food, energy, and shelter, humans have had to cope with limiting the impacts of these resource needs with preservation of the environment on which survival depends. Over time and even today, these air, water, and soil impacts are often addressed as separate issues by the technical community.

In the not too distant past, the nation's natural resources were exploited indiscriminately. Waterways served as industrial pollution sinks; skies dispersed smoke from factories and power plants; and the land proved to be a cheap and convenient place to indiscriminately dump industrial and municipal wastes. However, society is now more aware of human generated impacts on the environment and the need to protect it. The American people have been involved in a great social movement known broadly as "environmentalism." Society has come to be more aware of and concerned with the quality of the air one breathes, the water one drinks, and the land on which one lives and works. While economic growth and prosperity are still important goals, opinion polls show overwhelming public support for pollution controls and the pronounced willingness to pay for them.

Plastics discussed earlier are an example of the environmental impacts facing society due to the chemical process industry. Plastics themselves are relatively inert under ordinary conditions and are generally not easily degradable. Consequently, for many years, they were not usually considered an environmental problem, except in terms of the resources and energy used in their production and disposal. The major difficulty with plastics that has been known for many is with the chemicals used in their production and by-products generated when they are burned. The incineration of some plastics produces toxic gases that can cause serious problems if allowed to dissipate freely in the atmosphere. Likewise, the burning of plastics as a result of accidental fires can generate large amounts of smoke that contain these toxic gases. In addition, recent analysis of the fate of plastics in the environment through photochemical and physical degradation has identified the widespread distribution of microplastic particles throughout the environment. These microplastics are now of great concern as they are becoming incorporated within the food chain and because they could have associated with them a wide range of hazardous organic compounds that can adsorb to their surface and be transported large distances with the microplastic particles.

This chapter presents information on pollutants and categorizes their sources by the media they threaten. Topics include ambient and indoor air pollutants, water pollutants, land pollutants, and hazardous and toxic waste materials. Four illustrative examples and four problems are provided to highlight concepts presented throughout the chapter.

22.2 AIR POLLUTANTS

Since the Clean Air Act was passed in 1970, the United States has made impressive strides in improving and protecting air quality. As directed by this Act, the Environmental Protection Agency (EPA) set National Ambient Air Quality Standards (NAAQS) for those pollutants commonly found throughout the country that posed the greatest overall threats to air quality. These pollutants, termed "criteria pollutants" under the Act, include photochemical oxidants (i.e., ozone), carbon monoxide (CO), airborne particulate matter (PM), sulfur oxides, lead, and nitrogen oxides. Although the EPA has made considerable progress in controlling air pollution, all of the six criteria pollutants except lead and nitrogen dioxide are currently a major concern in a number of areas in the country (https://www.epa.gov/air-trends/air-quality-national-summary). The following subsections focus on a number of the most significant air quality challenges: ozone and CO, airborne particulates, sulfur dioxide, acid deposition, hazardous air pollutants (HAPs), and indoor air pollutants.

22.2.1 OZONE AND CARBON MONOXIDE

Ozone is one of the most intractable and widespread environmental problems. Chemically, ozone is a form of oxygen with three oxygen atoms instead of the two found in molecular oxygen. This makes it very reactive, so that it combines with practically every material with which it comes in contact, including human tissue. In the upper atmosphere, where ozone is needed to protect people from ultraviolet radiation, the ozone destruction by man-made chemicals is a concern, but at ground level, ozone can be a harmful pollutant.

Ozone is produced in the atmosphere when sunlight initiates chemical reactions between naturally occurring atmospheric gases and pollutants such as volatile organic compounds (VOCs) and nitrogen oxides. The main source of VOCs and nitrogen oxides is combustion sources such as motor vehicles, electric utilities, and industrial facilities, as well as gasoline vapor and chemical solvent releases.

Ozone can cause the muscles in the airways to constrict, trapping air in the alveoli. This leads to wheezing and shortness of breath. Depending on the level of exposure, ozone can cause human health effects including:

- Coughing and sore or scratchy throat
- Making it more difficult to breathe deeply and vigorously and cause pain when taking a deep breath
- Inflammation and damage the airways
- Making the lungs more susceptible to infection
- Aggravating lung diseases such as asthma, emphysema, and chronic bronchitis
- Increasing the frequency of asthma attacks

Some of these effects have been found even in healthy people, but effects can be more serious in people with lung diseases such as asthma. They may lead to increased school absences, medication use, visits to doctors and emergency rooms, and hospital admissions. Long-term exposure to ozone is linked to aggravation of asthma and is likely to be one of many causes of asthma development. Studies in locations with elevated concentrations also report associations of ozone with increased deaths from respiratory causes.

CO is a colorless, odorless gaseous product of incomplete fuel combustion. As with ozone, motor vehicles are the main contributor to CO formation. Other sources include wood-burning stoves, incinerators, and industrial processes. Breathing air with a high concentration of CO reduces the amount of oxygen that can be transported in the blood stream to critical organs like the heart and brain. At very high levels, which are possible indoors or in other enclosed environments, CO can cause dizziness, confusion, unconsciousness, and death. Very high levels of CO are not likely to occur outdoors. However, when CO levels are elevated outdoors, they can be of particular concern for people with some types of heart disease. These people already have a reduced ability for getting oxygenated blood to their hearts in situations where the heart needs more oxygen than usual. They are especially vulnerable to the effects of CO when exercising or under increased stress. In these situations, short-term exposure to elevated CO may result in reduced oxygen to the heart accompanied by chest pain also known as angina.

22.2.2 Airborne Particulates

PM contains microscopic solids or liquid droplets that are so small that they can be inhaled and cause serious health problems. Some particles less than 10 μm in diameter can get deep into the lungs and some may even get into the bloodstream. Of these, particles less than 2.5 μm in diameter, also known as fine particles or $PM_{2.5}$, pose the greatest risk to health. Fine particles are also the main cause of reduced visibility (haze) in parts of the United States, including many of our treasured national parks and wilderness areas. Particulate pollution regulated by the Clean Air Act includes these two types of PM: PM_{10} and $PM_{2.5}$.

Most fine particles form in the atmosphere as a result of complex reactions of chemicals such as sulfur dioxide, ammonia, and nitrogen oxides, which are pollutants emitted from power plants, industries, agricultural sources, and automobiles. Exposure to these fine, inhalable particles can affect both the lungs and heart. Numerous scientific studies have linked fine particulate pollution exposure to a variety of problems, including:

- Premature death in people with heart or lung disease
- Nonfatal heart attacks
- Irregular heartbeat
- Aggravated asthma
- Decreased lung function
- Increased respiratory symptoms, such as irritation of the airways, coughing or difficulty breathing

People with heart or lung diseases, children, and older adults are the most likely to be affected by particle pollution exposure.

Larger particulates in the air include dust, smoke, metals, and aerosols. Major sources of direct emissions of particulates include steel mills, power plants, cotton gins, cement plants, smelters, and diesel engines. Other sources are grain storage elevators, industrial haul roads, construction work, and demolition. Wood-burning stoves and fireplaces can also be significant sources of particulates. Urban areas are also likely to have windblown dust from roads, parking lots, and construction work ((Burke, Singh and Theodore 2000). Particles of sand and large dust, which are larger than 10 μm, are not regulated by EPA.

22.2.3 Sulfur Dioxide and Acid Rain

The largest source of SO_2 in the atmosphere is the burning of fossil fuels by power plants and other industrial facilities. Smaller sources of SO_2 emissions include industrial processes such as extracting metal from ore; natural sources such as volcanoes; and locomotives, ships, and other vehicles and heavy equipment that burn fuel with a high sulfur content. EPA's national ambient air quality standards for SO_2 are designed to protect against exposure to the entire group of sulfur oxides (SO_x). Emissions that lead to high concentrations of SO_2 generally also lead to the formation of other SO_x. SO_x can react with other compounds in the atmosphere to form small particles. These particles contribute to PM pollution. SO_2 is the component of greatest concern and is used as the indicator for the larger group of gaseous SO_x. Other gaseous SO_x (such as SO_3) are found in the atmosphere at concentrations much lower than SO_2. Control measures that reduce SO_2 can generally be expected to reduce people's exposures to all gaseous SO_x. This may have the important co-benefit of reducing the formation of particulate sulfur pollutants, such as fine sulfate particles.

Short-term exposures to SO_2 can harm the human respiratory system and make breathing difficult as described above for PM pollution. People with asthma, particularly children, are sensitive to these effects of SO_2. At high concentrations, gaseous SO_x can harm trees and plants by damaging foliage and decreasing plant growth. Finally, SO_2 and nitrogen oxides (NO_x) can react with water, oxygen, and other chemicals to form sulfuric and nitric acids. These then mix with water and other materials before falling to the ground as *acid rain*. As with SO_x, the major sources of NO_x are associated with fossil fuel combustion. Winds can blow SO_2 and NO_x over long distances and across borders making acid rain a problem for everyone and not just those who live close to these sources.

Acid rain can occur in two forms: *wet deposition* and *dry deposition*. Wet deposition is the term most commonly associated with acid rain, where sulfuric and nitric acids formed in the atmosphere fall to the ground mixed with rain, snow, fog, or hail. Dry deposition occurs when acidic particles and gases are deposited from the atmosphere in the absence of moisture. The amount of acidity in the atmosphere that deposits to earth through dry deposition depends on the amount of rainfall an area receives.

In addition to the human health effects associated with fine sulfate particulates, when acid rain enters lakes and streams it can cause some to turn acidic, significantly impacting water quality and aquatic habitats. When the accumulated dry deposition is washed off a surface by the next rain, this acidic water flows over and through the ground and can also harm plants and wildlife, such as insects and fish. Finally, this acid rain can have significant impacts on materials when nitric and sulfuric acid land on statues, buildings, and other man-made structures and damage their surfaces through metal corrosion, increased paint and stone deterioration, and material discoloration.

22.2.4 Hazardous Air Pollutants

HAPs, also known as toxic air pollutants or air toxics, are those pollutants that are known or suspected to cause cancer or other serious health effects or are responsible for adverse environmental effects. There are currently 188 regulated HAPs that include certain volatile organic chemicals, pesticides, herbicides, and radionuclides that present tangible hazards, based on scientific studies of exposure to humans and other mammals. Most HAPs originate from human-made sources, including mobile sources (e.g., cars, trucks, buses) and stationary sources (e.g., factories, refineries, power plants), as well as indoor sources (e.g., some building materials and cleaning solvents). Some air toxics are also released from natural sources such as volcanic eruptions and forest fires. Examples of HAPs include benzene, which is found in gasoline; perchloroethylene, which is emitted from some dry cleaning facilities; and methylene chloride, which is used as a solvent and paint stripper by a number of industries. Other examples of listed HAPs include dioxin, asbestos, toluene, and metals such as cadmium, mercury, chromium, and lead compounds.

People are exposed to toxic air pollutants in many ways that can pose health risks, such as by:

- Breathing contaminated air
- Eating contaminated food products, such as fish from contaminated waters; meat, milk, or eggs from animals that fed on contaminated plants; and fruits and vegetables grown in contaminated soil on which air toxics have been deposited
- Drinking water contaminated by toxic air pollutants
- Ingesting contaminated soil
- Making skin contact with contaminated soil, dust, or water (e.g., during recreational use of contaminated water bodies).

Once toxic air pollutants enter the body, some persistent toxic air pollutants accumulate in body tissues. Predators typically accumulate even greater pollutant concentrations than their contaminated prey. As a result, people and other animals at the top of the food chain who eat contaminated fish or meat are exposed to concentrations that are much higher than the concentrations in the water, air, or soil.

People exposed to toxic air pollutants at sufficient concentrations and durations may have an increased chance of getting cancer or experiencing other serious health effects including damage to the immune system, as well as neurological, reproductive (e.g., reduced fertility), developmental, respiratory, and other health problems. Detailed information about the health effects of HAPs is available in separate fact sheets through the EPA's web site (https://www.epa.gov/haps/health-effects-notebook-hazardous-air-pollutants) for nearly every HAP specified in the Clean Air Act Amendments of 1990. In addition to exposure from breathing air toxics, some toxic air pollutants such as mercury can deposit onto soils or surface waters, where they can be taken up by plants and ingested by animals and are eventually magnified up through the food chain. Like humans, animals may experience health problems if exposed to sufficient quantities of HAPs over time.

The Clean Air Act requires the EPA to regulate HAPs from categories of industrial facilities in two phases. The first phase is "technology-based," where EPA develops standards (maximum achievable control technology, MACT standards) for controlling the emissions of air toxics from sources in an industry group or "source category" based on emissions levels that are already being achieved by controlled, low-emitting sources. The second phase is a "risk-based" approach. Here, EPA must determine whether more health-protective standards are necessary within 8 years of setting the MACT standards. Within this time period, and at 8-year intervals after that, EPA must assess the remaining health risks from each source category to determine whether the MACT standards protect public health with an ample margin of safety and protect against adverse environmental effects. The regular 8-year process is used to account for improvements in air pollution controls and/or prevention. The first 8-year review, when combined with the residual risk review, is called the risk and technology review, RTR.

22.2.5 Indoor Air Pollutants

Indoor air quality (IAQ) refers to the air quality within and around buildings and structures, especially as it relates to the health and comfort of building occupants. Health effects from indoor air pollutants may be experienced soon after exposure or, possibly, years later. Indoor pollution sources that release gases or particles into the air are the primary cause of IAQ problems in homes. Inadequate ventilation can increase indoor pollutant levels by not bringing in enough outdoor air to dilute emissions from indoor sources and by not carrying indoor air pollutants out of the home. High temperature and humidity levels can also increase concentrations of some pollutants.

There are many sources of indoor air pollution in any home. These include:

- Combustion sources such as oil, gas, kerosene, coal, and wood
- Tobacco products
- Building materials and furnishings as diverse as deteriorated, asbestos-containing insulation, wet or damp carpet and cabinetry or furniture made of certain pressed wood products
- Products for household cleaning and maintenance, personal care or hobbies
- Central heating and cooling systems and humidification devices
- Outdoor sources such as radon, pesticides, and outdoor air pollution

The relative importance of any single source depends on how much of a given pollutant it emits and how hazardous those emissions are. In some cases, factors such as how old the source is and whether it is properly maintained are significant. For example, an improperly adjusted gas stove can emit significantly more CO than one that is properly adjusted.

Some sources, such as building materials, furnishings and household products like air fresheners, release pollutants more or less continuously. Other sources, related to activities carried out in the home, release pollutants intermittently. These include smoking, the use of unvented or malfunctioning stoves, furnaces, or space heaters, the use of solvents in cleaning and hobby activities, the use of paint strippers in redecorating activities and the use of cleaning products and pesticides in housekeeping. High pollutant concentrations can remain in the air for long periods after some of these activities, particularly if a ventilation system is insufficient to move pollutants out of an indoor space. Indoor air pollutants of special concern are briefly described below.

22.2.5.1 Radon

Radon is a naturally occurring odorless, colorless, and tasteless radioactive gas that is formed in the soil from the radioactive decay of radium-226. It can move indoors through cracks and openings in floors and walls that are in contact with the ground. A secondary source of radon is contaminated well water. Radon is the leading cause of lung cancer among nonsmokers and the second leading cause of lung cancer overall.

It is the emission of high-energy alpha particles during the radon decay process that increases the risk of lung cancer from radon exposure. These alpha emissions penetrate the cells of the epithelium lining the lung and are believed to initiate the process of carcinogenesis. Corrective steps include sealing foundation cracks and holes, and venting radon-laden air from beneath the foundation.

22.2.5.2 Secondhand Smoke

Environmental tobacco smoke (ETS) comes from burning tobacco products and is a major source of indoor air contaminants. The ubiquitous nature of ETS in indoor environments indicates that some unintentional inhalation of ETS by nonsmokers is unavoidable. ETS is a dynamic, complex mixture of more than 4,000 chemicals found in both vapor and particle phases. Many of these chemicals are known toxic or carcinogenic agents. Nonsmoker exposure to ETS-related toxic and carcinogenic substances will occur in indoor spaces where there is smoking.

All the compounds found in "mainstream" smoke, the smoke inhaled by the active smoker, are also found in "sidestream" smoke, the emission from the burning end of the cigarette, cigar, or pipe. ETS consists of both sidestream smoke and exhaled mainstream smoke. Inhalation of ETS is often termed "secondhand smoking," "passive smoking," or "involuntary smoking."

The EPA has classified ETS as a known human (Group A) carcinogen and estimates that it is responsible for approximately 3,000 lung cancer deaths per year among nonsmokers in the United States. The US Surgeon General, the National Research Council, and the National Institute for Occupational Safety and Health also concluded that passive smoking can cause lung cancer in otherwise healthy adults who never smoked.

Children's lungs are even more susceptible to harmful effects from ETS. In infants and young children up to 3 years, exposure to ETS causes an approximate doubling in the incidence of pneumonia, bronchitis, and bronchiolitis. There is also strong evidence of increased middle ear effusion, reduced lung function, and reduced lung growth. Several recent studies link ETS with increased incidence and prevalence of asthma and increased severity of asthmatic symptoms in children of mothers who smoke heavily, as well as increased risks of Sudden Infant Death Syndrome (SIDS). These respiratory illnesses in childhood may very well contribute to the small but significant lung function reductions associated with exposure to ETS in adults.

Airborne PM contained in ETS has been associated with impaired breathing, lung diseases, aggravation of existing respiratory and cardiovascular disease, changes to the body's immune system, and lowered defenses against inhaled particles. Acute cardiovascular effects of ETS include increased heart rate, blood pressure, and blood carboxyhemoglobin and related reduction in exercise capacity in those with stable angina and in healthy people.

Tobacco smoke in combination with radon exposure has a synergistic effect. Smokers and former smokers are believed to be at especially high risk. Scientists estimate that the increased risk of lung cancer to smokers from radon exposure is 10–20 times higher than to people who have never smoked.

While improved general ventilation of indoor spaces may decrease the odor of ETS, health risks cannot be eliminated by generally accepted ventilation methods. The most effective solution is to eliminate all smoking from an individual's environment, either through smoking prohibitions or by restricting smoking to properly designed smoking rooms.

22.2.5.3 Other Combustion Products

Combustion products are gases or particles that come from burning materials. In homes, the major source of combustion products are improperly vented or unvented fuel-burning appliances such as space heaters, woodstoves, gas stoves, water heaters, dryers, and fireplaces. The types and amounts of pollutants produced depend on the type of appliance, how well the appliance is installed, maintained, and vented, and the kind of fuel it uses. Common combustion pollutants include:

- CO which is a colorless, odorless gas that interferes with the delivery of oxygen throughout the body. CO causes headaches, dizziness, weakness, nausea, and even death at high exposure.
- Nitrogen dioxide which is a colorless, odorless gas that causes eye, nose, and throat irritation, shortness of breath, and an increased risk of respiratory infection.

Periodic professional inspection and maintenance of installed equipment such as furnaces, water heaters, and clothes dryers are recommended. Such equipment should be vented directly to the outdoors. Fireplace and wood or coal stove flues should be regularly cleaned and inspected before each heating season. Kitchen exhaust fans should be exhausted to outside air. Vented appliances should be used whenever possible. Individuals potentially exposed to combustion sources should consider installing CO detectors that meet the requirements of Underwriters Laboratory (UL) Standard 2034.

22.2.5.4 Volatile Organic Compounds

VOCs are chemicals found in personal items such as scents and hair sprays; graphics and craft materials including glues, adhesives, and permanent markers; paints and lacquers; paint strippers; cleaning supplies; varnishes and waxes; pesticides; building materials and furnishings; office equipment; moth repellents; air fresheners; and dry-cleaned clothing. VOCs evaporate into the air when these products are used or sometimes even when they are stored.

VOCs irritate the eyes, nose, and throat and cause headaches, nausea, and damage to the liver, kidneys, and central nervous system. Some of them can cause cancer. Two VOCs of particular concern are formaldehyde and those generated from the indoor use of pesticides.

Formaldehyde has been classified as a probable human carcinogen by the EPA. Urea-formaldehyde foam insulation (UFFI), one source of formaldehyde used in home construction until the early 1980s, is now seldom installed, but formaldehyde-based resins are components of finishes, plywood, paneling, fiberboard, and particleboard, all widely employed in mobile and conventional home construction as building materials (subflooring, paneling) and as components of furniture and cabinets, permanent press fabric, draperies, and mattress ticking.

Airborne formaldehyde acts as an irritant to the conjunctiva and upper and lower respiratory tract. Symptoms are temporary, depend upon the level and length of exposure, and may range from burning or tingling sensations in eyes, nose, and throat to chest tightness and wheezing. Acute, severe reactions to formaldehyde vapor – which has a distinctive, pungent odor – may be associated with hypersensitivity.

Pesticides sold for household use, notably impregnated strips, and foggers or "bombs," which are technically classed as semivolatile organic compounds, include a variety of chemicals in various forms. In addition to the active ingredient, pesticides are also made up of ingredients that are used to carry the active agent. These carrier agents are called "inerts" because they are not toxic to the targeted pest; nevertheless, some inerts are capable of causing health problems in exposed humans. Exposure to pesticides may cause harm if they are used improperly; however, exposure to pesticides via inhalation of spray mists may occur during normal use. Exposure can also occur via inhalation of vapors and contaminated dusts after use (particularly to children who may be in close contact with contaminated surfaces). Symptoms may include headache, dizziness, muscular weakness, and nausea. In addition, some pesticide active ingredients and inert components are considered possible human carcinogens, and these materials should only be used according to label directions.

22.2.5.5 Biologicals

Biological contaminants include bacteria, viruses, animal dander and cat saliva, house dust, mites, cockroaches, and pollen. There are many sources of these pollutants. Standing water, water-damaged materials, or wet surfaces also serve as a breeding ground for molds, mildews, bacteria, and insects. House dust mites, the source of one of the most powerful biological allergens, grow in damp, warm environments.

Some biological contaminants trigger allergic reactions, including hypersensitivity pneumonitis, allergic rhinitis, and some types of asthma. Infectious illnesses, such as influenza, measles, and chicken pox, are transmitted through the air. Molds and mildews release disease-causing toxins. Symptoms of health problems caused by biological pollutants include sneezing, watery eyes, coughing, shortness of breath, dizziness, lethargy, fever, and digestive problems. Some diseases, like humidifier fever, are associated with exposure to toxins from microorganisms that can grow in large building ventilation systems. However, these diseases can also be traced to microorganisms that grow in home heating and cooling systems and humidifiers.

Children, elderly people, and people with breathing problems, allergies, and lung diseases are particularly susceptible to disease-causing biological agents in the indoor air. Mold, dust mites, pet dander and pest droppings or body parts can trigger asthma attacks. Biological contaminants, including molds and pollens, can cause allergic reactions for a significant portion of the population. Tuberculosis, measles, staphylococcus infections, *Legionella*, and influenza are known to be transmitted by air.

General good housekeeping, and maintenance of heating and air conditioning equipment, are very important to control biological contaminants in indoor air. Adequate ventilation and good air distribution also help. The key to mold control is moisture control. Maintaining the relative humidity between 30% and 50% will help control mold, dust mites, and cockroaches. One can also employ integrated pest management to control insect and animal allergens. Cooling tower treatment procedures exist to reduce levels of *Legionella* and other organisms in indoor air ventilation systems.

22.3 WATER POLLUTANTS

The EPA, in partnership with state and local governments, is responsible for improving and maintaining water quality. These efforts are organized around three themes. The first is maintaining the quality of drinking water. This is addressed by monitoring and treating drinking water prior to consumption and by minimizing the contamination of the surface water and protecting against contamination of groundwater used for human consumption. The second is reducing the pollution of free-flowing surface waters and protecting their beneficial uses. The third is preventing the degradation and destruction of critical aquatic habitats, including wetlands, nearshore coastal waters, oceans, and lakes. The following is a discussion of various pollutants categorized by these themes.

22.3.1 DRINKING WATER SUPPLIES

The most severe and acute public health effects from contaminated drinking water, the transmission of waterborne diseases, such as cholera and typhoid, have been eliminated in America. However, other hazards remain in the nation's tap water. These hazards are associated with a number of specific contaminants in drinking water as defined by EPA's *National Primary and Secondary Drinking Water Standards* (https://www.epa.gov/ground-water-and-drinking-water/national-primary-drinking-water-regulations) that sets legal limits on over 90 contaminants of concern related to protection of public health and public water distribution systems. Contaminants of special concern to the EPA are lead, arsenic, radionuclides, microbiological contaminants, and disinfection by-products.

22.3.1.1 Lead

Lead can enter drinking water when plumbing materials that contain lead corrode, especially where the water has high acidity or low mineral content that corrodes pipes and fixtures. The most common sources of lead in drinking water are lead pipes, faucets, and fixtures. In homes with lead pipes that connect the home to the water main, also known as lead service lines, these pipes are typically the most significant source of lead in the water. Lead pipes are more likely to be found in older cities and homes built before 1986. Among homes without lead service lines, the most common problem is with brass or chrome-plated brass faucets and plumbing with lead solder.

The Safe Drinking Water Act requires EPA to determine the level of contaminants in drinking water at which no adverse health effects are likely to occur with an adequate margin of safety. These non-enforceable health goals, based solely on possible health risks, are called maximum contaminant level goals (MCLGs). EPA has set the MCGL for lead in drinking water at 0.0 because lead is a toxic metal that can be harmful to human health even at low exposure levels. For most contaminants, EPA sets an enforceable regulation called a maximum contaminant level (MCL) based on the MCLG. MCLs are set as close to the MCLGs as possible, considering cost, benefits, and the ability of public water systems (PWSs) to detect and remove contaminants using suitable treatment technologies. However, because lead contamination of drinking water generally results from corrosion of the plumbing materials belonging to water system customers, EPA established a treatment technique rather than an MCL for lead. A treatment technique is an enforceable procedure or level of technological performance which water systems must follow to ensure control of a contaminant.

The treatment technique regulation for lead (referred to as the Lead and Copper Rule, EPA 2008) requires water systems to control the corrosivity of the water. The regulation also requires systems to collect tap samples from sites served by the system that are more likely to have plumbing materials containing lead. If more than 10% of tap water samples exceed the lead action level of 15 parts per billion (ppb), then water systems are required to take additional actions including taking further steps to optimize their corrosion control treatment (for water systems serving 50,000 people that have not fully optimized their corrosion control); educating the public about lead in drinking water and actions consumers can take to reduce their exposure to lead; and replacing the portions of lead service lines (lines that connect distribution mains to customers) under the water system's control.

Young children, infants, and fetuses are particularly vulnerable to lead because the physical and behavioral effects of lead occur at lower exposure levels in children than in adults. A dose of lead that would have little effect on an adult can have a significant effect on a child. In children, low levels of exposure have been linked to damage to the central and peripheral nervous system, learning disabilities, shorter stature, impaired hearing, and impaired formation and function of blood cells.

Lead is also harmful to adults. Adults exposed to lead can suffer from: cardiovascular effects, increased blood pressure and incidence of hypertension; decreased kidney function; and reproductive problems (in both men and women).

22.3.1.2 Arsenic

Arsenic is a semi-metal element in the periodic table. It is odorless and tasteless. The contamination of a drinking water source by arsenic can result from either natural or human activities. Arsenic is an element that occurs naturally in rocks and soil, water, air, plants, and animals. Volcanic activity, the erosion of rocks and minerals, and forest fires are natural sources that can release arsenic into the environment. Although about 90% of the arsenic used by industry in the United States is currently used for wood preservative purposes, arsenic is also used in paints, drugs, dyes, soaps, metals, and semi-conductors. Agricultural applications, mining, and smelting also contribute to arsenic releases.

Non-cancer effects of arsenic can include thickening and discoloration of the skin, stomach pain, nausea, vomiting, diarrhea, numbness in hands and feet, partial paralysis, and blindness. Arsenic has also been linked to a number of cancers including cancer of the bladder, lungs, skin, kidney, nasal passages, liver, and prostate.

Arsenic is one of the inorganic contaminants regulated under the Phase II/V Chemical Contaminants Rule (https://www.epa.gov/dwreginfo/chemical-contaminant-rules). In 2001, under the Arsenic Rule, EPA adopted a lower standard for arsenic in drinking water. The lower standard of 10 ppb replaced the prior standard of 50 ppb.

22.3.1.3 Radionuclides

Radionuclides are radioactive isotopes that emit radiation as they decay. The most significant radionuclides in drinking water are radium, uranium, and radon, all of which occur naturally in nature. While radium and uranium enter the body by ingestion, radon is usually inhaled after being released into the air during showers, baths, and other activities, such as washing clothes or dishes. Radionuclides in drinking water occur primarily in those systems that use groundwater. Naturally occurring radionuclides seldom are found in surface waters (such as rivers, lakes, and streams).

Ionizing radiation has sufficient energy to affect the atoms in living cells and thereby damage their genetic material (DNA). Fortunately, the cells in the body are extremely efficient at repairing this damage. However, if the damage is not repaired correctly, a cell may die or eventually become cancerous.

Exposure to low levels of radiation does not cause immediate health effects but can cause a small increase in the risk of cancer over a lifetime. There are studies that keep track of groups of people who have been exposed to radiation, including atomic bomb survivors and radiation industry workers. These studies show that radiation exposure increases the chance of getting cancer, and the risk increases as the dose increases: the higher the dose, the greater the risk. Conversely, cancer risk from radiation exposure declines as the dose falls: the lower the dose, the lower the risk.

Radiation doses are commonly expressed in millisieverts (international units) or rem (US units). A dose can be determined from a one-time radiation exposure or from accumulated exposures over time. About 99% of individuals would not get cancer as a result of a one-time uniform whole-body exposure of 100 millisieverts (10 rem) or lower. At this dose, it would be extremely difficult to identify an increase in cancer rates caused by radiation when about 40% of men and women in the United States will be diagnosed with cancer at some point during their lifetime. The EPA sets regulatory limits and recommends emergency response guidelines well below

Environmental Considerations

100 millisieverts (10 rem) to protect the US population, including sensitive groups such as children, from increased cancer risks from accumulated radiation dose over a lifetime.

On December 7, 2000, EPA published the Radionuclides Final Rule (U.S. EPA 2001). The new rule revised the radionuclides regulation, which had been in effect since 1977. The revisions set new monitoring requirements for community water systems (CWS). This ensured customers receive water meeting MCLs for radionuclides in drinking water. The MCLG for radionuclides is set to 0, with the MCL values for regulated radionuclides as follows:

- Beta/photon emitters – 4 mrem/yr
- Gross alpha particles – 15 pCi/L
- Combined radium – 226/228 – 5 pCi/L
- Uranium – 30 µg/L

22.3.1.4 Microbiological Contaminants

Water contains many microbes, i.e., bacteria, viruses, and protozoa, and although most organisms are harmless, others can cause serious disease in the exposed human population. Microbiological contamination continues to be a national concern because contaminated drinking water systems can rapidly spread disease throughout their service population. The Surface Water Treatment Rules (SWTRs) (https://www.epa.gov/dwreginfo/surface-water-treatment-rules) established by EPA are designed to reduce illnesses caused by these pathogens in drinking water. The disease-causing pathogens of most concern include *Legionella*, *Giardia lamblia*, and *Cryptosporidium*.

The SWTRs were developed in phases between 1989 and 2006 and require water systems to filter and disinfect surface water sources. Some water systems are allowed to use disinfection only for surface water sources that meet criteria for water quality and watershed protection. The various phased rules include the following.

- *Surface Water Treatment Rule* – June 1989: Applies to all PWSs using surface water sources or ground water sources under the direct influence of surface water (GWUDI); requires most water systems to filter and disinfect water from surface water sources or GWUDI; established MCLGs for viruses, bacteria and *Giardia lamblia*; and included treatment technique (TT) requirements for filtered and unfiltered systems to protect against adverse health effects of exposure to pathogens.
- *Interim Enhanced Surface Water Treatment Rule* – December 1998: Applies to all PWSs using surface water, or GWUDI, that serve 10,000 or more persons; set an MCLG of 0 for *Cryptosporidium*; set a 2-log *Cryptosporidium* removal requirements for systems that provide filtration; required that watershed protection programs address *Cryptosporidium* for system that are not required to provide filtration; required certain PWSs to meet strengthened filtration requirements; established requirements for covers on new finished water reservoirs; required sanitary surveys, conducted by states, for all surface water systems regardless of size; and requires systems to calculate levels of microbial inactivation to address risk trade-offs with disinfection by-products.

- *Filter Backwash Recycling Rule* – June 2001: Applies to all PWSs using conventional or direct filtration to treat surface water, or GWUDI, regardless of size; required PWSs to review their backwash water recycling practices to ensure that they do not compromise microbial control; required recycled filter backwash water to go through all processes of a system's conventional or direct filtration treatment.
- *Long-Term 1 Enhanced Surface Water Treatment Rule* – January 2002: Set Interim Enhanced Surface Water Treatment Rules for all PWSs using surface water, or GWUDI, serving **fewer** than 10,000 persons.
- *Long-Term 2 Enhanced Surface Water Treatment Rule* – January 2006: Applies to **all** PWSs that use surface water or GWUDI; targets additional *Cryptosporidium* treatment requirements to higher risk systems; requires provisions to reduce risks from uncovered finished water storage facilities; provides provisions to ensure that systems maintain microbial protection as they take steps to reduce the formation of disinfection by-products.

Two additional drinking water regulations of interest are those that apply specifically to groundwater sources and those that apply to aircraft drinking water supplies.

EPA issued the Ground Water Rule (GWR) on November 8, 2006, to improve drinking water quality and reduce disease incidence associated with harmful microorganisms in drinking water by establishing a risk-based approach to target groundwater systems vulnerable to fecal contamination. In many cases, fecal contamination can contain disease-causing pathogens. The GWR applies to PWSs that use groundwater as a source of drinking water. The rule also applies to any system that delivers surface and groundwater to consumers where the groundwater is added to the distribution system without treatment.

The GWR's targeted, risk-based strategy addresses risks through an approach that relies on four major components:

- Routine sanitary surveys of systems that require the evaluation of eight critical elements of a PWS and the identification of significant deficiencies (e.g., a well located near a leaking septic system).
- Source water monitoring for a system that (not treating drinking water to remove 99.99% (4-log) of viruses) identifies a positive sample during regular Total Coliform monitoring or assessment monitoring (at the option of the state) targeted at high-risk systems.
- Corrective action is required for any system with a significant deficiency or source water fecal contamination.
- Compliance monitoring to ensure that treatment technology installed to treat drinking water reliably achieves 99.99% (4-log) inactivation or removal of viruses.

In 2004, EPA found all aircraft PWSs to be out of compliance with the National Primary Drinking Water Regulations (NPDWRs). The existing NPDWRs were designed for traditional, stationary PWSs, not mobile aircraft water systems that are operationally very different. The primary purpose of the Aircraft Drinking Water Rule (ADWR) is to ensure that safe and reliable drinking water is provided to aircraft passengers and crew. Using a collaborative rulemaking process among the EPA (regulates systems that supply water to airports and onboard aircraft), Food and Drug Administration (regulates water used in food and drink preparation and water supply lines for the aircraft), and Federal Aviation Administration (FAA) (FAA oversees airline operation and maintenance programs, including the potable water system), EPA developed ADWR to address aircraft PWSs. The ADWR establishes barriers of protection from disease-causing organisms targeted to the air carrier industry and applies to aircraft with onboard water systems that provide water for human consumption through pipes and regularly serve an average of at least 25 individuals daily, at least 60 days out of the year, and that board only finished water for human consumption.

The ADWR protect against disease-causing microbiological contaminants through requiring the development and implementation of *aircraft water system operations and maintenance plans*. The plans include routine disinfection and flushing of the water system, air carrier training requirements for key personnel, and periodic sampling of the onboard drinking water, as well as self-inspections of each aircraft water system and immediate notification of passengers and crew when violations or specific situations occur.

22.3.1.5 Disinfection By-products

Disinfection by-products (DBPs) can be produced during water treatment by the chemical reactions of disinfectants (used for microbial contaminant reduction) with naturally occurring or synthetic organic materials present in untreated water. These DBPs include trihalomethanes, haloacetic acids (HAA), chlorite, and bromate, and if consumed in excess of EPA's standards (https://www.epa.gov/ground-water-and-drinking-water/national-primary-drinking-water-regulations#Byproducts) over many years may increase health risks.

Since disinfectants are essential to safe drinking water, the EPA developed Stage 1 and Stage 2 Disinfectants and Disinfection Byproducts Rules (DBPRs) (https://nepis.epa.gov/Exe/ZyPDF.cgi?Dockey=P100C8XW.txt) to minimize the risks from by-products generated from the use of these materials. DBPs can form in water when disinfectants used to control microbial pathogens combine with naturally occurring materials found in source water. The DBPRs apply to all Community Water Systems (CWS) and Non-Transient Non-Community Water Systems (NTNCWS), including those serving fewer than 10,000 people, that add/deliver a primary or residual disinfectant, and TNCWs that use chlorine dioxide. The Rules do not apply to water systems that use ultraviolet (UV) light for disinfection.

Results from toxicology studies have shown several DBPs (e.g., bromodichloromethane, bromoform, chloroform, dichloroacetic acid, and bromate) to be carcinogenic in laboratory animals. Other DBPs (e.g., chlorite, bromodichloromethane, and certain haloacetic acids) have also been shown to cause adverse reproductive or developmental effects in laboratory animals. Several epidemiology studies have suggested a weak association between certain cancers (e.g., bladder) or reproductive and developmental effects, and exposure to chlorinated surface water. More than 200 million people consume water that has been disinfected. Because of the large population exposed, health risks associated with DBPs, even if small, need to be taken seriously.

22.3.2 Surface Water Pollutants

Pollutants in waterways come from industries or treatment plants discharging wastewater into streams or from waters running across urban and agricultural areas, carrying surface pollution with them (nonpoint sources, NPSs). The following is a discussion of surface water pollutants categorized by their main sources along with the EPA programs designed to control their impact on surface and groundwater quality.

22.3.2.1 Point Sources

The term point source is defined very broadly to mean any discernible, confined and discrete conveyance, such as a pipe, ditch, channel, tunnel, conduit, discrete fissure, or container. It also includes vessels or other floating craft from which pollutants are or may be discharged. By law, the term "point source" also includes concentrated animal feeding operations (CAFOs), which are places where animals are confined and fed. Also, by law, agricultural stormwater discharges and return flows from irrigated agriculture are not considered point sources.

Raw or insufficiently treated wastewater from municipal and industrial treatment plants (point sources) still threatens water resources in many parts of the country. In addition to harmful nutrients, poorly treated wastewater may contain bacteria and toxic or hazardous chemicals. To address the control of pollutants from point sources into surface water bodies, the Clean Water Act (CWA) established the National Pollutant Discharge Elimination System (NPDES) permit program in 1972. The NPDES permit provides two levels of control: technology-based limits (Federal minimum standards) and state-established water quality-based limits (if technology-based limits are not sufficient to provide protection of the water body). Under the CWA, EPA authorizes the NPDES permit program to state, tribal, and territorial governments, enabling them to perform many of the permitting, administrative, and enforcement aspects of the NPDES program. In states authorized to implement CWA programs, EPA retains oversight responsibilities. Currently 47 states and one territory are authorized to implement the NPDES program at that state level.

An NPDES permit can be considered a license for a facility to discharge a specified amount of a pollutant under certain conditions and at a specified location in order to protect the beneficial use of the receiving water into which the facility discharges. Sludge or biosolids, the residue left from wastewater treatment plants, is a growing problem. Although some sludges are relatively "clean," or free from toxic substances, other sludges may contain organic, inorganic, or toxic pollutants and pathogens. Permits may also authorize facilities to process, incinerate, landfill, or beneficially use sewage sludge under specified conditions, again, to minimize the impact of this sludge management on human health and the environment.

The two basic types of NPDES permits issued are individual and general permits.

- An individual permit which is a specifically tailored to an individual facility. Once a facility submits the appropriate application(s), the permitting authority develops a permit for that particular facility based on the information contained in the permit application (e.g., type of activity, nature of discharge, receiving water quality). The authority issues the permit to the

facility for a specific time period (not to exceed 5 years) with a requirement that the facility reapply for a new permit prior to the expiration date.
- A general permit which covers a group of dischargers with similar qualities within a given geographical location. General permits may offer a cost-effective option for permitting agencies because of the large number of facilities that can be covered under a single permit.

EPA has two general components of the NPDES program, one for municipal wastewater treatment plants, i.e., Publicly Owned Treatment Works (POTWs), and one for industrial wastewater treatment plants. While POTWs have similar characteristics across the country if primarily treating residential and commercial wastewater, characteristics of some commercial and particularly industrial wastewaters can vary widely depending on the nature of the facility generating the wastewater. Wastewater discharges from these commercial and industrial sources may contain pollutants at levels that could affect the quality of receiving waters or interfere with POTWs that receive those discharges. The CWA has established specific effluent limitations for more than 50 different categories of industrial and commercial activities (https://www.epa.gov/eg/learn-about-effluent-guidelines) and developed the National Pretreatment Program (https://www.epa.gov/npdes/national-pretreatment-program) to protect POTWs' infrastructure and reduce conventional and toxic pollutant levels discharged by industries and other nondomestic wastewater sources into municipal sewer systems and into the environment.

Animal feeding operations (AFOs) are a category of point sources that are agricultural operations where animals are kept and raised in confined spaces. An AFO is a lot or facility (other than an aquatic animal production facility) where the following conditions are met:

- Animals have been, are, or will be stabled or confined and fed or maintained for a total of 45 days or more in any 12-month period.
- Where crops, vegetation, forage growth, or post-harvest residues are not sustained in the normal growing season over any portion of the lot or facility.

AFOs that meet the regulatory definition of a CAFO are regulated under the NPDES permitting program. Manure and wastewater from AFOs have the potential to contribute pollutants such as nitrogen and phosphorus, organic matter, sediments, pathogens, hormones, and antibiotics to the environment. The federal CAFO program is designed to support and complement an array of voluntary and regulatory programs administered by USDA, EPA, and states. The CAFO regulations are an integral part of an overall federal strategy to support a vibrant agricultural economy while simultaneously ensuring that all AFOs manage their manure and develop effective nutrient management plans that are protective of the environment. These nutrient management plans include the following best management practices for controlling pollutant releases from AFOs (US EPA 2012):

- Provisions for adequate manure, wastewater, and stormwater storage
- Provisions for adequate management of animal mortality and disposal
- Provisions for diversion of clean water from AFO facilities

- Prevention of direct animal contact with waters of the United States
- Best practices in handling, storage, and disposal of agrochemicals
- Site-specific conservation practices (i.e., crop rotation, buffer strips, irrigation water management)

22.3.2.2 Nonpoint Sources

The term NPS is defined to mean any source of water pollution that does not meet the legal definition of point source as given above. NPS pollution generally results from surface runoff, precipitation, atmospheric deposition, drainage, seepage or hydrologic modification. NPS pollution, unlike pollution from industrial and sewage treatment plants, comes from many diffuse sources and is caused by rainfall or snowmelt moving over and through the ground, carrying with it a variety of natural and man-made water pollutants that can be deposited into lakes, rivers, wetlands, coastal waters, and groundwater.

NPS pollution can include:

- Excess fertilizers, herbicides, and insecticides from agricultural lands and residential areas
- Oil, grease, and toxic chemicals from urban runoff and energy production
- Sediment from improperly managed construction sites, crop and forest lands, and eroding streambanks
- Salt from irrigation practices and acid drainage from abandoned mines
- Bacteria and nutrients from livestock, pet wastes, and faulty septic systems
- Atmospheric deposition and channelization and channel modification, dams, and streambank and shoreline erosion

NPS pollution presents continuing problems for achieving national water quality in many parts of the country. Sediment and nutrients are the two largest contributors to NPS problems. NPS pollution is also a major source of pesticide runoff from agricultural areas, metals from active or abandoned mines, gasoline, and asbestos from urban areas. In addition, the atmosphere is an NPS of toxics since many toxics can attach themselves to dust, later to be deposited in surface waters hundreds of miles away through precipitation. The effects of NPS pollutants on specific waters vary and may not always be fully assessed. However, it is known that pollutants in NPS pollution have harmful effects on drinking water supplies, recreation, fisheries, and wildlife.

To address these water quality problems, federal, state, tribal, territorial, and local governments provide technical assistance and funding programs to implement NPS controls. A number of federal programs exist to support efforts at the local level for NPS pollution control that include:

- US EPA's Nonpoint Source Pollution Management Program which implements Section 319 of the CWA
- Other US EPA programs implemented through the CWA to support NPS pollution control such as Section 104(b)(3) Water Quality Cooperative Agreements; Section 104(g) Small Community Outreach program; Section 106 Grants for Pollution Control program; Section 314 Clean Lakes program; and Section 320 National Esturary program

Environmental Considerations

- USDA incentive-based conservation programs through the Consolidated Farm Services Agency, the Natural Resources Conservation Service, and the US Forest Service
- Federal Highway Administration's erosion control guidelines for federally funded transportation-related construction projects
- US Department of Interior's technical assistance and financial support programs for NPS pollution control through the Bureau of Reclamation, Bureau of Land Management, and the Fish and Wildlife Service

The primary federal NPS pollution control program is the Section 319 program implemented by the US EPA titled the Nonpoint Source Pollution Management Program. This program provides states, territories, and tribes with grants to implement NPS pollution controls described in approved NPS pollution management programs. All states, territories, and some tribes must meet two basic requirements to be eligible for a Section 319 grant, the first of which is to develop and gain EPA approval of an NPS pollution assessment report. In the assessment report, the state, territory, or tribe identifies waters impacted or threatened by NPS pollution. They also describe the categories of NPS pollution, such as agriculture, urban runoff, or forestry, that are causing water quality problems in their jurisdiction. To meet the second requirement a state, territory, or tribe must develop and obtain EPA approval of an NPS pollution management program. This program becomes the framework for controlling NPS pollution in their jurisdiction, given the existing and potential water quality problems described in the NPS pollution assessment report. A well-developed management program supports activities with the greatest potential to produce early, demonstrable water quality improvement results; assists in the building of long-term institutional capacity to address NPS pollution problems; and encourages strong interagency coordination and ample opportunity for public involvement in the NPS pollution control strategy decision-making process.

22.4 SOLID WASTE

The Resource Conservation and Recovery Act (RCRA) gives the US EPA the authority to control non-hazardous municipal, commercial and industrial waste, as well as hazardous waste from the "cradle-to-grave." This includes the generation, transportation, treatment, storage, and disposal of hazardous waste. To achieve this, EPA has developed regulations, guidance, and policies that ensure the safe management and cleanup of solid and hazardous waste, and programs that encourage source reduction and beneficial reuse. RCRA establishes the framework for a national system of solid waste control. Subtitle D of the Act is dedicated to non-hazardous solid waste requirements, while Subtitle C focuses on hazardous solid waste.

22.4.1 Non-Hazardous Waste

Non-hazardous solid waste is regulated under Subtitle D of RCRA. Regulations established under Subtitle D ban open dumping of waste and set minimum federal criteria for the operation of municipal waste and industrial waste landfills, including design criteria, location restrictions, financial assurance, corrective action (cleanup), and closure requirement. States play a lead role in implementing these regulations and must meet minimum federal standards but may set more stringent requirements if they are so inclined. In the absence of an approved state program, the federal requirements must be met by waste facilities and the EPA administers the program in these "non-primacy" states.

RCRA's Subtitle D program established the framework for states to implement effective municipal solid waste (MSW) and non-hazardous secondary material management programs to prevent contamination from adversely impacting communities and to prevent future hazardous waste Superfund sites. A total of 18 million acres of contaminated lands have been restored for productive reuse through the RCRA Corrective Action program, and various partnerships and award programs have been created to encourage companies to modify manufacturing practices to generate less waste and reuse materials safely. A major focus of the RCRA Subtitle D program has been to change the public's perceptions of wastes to vision it as valuable commodities that can be part of new products through sustainable materials management efforts and recycling. Through this effort, the nation's recycling infrastructure has been strengthened and the MSW recycling/composting rate has increased from less than 10% in 1980 to more than 32% as of 2018, the most recent year for which national data are available.

22.4.2 Hazardous Waste

Hazardous waste is regulated under Subtitle C of RCRA. EPA has developed a comprehensive program to ensure that hazardous waste is managed safely from the moment it is generated to its final disposal (cradle-to-grave). Under Subtitle C, EPA may authorize states to implement key provisions of hazardous waste requirements in lieu of the federal government. If a state program does not exist, EPA directly implements the hazardous waste requirements in that state. Subtitle C regulations set criteria for hazardous waste generators, transporters, and treatment, storage, and disposal facilities. This includes permitting requirements, enforcement and corrective action or cleanup.

Simply defined, a hazardous waste is a waste with properties that make it dangerous or capable of having a harmful effect on human health or the environment. Hazardous waste is generated from many sources, ranging from industrial manufacturing process wastes to batteries and may come in many forms, including liquids, solids gases, and sludges. In order for a material to be classified as a hazardous waste, it must first be a solid waste (any material that is discarded and in the form of refuse, sludge from a wastewater treatment plant, water supply treatment plant, or air pollution control facility and other discarded material, resulting from industrial, commercial, mining, and agricultural operations, and from community activities). Once a generator determines that their waste meets the definition of a solid waste, they investigate whether or not the waste is a listed or characteristic hazardous waste and whether it fits into EPA's hazardous waste management (Subtitle C) regulatory program.

EPA established a comprehensive regulatory program to ensure that hazardous waste is managed safely from "cradle to grave" meaning from the time it is created, while it is transported, treated, and stored, and until it is disposed. Under RCRA, hazardous waste generators are the first link in the hazardous waste management

Environmental Considerations

system. All generators must determine if their waste is hazardous and must oversee the ultimate fate of the waste. Furthermore, generators must ensure and fully document that the hazardous waste that they produce is properly identified, managed, and treated prior to recycling or disposal through a hazardous waste manifest system. The degree of regulation that applies to each generator depends on the amount of waste that a generator produces (US EPA 2022).

After generators produce a hazardous waste, transporters may move the waste to a facility that can recycle, treat, store, or dispose of the waste. Since such transporters are moving regulated wastes on public roads, highways, rails, and waterways, United States Department of Transportation hazardous materials regulations, as well as EPA's hazardous waste regulations, apply.

To the extent possible, EPA tried to develop hazardous waste regulations that balance the conservation of resources, while ensuring the protection of human health and environment. Many hazardous wastes can be recycled safely and effectively, while other wastes will be treated and disposed of in landfills or incinerators. Recycling hazardous waste has a variety of benefits including reducing the consumption of raw materials and the volume of waste materials that must be treated and disposed. However, improper storage of those materials might cause spills, leaks, fires, and contamination of soil and drinking water. To encourage hazardous waste recycling while protecting health and the environment, EPA developed regulations to ensure recycling would be performed in a safe manner.

Treatment Storage and Disposal Facilities (TSDFs) provide temporary storage and final treatment or disposal for hazardous wastes. Since they manage large volumes of waste and conduct activities that may present a higher degree of risk, TSDFs are stringently regulated. The TSDF requirements establish generic facility management standards, specific provisions governing hazardous waste management units and additional precautions designed to protect soil, groundwater, and air resources.

22.5 TOXIC SUBSTANCES

Under a broad range of federal statutes, EPA gathers health, safety, and exposure data; requires necessary testing; and controls human and environmental exposures for numerous chemical substances and mixtures. EPA regulates the production and distribution of commercial and industrial chemicals to ensure that chemicals for sale and use in the United States do not harm human health or the environment. The Toxic Substances Control Act (TSCA) addresses the manufacturing, processing, distribution, use, and disposal of commercial and industrial chemicals. The Pollution Prevention Act (PPA) establishes pollution prevention as the national policy for controlling industrial pollution at its source, and The Department of Transportation's Pipeline and Hazardous Materials Safety Administration regulates the transport of hazardous materials used in commercial and industrial settings.

Under TSCA, EPA's Compliance Monitoring Strategy for the TSCA (US EPA 2016) provides guidance to EPA and authorized states with respect to administering and implementing the Agency's national compliance program for New and Existing Chemicals, also known as "Core TSCA;" Polychlorinated Biphenyls (PCBs); Asbestos, including the Asbestos Hazard Emergency Response Act (AHERA),

Worker Protection Rule (WPR), and Model Accreditation Program (MAP); and Lead-based Paint (LBP).

EPA enforces the proper management of hazardous materials in the manufacturing and industrial process sectors via requirements under the Emergency Planning and Community Right to Know Act (EPCRA). EPCRA is used to ensure that facilities are prepared for chemical emergencies and report any releases of hazardous and toxic chemicals to both EPA and surrounding communities. EPA and the states verify EPCRA compliance through a comprehensive EPCRA compliance monitoring program which includes inspecting facilities, reviewing records, and taking enforcement action where necessary. The EPCRA compliance assistance program provides businesses, federal facilities, local governments, and tribes with tools to help meet environmental regulatory requirements. EPCRA requires that citizens be informed of toxic chemical releases in their area. Industrial facilities must annually report releases and transfers of certain toxic chemicals. This information is publicly available in the Toxics Release Inventory (TRI) database (https://www.epa.gov/toxics-release-inventory-tri-program/tri-data-and-tools).

22.5.1 New and Existing Chemicals Program

Section 8(b) of TSCA requires EPA to compile, keep current and publish a list of each chemical substance that is manufactured or processed, including imports, in the United States for uses under TSCA. TSCA defines a "chemical substance" as any organic or inorganic substance of a particular molecular identity, including any combination of these substances occurring in whole or in part as a result of a chemical reaction or occurring in nature, and any element or uncombined radical. Also called the "TSCA Inventory" or simply "the Inventory," this database of commercial chemicals plays a central role in the regulation of most industrial chemicals in the United States.

The initial reporting period by manufacturers, processors, and importers was January to May of 1978 for chemical substances that had been in commerce since January of 1975. The Inventory was initially published in 1979, and a second version, containing about 62,000 chemical substances, was published in 1982. The TSCA Inventory has continued to grow since then, and now lists more than 86,000 chemicals.

The Chemical Data Reporting (CDR) rule, under TSCA, requires manufacturers (including importers) to provide EPA with information on the production and use of chemicals in commerce. Under the CDR rule, EPA collects basic exposure-related information including information on the types, quantities, and uses of chemical substances produced domestically and imported into the United States. The CDR database constitutes the most comprehensive source of basic screening-level, exposure-related information on chemicals available to EPA and is used by the Agency to protect the public from potential chemical risks.

Chemicals substances in the Inventory include:

- Organics
- Inorganics
- Polymers
- Chemical substances of unknown or variable composition, complex reaction products, and biological materials (UVCBs)

Environmental Considerations

Chemical substances not in the Inventory are those with uses not regulated under TSCA. The use of these chemical substances is governed by other US statutes on, for example:

- Pesticides
- Foods and food additives
- Drugs
- Cosmetics
- Tobacco and tobacco products
- Nuclear materials
- Munitions

For purposes of regulation under TSCA, if a chemical is on the Inventory, the substance is considered an "existing" chemical substance in US commerce. Any chemical that is not on the Inventory is considered a "new chemical substance." In addition to defining whether a specific substance is "new" or "existing," the Inventory also contains "flags" for those existing chemical substances that are subject to manufacturing or use restrictions.

Determining if a chemical is on the Inventory is a critical step before beginning to manufacture (which includes importing) a chemical substance. Mandated by Section 5 of TSCA, EPA's New Chemicals program helps manage the potential risk to human health and the environment from chemicals new to the marketplace. The program functions as a "gatekeeper" that can identify conditions, up to and including a ban on production, to be placed on the use of a new chemical before it is entered into commerce. The New Chemicals program requires anyone who plans to manufacture a new chemical substance for a non-exempt commercial purpose to provide EPA with a Premanufacture Notice (PMN) at least 90 days before initiating the activity. After PMN review has been completed, the company that submitted the PMN must provide a Notice of Commencement of Manufacture or Import (NOC, EPA Form 7710-56) to EPA within 30 calendar days of the date the substance is first manufactured or imported for nonexempt commercial purposes.

Once a complete NOC is received by EPA, the reported substance is considered to be on the Inventory and becomes an "existing chemical." The Agency receives approximately 400 NOCs each year, thus the TSCA Inventory changes almost daily.

Substances reported through exemption submissions and exempt uses that are not subject to reporting do not require an NOC and are not added to the Inventory. Examples include:

- Low Volume Exemptions (LVEs)
- Low Release/Low Exposure Exemptions (LoREXs)
- Test Market Exemptions (TMEs)
- Substances used for research and development
- Polymers that meet the 1995 Polymer Exemption Rule Amendments

The CDR information is collected every 4 years from manufacturers when production volumes for a chemical are 25,000 lb or greater for a specific reporting year.

Collecting the information every 4 years assures that EPA and (for non-confidential data) the public have access to up-to-date information on chemicals.

As part of EPA's commitment to strengthen the management of chemicals and increase information on chemicals, the Agency provides free access to the inventory online (https://www.epa.gov/tsca-inventory/how-access-tsca-inventory).

22.5.2 Polychlorinated Biphenyls Program

PCBs are regulated under TSCA Section 6(e) and related regulations found at 40 CFR Part 761. In 1979, PCBs were banned from manufacture in the United States. However, some products and equipment that used PCBs were allowed to continue to use them, such as electrical transformers, coatings, and pigments. Compliance activities monitor:

- Manufacture (including import), processing, distribution in commerce, and use of PCBs
- Storage or disposal of waste PCBs and PCB items (e.g., articles, containers, equipment), including the proper management of PCBs through prescribed or approved handling, marking, and storage and disposal methods; and cleanup of PCB spills

Compliance monitoring involves reviewing compliance with 40 CFR Part 761 by operations that have or use equipment or other items containing PCBs, such as transformers, capacitors, voltage regulators, hydraulic systems, small capacitors in fluorescent light ballasts, and caulking compounds.

22.5.3 Asbestos Program

The Title II Asbestos Hazard Emergency Response Act (AHERA) program (also called the Asbestos in Schools Program) governs the management of asbestos in kindergarten through Grade 12 schools. The objective of AHERA compliance monitoring is to ensure regulatory compliance and, thereby, minimize the risk of exposure to asbestos in schools.

The Section 6 Worker Protection Rule protects state and local government employees who are not protected by the federal OSHA asbestos standard. The Model Accreditation Program supports the AHERA requirement of training for asbestos abatement professionals.

Compliance monitoring involves reviewing local education agencies' management of asbestos under AHERA; state and local government employers' compliance with the federal OSHA asbestos standard under the WPR; and the adequacy of state asbestos accreditation programs under the MAP.

22.5.4 Lead-Based Paint Program

EPA monitors compliance with three major Lead-based Paint Program regulations under TSCA Subchapter IV and Residential Lead-Based Paint Hazard Reduction

Environmental Considerations

Act of 1992 (enacted as Title X of the Housing and Community Development Act of 1992).

The Lead-based Paint Real Estate Notification and Disclosure Rule promulgated under Section 1018 of the Housing and Community Development Act of 1992 requires sellers and lessors of pre-1978 housing to provide purchasers and lessees with a lead hazard information pamphlet and any lead hazard evaluation reports available to the seller or lessor. Receipt must be acknowledged. The Department of Housing and Urban Development shares compliance responsibilities with EPA for the Lead Disclosure Rule.

The Lead-Based Paint Activities, Certification, and Training Rule (Abatement Rule) requires individuals and firms performing abatements to be trained and certified by accredited training providers; to give notice to EPA prior to the abatement work; and to follow work practice standards. A lead abatement is intended to permanently eliminate lead-based paint.

The Renovation, Repair, and Painting Rule (RRP Rule) requires firms and workers performing renovations to be trained and certified by accredited training providers and to follow work practice standards. In addition, prior to starting a renovation the firm must provide a lead hazard information pamphlet to the owner and occupant of pre-1978 housing or childcare facilities, and to parents and guardians of children under age 6 that attend a childcare facility.

Compliance monitoring involves reviewing sellers', landlords', and property managers' compliance with lead disclosure requirements; reviewing compliance with the requirements of the Abatement Rule by training providers and by firms and individuals performing abatements; and reviewing compliance with the requirements of the RRP Rule by training providers and by firms and individuals performing renovations.

22.5.5 FORMALDEHYDE

The Formaldehyde Standards for Composite Wood Products Act added TSCA Subchapter VI to reduce emissions of formaldehyde from composite wood products by establishing formaldehyde emissions limits for domestic or imported hardwood plywood, particleboard, and medium-density fiberboard sold, supplied, offered for sale, or manufactured in the United States, whether in the form of an unfinished panel or incorporated into a finished good. EPA is in the process of developing regulations which include a third-party certification (TPC) component and implement statutory emission standards.

22.6 ILLUSTRATIVE EXAMPLES

Four illustrative examples complement the material presented above regarding environmental considerations related to health and hazard risk assessment.

22.6.1 ILLUSTRATIVE EXAMPLE 1

An electronic environmental monitoring system in a refinery consists of three components (A, B, and C) connected in parallel. If the time to failure for each component is exponentially distributed with mean times of failures for components A, B, and C of 600, 300, and 200 weeks, respectively, determine the probability the monitoring system will fail before 365 weeks.

Solution: For this series system, the probability of failure is (see Chapter 5):

$$P(F) = P(\text{all components fail}) = (1 - P_A)(1 - P_B)(1 - P_C)$$

where $P(F)$ is the probability of surviving 365 weeks. The system reliability is:

$$R = 1 - P(F) = 1 - (1 - P_A)(1 - P_B)(1 - P_C)$$

Since the time to failure for each component is exponentially distributed,

$$P_A = e^{-\frac{365}{600}} = 0.544$$

$$P_B = e^{-\frac{365}{300}} = 0.296$$

$$P_C = e^{-\frac{365}{200}} = 0.161$$

Therefore, the system reliability is:

$$R = 1 - (1 - 0.544)(1 - 0.296)(1 - 0.161) = 1 - (0.456)(0.704)(0.839) = 1 - 0.269 = 0.731$$

$$R = 73.1\%$$

The probability the system will fail within that same period is simply:

$$P(F) = 1 - R(\text{fractional basis}) = 1 - 0.731 = 0.269 = 26.9\%$$

22.6.2 ILLUSTRATIVE EXAMPLE 2

The probability that a fan in a refinery controlling toxic emissions within a process building will not survive for more than 5 years is 0.90. How often should the fan be replaced? Assume that the time to failure is exponentially distributed and that the replacement time should be based on the fan's expected life.

Solution: This requires the calculation of μ in the exponential model with units of 1/years, i.e.,

$$F(t) = 1 - e^{-\lambda t} \tag{5.36}$$

Based on the information provided,

$$P(t \leq 5) = 0.90$$

or

$$0.90 = 1 - e^{-\lambda(5)}$$

Solving for λ yields:

$$1 - 0.90 = e^{-\lambda(5)}; \ln(0.10) = -5\lambda; \lambda = -(-2.303)/5 = 0.4605$$

Therefore, the expected time (or life), $E(t)$, is

$$E(t) = \frac{1}{\lambda} = \frac{1}{0.4605} = 2.17 \text{ years}$$

Therefore, the fans should be replaced in approximately 2 years to avoid fan failure and development of a potentially hazardous atmosphere within the process building.

22.6.3 ILLUSTRATIVE EXAMPLE 3

The number of environmental control patents issued per month has a Poisson distribution with a mean of 4. Find the probability that at most two patents will be issued in a given month.

Solution: The probability of at most two patents issued in a given month is expressed as:

$$P(X \leq 2) = P(X = 0) + P(X = 1) + P(X = 2)$$

Using the equation for the Poisson distribution (5.24) and solving yield:

$$f(x) = \frac{e^{-\mu} \mu^x}{x!}, x = 0, 1, 2, \ldots \quad (5.24)$$

$$P(X \leq 2) = e^{-4}\left[\left(4^0/0!\right) + \left(4^1/1!\right) + \left(4^2/2!\right)\right] = (0.0183)[1 + 4 + 8] = 0.238 = 23.8\%$$

Thus, the probability that at most two environmental control patents will be issued in a given month is approximately 24%.

22.6.4 ILLUSTRATIVE EXAMPLE 4

A sample of 27 ambient hazardous particulate readings at an industrial site has a mean concentration of 92 µg/m³ and a standard deviation of 14 µg/m³. Assuming that the measurements are approximately normally distributed, obtain the 98% confidence limits for the true mean concentration.

Solution: Based on the problem statement, $\mu = 92$, $\sigma = 14$, and $n = 27$. Assume that $s \approx \sigma/\sqrt{n}$ since the readings are approximately normally distributed. Using Equation 5.35,

$$P(T_1 < T < T_2) = P\left(\frac{T_1 - \mu}{\sigma} < \frac{T - \mu}{\sigma} < \frac{T_2 - \mu}{\sigma}\right) \quad (5.35)$$

Substituting knowing that for a two-tailed test with Z normally distributed, using the standard normal cumulative probability values from Table 5.4, for a probability of 0.01 on each tail, T_1 and T_2 would both be 2.33. Therefore,

$$P\left(-2.33 < \frac{T-\mu}{\sigma/n} < 2.33\right) = 0.98; P\left(-2.33 < \frac{T-92}{14/\sqrt{27}} < 2.33\right) = 0.98$$

$$P\left(-2.33 < \frac{T-92}{2.694} < 2.33\right) = 0.98$$

Solving yields:

$$2.33 = \frac{T-92}{2.694}; (2.33)(2.698) = T - 92; T = 92 + 6.29$$

$$-2.33 = \frac{T-92}{2.694}; (-2.33)(2.698) = T - 92; T = 92 - 6.29$$

Finally, the 98% confidence interval can be described as ±6.29 and the mean concentration can be expressed as:

$$\mu = 92 \pm 6.29 \, \mu g/m^3$$

PROBLEMS

22.1 The number of hazardous waste trucks arriving daily at a certain hazardous waste incineration facility has a Poisson distribution with parameter n. Present facilities can accommodate three trucks a day. If more than three trucks arrive in a day, the trucks in excess of three must be sent elsewhere. How much must the present waste facilities be increased to permit handling of all trucks for 95% of the days?

22.2 The time between outages for an electrostatic precipitator controlling emissions of hazardous particulates generated from a chemical process is exponentially distributed with an estimated mean time between outages of 50 months. Calculate the probability that the precipitator will survive 20 months before an outage and a hazardous particulate release.

22.3 Over the last 10 years, a local hospital reported that the number of deaths per year due to excess $PM_{2.5}$ air pollution due to temperature inversions was 0.5. What is the probability of exactly three deaths in a given year? Assume the death rate due to $PM_{2.5}$ pollution can be described by a Poisson distribution. Also calculate the annual probability of three or more deaths being attributed to $PM_{2.5}$ pollution due to temperature inversions.

22.4 A sample of 82 PCB concentrations at a Superfund site has a mean of 650 ppm and a standard deviation of 42 ppm. Assuming that the measurements are approximately normally distributed, obtain the 98% confidence limits for the mean PCB concentration at this site.

REFERENCES

Burke, G., Singh, B., and Theodore, L. 2000. *Handbook of Environmental Management and Technology*, 2nd Edition. Hoboken, NJ: John Wiley and Sons.

US Environmental Protection Agency 2001. *Radionuclides Rule: A Quick Reference Guide.* EPA 816-F-01-003. Washington, DC: Office of Water, US EPA. http://nepis.epa.gov/Exe/ZyPDF.cgi?Dockey=30006644.txt.

US Environmental Protection Agency 2008. *Lead and Copper Rule: A Quick Reference Guide.* EPA 816-F-08-018. Washington, DC: Office of Water, US EPA. http://nepis.epa.gov/Exe/ZyPDF.cgi?Dockey=60001N8P.txt.

US Environmental Protection Agency 2012. *NPDES Permit Writers' Manual for Concentrated Animal Feeding Operations.* EPA 833-F-12-001. Washington, DC: Office of Water, US EPA. https://www.epa.gov/sites/default/files/2015-10/documents/cafo_permitmanual_entire.pdf.

US Environmental Protection Agency 2016. *Compliance Monitoring Strategy for the Toxic Substances Control Act (TSCA).* Washington, DC: Office of Enforcement and Compliance Assurance, US EPA. https://www.epa.gov/sites/default/files/2014-01/documents/tsca-cms.pdf.

US Environmental Protection Agency 2022. *Learn the Basics of Hazardous Waste.* Washington, DC: US EPA Office of Resource Conservation and Recovery. https://www.epa.gov/hw/learn-basics-hazardous-waste (accessed April 11, 2022).

Index

A

Absorbate 102
Absorbent 102, 109, 303
Absorber 102, 167–168
Absorption 26–27, 102, 107, 114, 117, 135, 138, 142, 172, 227, 263
Accidents 4–10, 14, 22–23, 28, 30–32, 36, 40–41, 107–108, 110, 115, 119, 129, 134, 165, 167, 174–175, 201–202, 204–205, 208–210, 212, 214, 216, 220–221, 261, 294, 306–307, 314, 316–317, 319, 327, 330
Accidental release 32, 212, 216, 218
Accuracy 16, 20, 33–34, 37, 104, 132, 212, 343
Acute 3, 10, 26–27, 32, 92, 102, 305–306, 368–370
Acute (risk) 3, 102
Acute toxicity 26
Adiabatic 102, 111
Adsorbent 102, 141, 274, 301, 304
Adsorber 102
Adsorption 27, 52, 102, 135, 141–142, 172, 278
Adverse effects 6, 19, 22, 24, 27, 33, 35, 38, 291, 306, 308
Adverse health effect 21, 25–28, 102, 110, 119, 205, 371, 373
Agriculture 286, 322, 341–342, 376, 379
AICHE 32, 130
Air toxics 365–366
Aircraft 337, 343, 374–375
Airports 353, 375
Alarms 154, 212–213, 215, 264, 329
Algebraic 179–180, 188
Algorithm 30, 112, 307, 340
Ammonium nitrate 218, 225–226, 228–229, 323–324
Analysis of variance (ANOVA) 56
Animals 5, 25–26, 28, 35, 112, 230, 273, 291, 351, 365–366, 372, 375–377
Arithmetic 49–50
Asbestos 39, 144, 298, 307, 352, 365, 378, 381, 384
Asthma 362–364, 368–370
Atmospheric dispersion 26, 38, 103, 106, 209, 314, 320, 337, 339–341, 347. *See also* Gaussian model; Stability class
Autoignition temperature 3, 103, 109
Automobile 10, 157, 172–173, 250, 363
Average annual individual risk (AAIR) 314, 317–318
Average annual risk (AAR) 317–318
Average individual risk 15

B

Bacterium 277, 287–289, 355, 369, 373, 376, 378
Baghouse 98
Barometer 243, 336, 342
Basic event 103, 165
Bathtub curve 76–77. *See also* Weibull distribution
Benzene 238–239, 247, 251, 302, 310, 365
Bhopal 202, 216, 327
Binomial distribution 64–65, 69–70, 72, 93–94, 244, 257, 268, 283, 294, 311, 331–333, 357, 359
Bioterrorism 326, 328, 333
Biphenyls 33, 381, 384
Blower 103, 109, 111, 146–147
Boiler 107, 118, 138, 149–150, 167, 230, 264–265, 279–280, 358
Buoyancy 249, 338–339
Butadiene 252–253
Butane 184, 247, 250–251
Butylene 250–251

C

Cadmium 256, 302, 310, 365
Cancer 3, 18–19, 21, 25, 33, 41, 61, 104, 128, 275, 302, 304, 310, 354, 365, 367–369, 372–373, 375. *See also* Carcinogenesis; Carcinogens
Carboxyhemoglobin 368
Carcinogenesis 367. *See also* Cancer
Carcinogens 350, 369. *See also* Cancer
Catastrophic 9–10, 32, 115, 123, 345
Cause-consequence analysis 3
Centrifugal force 104, 147
Centrifuges 274, 277–278
Checklists 30
Chemical abstract service (CAS) 3, 104
Chemical accidents 36, 165, 208, 221
Chemical agents 197
Chemical engineers 41, 129–130, 134, 155, 173, 194, 206
Chemical formula 118, 227, 229
Chemical plant 28–29, 57, 59, 155, 221, 233, 262, 265, 306–307, 349
Chemical process quantitative risk assessment (CPQRA) 4, 105
Chemical release 30, 32, 39, 214, 217, 330, 343, 382
Chemophobia 202, 327

391

Chromium 365
Chronic 3, 10, 26–27, 32, 40, 105, 305–306, 328, 362
Circulation 262, 337, 341, 343, 350
Civilization 126, 242, 285–286, 321
Clean water act (CWA) 376–378
Cleanup 39, 133, 220, 304, 379–380, 384
Climate 121, 223, 261, 287, 335–337, 339, 341, 343, 345–347, 351, 354
Climatology 335
Coastal 149, 340, 343, 351, 370, 378
Code of Federal Regulations (CFR) 174, 217, 384
Colloidal 243, 298, 301
Combustion 102, 105, 107–108, 110, 114, 126–127, 129, 138, 145, 169–170, 184, 192–194, 227, 248, 250, 261, 279, 362–364, 366, 368
Community map 208
Completely stirred tank reactor (CSTR) 137, 170–172
Compliance 107, 287, 302, 310, 374, 381–382, 384–385, 389
Componential 155, 159, 180
Composting 38, 355, 380
Condensate 105, 107, 140
Condensation 129, 148, 193, 243, 250, 252, 265–266, 300–301
Condense 102, 198, 262, 290, 300
Condensed 116, 140–141, 150–151, 229, 300
Condenser 140–141, 149, 164, 229, 290
Condensing 126, 300
Conduction 138–139
Conductivity 229
Conductor 228, 372
Conduit 107, 109–110, 112, 143–144, 376
Consequence analysis 3
Consequences 3–4, 6–7, 9, 14–15, 17, 22–23, 30, 32–33, 37, 39–40, 99, 104–105, 116, 165, 169–172, 190, 210, 219, 302, 306–307, 313–314, 328–329
Construction 11, 133, 139, 142–144, 149, 157, 194, 207, 230, 248, 262, 264, 267, 302, 349–350, 352, 354–355, 364, 369, 378–379
Contaminant 24–25, 34–35, 106, 245, 256, 266, 302, 304, 325, 338, 355, 367, 369–373, 375
Contaminated 56–57, 62, 106, 149, 243–244, 248, 256, 280, 283, 311, 365, 367, 369–370, 373, 380
Contamination 118, 132, 144, 260, 274, 289, 301, 311, 328, 330, 370–374, 380–381
Continuous random variable 50–51, 57, 61–63, 74–76
Convection 105, 109, 138–139, 142, 262, 337–338
Coolant 97, 140, 235, 261
Coordinate system 56, 180
Correlation 55–56, 339

Corrosion 28, 105, 144–145, 148–149, 154, 194, 229, 254, 264–265, 306, 365, 371
Corrosive 32, 121, 144, 146, 227
Corrosiveness 152, 267
Corrosivity 208, 371
Council for Environmental Quality (CEQ) 37
Counterterrorism 287
COVID 21, 276
Criticality 123, 328
Cryptosporidium 373–374
Crystal 106, 113, 173–174, 230–231, 238, 288, 300
Crystalline 113, 229, 232
Crystallization 106, 129, 173, 228, 231–232
Crystallizers 274, 277
Cubic model 55
Cumulative distribution function (CDF) 57, 61–63, 75–76, 78, 87, 234, 312, 345
Cyclohexane 170–173, 247
Cyclones 336, 346

D

Damper 104, 106, 143, 145–146
De minimus risk 15, 19
Decision-making process 22, 379
Decomposition 105, 111, 113, 115, 120, 225, 247, 250, 322–323
Defect 16, 25, 43, 64, 333
Defective 51, 64, 71, 268, 286, 293, 332–333
Degradation 12, 132, 241, 260, 263, 287–288, 361, 370
Dehydration 129, 229, 288–289
Dehydrogenation 129, 185, 252
Delphi method 106
Demineralized 150
Demister 106
Demolition 322–323, 350, 364
Deposition 111, 299–300, 362, 364, 378
Derivation 63, 75, 180
Dermal 4, 27, 106
Desalination 109, 142, 235, 310
Desorption 106
Deterioration 105, 241, 263, 340, 365
Detonation 4, 313, 322–324. *See also* Explosion
Deviations 5, 31, 85, 87, 110
Devices 16, 114, 131, 143, 148, 208, 218, 265, 274, 278, 286–287, 289, 302, 310, 342, 354, 366
Dewatering 142
Diffusion 136, 142, 340
Dike 106, 208
Dilution 106, 169, 227, 265–266, 337
Dioxin 233–234, 365
Disaster 13, 30, 202, 207–209, 211, 214, 216, 218, 221, 326–327, 333, 345–346

Index

Discharge 38, 103, 105, 108–110, 112, 145, 147–148, 154, 162, 212, 228, 250, 257, 260–261, 279, 314, 338, 376–377
Discrete random variable 51, 62–63, 76
Disease 4–5, 21, 38, 40, 102, 121, 128, 132, 273–275, 303, 305, 320, 336, 362–364, 368, 370, 373–375
Disinfection 370, 373–375
Dispersion 26, 36, 38, 50–51, 103, 106, 110, 148, 209, 314–315, 320, 337–341, 347
Disposal 38–39, 112, 118, 121, 132–133, 211, 248, 260–261, 278–279, 308, 319, 361, 377–381, 384
Distillation 106–107, 109, 114, 116, 126, 129, 135, 140–141, 178, 197, 199, 234, 239, 241, 249–252, 274, 277
Distributions 16, 52, 61–63, 65, 67, 69–77, 79, 81, 83, 85, 87, 89–93, 95, 97
Domino effects 4
Dose 4–5, 7, 21, 23–26, 28, 34–35, 37, 92, 107, 110, 112, 119, 278, 305, 371–373
Dose-response 23, 25, 28, 34–35, 92, 110, 305
Downwind concentration 245, 283, 315–316. *See also* Gaussian model
Dread 6, 117
Dusts 266, 279, 369

E

Earthquakes 10, 209, 343
Ecological impacts 28
Ecological risk 15
Ecology 133, 353
Economic analysis 12, 186, 226
Economizer 107, 145
Ecosystems 21, 24, 28, 302, 310, 353–354
Effective height 339
Electrolysis 129, 230
Electrostatic precipitators 105
Emergency action 210
Emergency planning and community right-to-know act (EPCRA) 203, 216–218, 221, 382
Emergency response plan (ERP) 326
Emissions 5, 9, 92, 98–99, 105, 109, 111, 116, 138, 151–152, 201, 212, 248, 274, 279, 302, 306, 338–340, 364, 366–367, 385–386, 388
Emitters 373
Endothermic 107
Enforcement 133, 204, 376, 380, 382, 389
Engineering approach 25, 194
Enthalpy 266
Entrainment 104, 107, 338
Entropy 291
Environmental control 133, 255, 304, 387
Environmental damage 22–23, 305
Environmental management 8, 41, 92, 123, 319, 389

Environmental Protection Agency (EPA) 19, 22, 32, 34–35, 37, 40–41, 59, 98, 134, 202–204, 209–210, 214, 216–217, 221, 304, 307–309, 319–320, 325–326, 328, 330, 333, 340–341, 347, 352, 362, 364–367, 369–385, 389
Environmental risk assessment 22, 304
Environmental tobacco smoke (ETS) 367–368
Environmentalism 361
Epidemiology 19, 25, 41, 92, 305, 375
Episodic release 4, 107
Equilibrium 52, 117, 141, 183, 193, 240–241
Equipment design 31, 132, 157
Equipment reliability 4, 245
Estimation 6–7, 17, 26, 32, 37, 79, 108, 116, 118, 121, 347
Ethics 9
Ethyl alcohol 238, 240–241, 253
Ethylene 125, 238, 240, 251, 253, 266, 290, 311
Evacuation 205, 210, 212–215, 219, 283, 327
Evaporation 105, 107, 135, 142–143, 151, 229–231, 265–266, 289, 300, 337
Evaporator 142, 173, 278, 291
Event tree analysis (ETA) 4, 36, 108, 165, 171, 173
Event trees 170, 316, 329
Explosion 4–5, 7, 9–10, 21, 28, 30–32, 99, 106, 108, 116, 119, 128, 141, 170–171, 201, 209, 218, 220, 256, 264, 267, 280–281, 306, 311, 313–314, 317–318, 324. *See also* Detonation
Exponential distribution 76, 78–79, 87–89, 96, 233, 244, 318, 358
Exposure assessment 23–24, 26–28, 34, 305
Exposure period 4, 7, 108
External event 4, 108
Extraction 50, 108, 112, 135, 143, 178, 260–261, 266, 274, 276–279, 356
Extrapolation 25, 35–36, 108

F

Fabrication 31, 125, 139, 242–243
Failure modes 123
Fans and blowers 146
Fatal accident rate (FAR) 4, 37, 56, 108, 139, 177, 261, 285
Fatalities 3–4, 93, 103, 108
Fault tree 4, 36, 103, 108, 119, 165, 173
Fault tree analysis (FTA) 4, 108, 165
Federal aviation administration (FAA) 375
Federal register 4, 108
Fermentation 118, 125, 129, 172, 197, 199, 274, 276–279
Fertilizer 196–197, 218, 227–228, 247, 252, 354, 378
Fireballs 32
Fires 30, 209
First aid 204

Flammability 4, 109
Flammable 211
Flash point 109
Floods 209
Fluidization 109
Formaldehyde 127, 185, 237–238, 241, 253, 289, 355, 369, 385
Formulation 13, 18, 133, 197, 274, 276–278, 280, 338
Fractionation 140
Fuel 105, 110–111, 114, 118, 149, 168–170, 197, 247, 249–251, 254–255, 257, 259–262, 264, 266, 269, 279, 311, 322–323, 351–352, 354, 356, 363–364, 368
Furnace 107, 110, 118, 126, 138, 145, 167, 231–232, 239, 250, 264, 367–368

G

Gases 102, 104, 106–110, 113–114, 117, 142–144, 148, 151–152, 191, 197, 227, 250–251, 266, 300, 324, 338–340, 361–362, 364, 366, 368, 380
Gasoline 111, 172–174, 178, 195, 197, 247, 249–251, 255, 260, 362, 365, 378
Gaussian model 110. *See also* Atmospheric dispersion; Downwind concentration; Plume; Plume rise; Stability class
Generator 39, 215, 279, 306, 380–381
Glycol 238, 253
Government 4, 20, 34, 37, 39, 108, 119, 131, 201, 203, 205, 207, 212, 214, 216–217, 221, 285, 307, 310, 321, 324–328, 340–341, 343, 352, 370, 376, 378, 380, 382, 384
Groundwater 26–27, 302, 310, 354, 370, 372, 374, 376, 378, 381
Guide words 5, 110
Gypsum 126, 197, 230–231

H

Half-life 5, 110
Hazard 4–5, 8–10, 18–23, 25, 28, 30–33, 35–36, 40–41, 61, 87, 98, 101–102, 105–108, 110, 112, 117, 120–121, 123, 154, 165, 170–171, 174, 201–209, 212–213, 215–219, 221, 223, 242, 246, 260, 262, 264, 267, 287, 291, 297–298, 302, 304–306, 308–311, 317–318, 320, 327–330, 332–333, 335, 343–344, 350, 353, 356, 365, 370, 381, 384–385
Hazard and operability study (HAZOP) 5, 30, 110, 306, 329
Hazard risk assessment (HZRA) 4–5, 7, 22, 27, 29–30, 40, 105, 110, 306–307
Hazard risk characterization 32
Hazardous air pollutant (HAP) 309, 362, 365–366
Hazardous chemicals 25, 117, 202–203, 205, 214, 327–328, 376
Hazardous materials 32, 39–40, 108, 204, 220–221, 350, 352, 381–382
Hazardous spills 30
Hazardous waste 38–39, 41, 133, 271, 284, 302, 309–310, 379–381, 388–389
Hazardous waste management 133, 380–381
Hazards survey 201
Health effects 5, 15, 21, 27–28, 32, 102, 110, 112, 119, 205, 319, 362, 364–366, 370–373
Health risk 1, 3, 5, 7–9, 11, 13–15, 17, 19–25, 27, 29, 31, 33–35, 37, 39–41, 43, 51–52, 61, 76, 102, 110, 123, 154, 174, 205, 214, 219–220, 223, 246, 304–306, 317, 320, 333, 365–366, 368, 371, 375
Health risk assessment (HRA) 5, 21, 23–24, 26–27, 34–35, 110
Heat exchangers 135, 138–140, 145, 149, 163, 172–173, 250, 268, 280
Herbicides 365, 378
Homogenous 146
Human error 5, 31, 40, 110, 165
Human factors 5, 110, 129
Human reliability 5, 110
Hurricanes 208–209, 344–345
Hydration 231
Hydrazine 253
Hydrocarbon 111, 172, 219–220, 225, 239, 243, 247–248, 251, 290, 311
Hydrochloric acid 126, 197, 225–230
Hydrodynamics 335, 342
Hydrogenation 129
Hydrology 133, 353
Hydrolysis 129, 198, 299–301
Hygroscopic 106

I

Icebergs 254
Ignitability 208
Ignition 3, 30, 37, 103, 121, 311
Ignition sources 37, 121
Illness 107, 121, 276, 304–305, 328, 368–369, 373
Immediately dangerous to life and health (IDLH) 246, 283
Incident 3–6, 17–18, 32, 37, 61, 103, 105, 107, 111, 115–116, 119, 204, 209, 212, 214, 218, 220, 260, 313, 324–325, 329
Incineration 38, 41, 271, 279, 284, 361, 388
Individual risk 5–6, 9, 15, 17, 35, 111–112, 314
Industrial revolution 286, 297, 302
Infection 362, 368, 370
Infrastructure 132, 324–325, 328–329, 377, 380
Ingestion 5, 27, 111, 372
Inhalation 27, 367, 369
Initiating event 4–6, 107–108, 111, 165

Index

Injury 3, 6–7, 13–15, 19, 21–22, 32, 104, 116, 169–172, 202, 216–217, 308
Insecticides 239, 378
Insulation 144, 151, 227, 232, 241, 354, 356, 366, 369
Intakes 355
Intermediate event 4–5, 107, 111, 165
Inversion 94–95, 107, 338–339, 388. *See also* Stability class
Ionization 105, 299–300
Irradiation 172–173
Irrigation 354–355, 378
Irritant 216, 369
Isobaric 111, 194
Isobutane 251
Isobutyl 240
Isocyanate 216
Isokinetic 111
Isomerization 129, 252
Isooctane 251
Isopleth 5, 111
Isoprene 243
Isotherm 111

K

Kerosene 111, 249–250, 366
Ketone 240, 252–253
Kinetic 136, 147, 181–182, 192–193

L

Lamblia 373
Latency 303
Leaching 111, 143
Legionella 370, 373
Lethal concentration 50 (LC50) 5, 112
Lethal dose 50 (LD50) 5, 112
Liability 9, 39, 128, 201, 309
Lightning 264
Linear equations 55–56
Liquefaction 150
Local emergency planning committee (LEPC) 214, 218
Log-normal distribution 76, 89–92, 97, 235, 245, 346
Lower Explosive Limit (LEL) 5
Lower flammability limit (LFL) 5
Lubrication 248

M

Maintenance 31, 107, 117, 125, 135, 147, 154, 264, 279, 301, 324, 329, 353–355, 366, 368, 370, 375
Malignant 5, 112
Manifold 112, 137, 145–146, 274

Mass transfer 20, 98, 135–137, 140, 174, 177, 193, 246, 257, 284
Mathematical models 26
Maximum contaminant level (MCL) 371, 373
Mercury 256, 302, 310, 323, 342, 352, 355, 365–366
Meteorology 335–336, 338, 343
Methacrylate 240
Methane 185, 240, 251, 266
Methanol 185, 195, 238, 253
Methylstyrene 239
Microbiology 133
Micrometeorology 335
Microorganisms 38, 111, 278, 287–288, 370, 374
Microplastic 361
Mobile sources 365
Model uncertainty 17–18, 35, 39
Modeling 18, 36–37, 39, 101, 133, 137, 338, 340, 343
Momentum 138, 177, 179–180, 182, 338–339
Monomer 113, 241–243
Morbidity 303
Multinomial 65, 69–70, 257, 269

N

Nanochemical 313, 318
Nanofacility 318
Nanomaterial 297–304, 306, 308–311, 313, 316, 319–320
Nanoparticle 94, 297, 299–301, 303, 305–306, 311–313, 318–319
Nanoporous 303
Nanorobots 306
Nanoscale 298–300, 308–309, 319
Nanosensors 302, 310
Nanosized 301, 304
Nanotechnology 198, 223, 297, 299, 301–311, 313, 315, 317, 319–320
Nanotoxicology 320
Nanotubes 298–299, 308
Nanowires 299
Naphthenes 247
National Ambient Air Quality Standards (NAAQS) 362
National Environmental Policy Act (NEPA) 37
National Response Team (NRT) 190
Natural disasters 208, 326
Neutralize 104, 228
Nitrate 174, 218, 225–226, 228–229, 241, 322–324
Nitration 129, 197, 241
Nitric 126–127, 226–229, 364–365
Nitrocellulose 127, 322–324
Nitrocompounds 322
Nitroglycerin 322–324
Noncarcinogenic 25

Nonpoint 376, 378–379
Nonrenewable 247, 262, 310, 350
Nozzle 111, 143, 249
Nucleation 106
Nucleus 102–103, 109, 113
Nutrients 277, 285–286, 291, 376, 378

O

Occupational Safety and Health Administration (OSHA) 6, 35, 114, 117, 219, 304, 307–308, 384
Octane 247, 250–251
Odor 102, 106, 108, 142, 265, 368–369
Offshore 13, 248–249, 254, 260, 340
Oil 13, 111–112, 115, 125–126, 140, 146, 150, 157, 196–197, 231, 247–250, 253–257, 259–262, 267, 269, 278, 311, 323, 343, 353, 366, 378
Optimization 113, 279, 311
Optimum 15, 153–154, 178, 185–186
Osmosis 113, 116
Outbreak 5
Oxidation 126–127, 129, 185, 228, 239–240, 299
Ozone 105, 172, 241, 290, 362–363

P

Packed columns 140, 142
Paleoclimatology 336
Pandemic 5, 8, 19, 21, 61, 276
Particulates 110, 117, 232, 302, 310, 362–364, 388
Pasquill-Gifford model 315. *See also* Gaussian model
Pathogens 302, 310, 373–377
Permissible exposure limit (PEL) 6, 114
Permutation 65–69
Personal protection equipment (PPE) 6, 114, 207, 217–218
Pesticides 129, 237, 354–355, 365–367, 369, 383
Petrochemicals 195, 197, 238, 248, 251–253
Pharmaceuticals 136, 195, 198–199, 223, 273–275, 277–279, 281, 283
Phenol 127, 237–241, 252–253
Phosphate 226–227, 230–231
Photochemical 361–362
PHRMA 273, 275
Phthalate 238
Physiological processes 336
Pipes and tubing 144
Piping and instrumentation diagram (P&ID) 155–157, 159–160
Planning committee 201, 203–204, 207–208, 212, 214
Plant hazard risk assessment 318
Plume 106, 110, 317–318, 338–340. *See also* Gaussian model
Plume rise 338–340. *See also* Gaussian model
Point source 37, 314, 347, 376–378
Poisons 297, 305
Poisson distribution 72–73, 87, 95, 97, 232–233, 269, 295, 332–333, 357–359, 387–388
Pollution Prevention Act 39, 217, 309, 381
Polychlorinated biphenyls (PCBs) 33, 307, 381, 384
Polyethylene 125, 241–242, 253
Polymer 113, 127, 136, 241, 243, 298, 382–383
Polymerization 114, 129, 241–243, 251, 301
Polystyrene 242
Power outage 215
Precipitation 129, 278, 335, 337, 342, 378
Precision 6, 16, 20, 33–34, 114
Preliminary hazard analysis (PHA) 30
Preparedness 115, 203, 214, 221, 326, 328, 333
Prevention 14, 39, 41, 115, 143, 150, 167, 201, 203–204, 206, 208, 210, 213, 217, 273, 299, 303, 309–310, 319–320, 328, 340, 356, 366, 378, 381
Probability density function (PDF) 50–51, 57, 61–65, 71–78, 82–83, 87, 90–91, 97, 157, 221, 233–234, 257, 267, 270, 293, 311, 319–320, 333, 344, 389
Probability distributions 61–63, 65, 67, 69, 71, 73–77, 79, 81, 83, 85, 87, 89, 91–93, 95, 97
Probit model 92
Propagating factors 6
Propylene 239, 250–251, 290
Protozoa 373
Psychrometric 115
Public perception 33
Pumps 45, 70, 92, 114, 143, 147, 151, 161, 164, 255, 264
Pyrolysis 115, 129, 199, 300

Q

Quadratic 52

R

Radiation 37, 109, 114, 116, 139, 209, 230, 232, 261–263, 286–287, 335–336, 342, 362, 372–373
Radioactive elements 102
Radioactive materials 128
Radioactivity 261–262
Radionuclides 365, 370, 372–373, 389
Radon 352, 354–355, 366–368, 372
Random numbers 113
Random variable 50–51, 57, 61–64, 69, 71–72, 74–76, 83–87, 90–91, 97, 234, 293, 311, 344
Reactors 129, 135–138, 140, 160, 164, 170–171, 193, 197, 274, 277, 300
Reboiler 116, 140, 164

Index

Receptor 21, 23, 25–26, 28, 208, 246, 315, 338
Recharge 354
Redundancy system 269
Refractory 116, 148, 197, 230, 232
Refrigeration 116, 148, 150–151, 197, 259, 263, 265–266, 285, 288–290
Regression 51–56, 58–59, 116
Reliability 4–5, 7, 16, 43, 65, 76, 78, 87–89, 95, 110, 117, 120–121, 233, 235, 244–245, 255–256, 267–269, 282–283, 292, 295, 321, 330, 386
Remediation 299, 302–303, 319–320
Reservoir 247–249, 263–265, 373
Resource conservation and recovery act (RCRA) 38, 309, 379–380
Risk analysis 4, 6, 10–11, 15, 20–21, 23, 25, 27, 29, 31, 33, 35, 37, 39, 41, 43, 51, 61, 76, 92, 105, 116
Risk assessment 4–9, 15–16, 20–23, 28, 33–35, 41, 61, 76, 105, 110, 116, 123, 174, 207–208, 246, 297, 304–306, 313, 315, 317–318, 320, 328–329, 333, 335, 350, 385
Risk communication 15, 22
Runoff 38, 118, 120, 208, 280, 354, 378–379

S

Safe Drinking Water Act (SDWA) 326
Safety audits 115, 123
Scale-up 117, 171–172
Security 287, 310, 325–326, 329–330, 333
Sedimentation 118
Segregation 219, 279, 357
Seismic 119, 207–208
Severity 14, 22, 28, 30, 32–33, 117, 306–307, 337, 368
Shock 231, 323
Short term exposure limit (STEL) 6, 119
SO2 emissions 364
Societal risk 6, 9, 15, 118
Soluble 111, 228–229, 238
Solute 118, 143, 300
Solvent 108, 112, 118, 128, 141, 143, 168, 238, 240–241, 274, 278–279, 300–302, 310, 362, 365, 367
Sorbents 303
Source reduction 217, 319, 379
Stability class 313, 315. *See also* Atmospheric dispersion; Gaussian model; Inversion
Stakeholder 204, 325
Standard normal curve 83, 85–86
Standard normal variable 83–85, 96, 233, 270, 294, 312, 358
Steam engine 265
Stoichiometry 177–178, 182, 185, 187
Storage tank 39, 57, 164, 228, 274, 278
Stormwater 38, 280, 376–377

Stratosphere 337
Styrene 237–238, 252–253
Sulfate 125–126, 226–227, 229–230, 364
Sulfonation 129, 239
Sulfuric acid 125–127, 178, 195, 197, 225–227, 239–241, 251, 365
Superfund Amendments and Reauthorization Act (SARA) 203–204, 207, 210, 212, 214, 221, 327
Superheated 118, 263, 288, 291, 300
Sustainability 12, 350, 352–354, 359
Symbol 105, 155, 157–163, 181, 188, 229
Synthesis 118, 127, 237, 243–246, 274, 276–277, 279, 300
Synthetic 113, 120, 127, 195–197, 199, 237–238, 241, 243, 252, 273–275, 291, 356, 375
System checklists 30

T

Tanker 150, 254, 260, 279
Terrorism 195, 198, 223, 321, 323–331, 333
Terrorist attack 10, 326
Thermodynamics 178, 182, 187, 192–194, 271, 295, 300, 335, 342, 347
Thickening 142, 372
Threat 204, 207, 287, 325–326, 329, 354–355, 362
Threshold limit value (TLV) 6–7, 119, 305, 308
Thunderstorms 337
Time to failure 43, 51, 77–79, 82–84, 86, 91, 95, 97, 244, 256–257, 270, 318, 385–386
Time weighted average (TWA) 7, 119
Toluene 252, 302, 310, 365
Top event 4, 108, 119, 165
Tornadoes 208–209, 343
Toxic dose 7, 119
Toxic emissions 138, 386
Toxic materials 27, 121, 146, 266
Toxicity 21, 23–27, 30, 32, 37, 208, 305–306, 308
Toxicology 5, 15, 24–25, 110, 208, 305, 375
Toxins 288, 370
Training programs 154, 212
Transformations 223
Transportation 112, 135, 151, 203–204, 206–210, 214, 218, 220, 226, 229, 248, 252–253, 260, 328, 350, 353, 356–357, 379, 381
Trihalomethanes 375
Trinitrotoluene (TNT) 252, 323
Troposphere 337
Tsunami 10, 119, 209, 343
Turbulence 103, 110, 116, 119, 140, 337–339

U

Uncertainty 7, 13, 17–18, 22–23, 25–26, 28, 33–37, 39–40, 106, 119, 304–305
Utilities 31, 106, 121, 148, 152, 269, 325, 328–329, 362

V

Vaccine 56, 62, 273, 275–276, 286
Valves 6, 44, 115–116, 143, 145–147, 154–155, 157, 160, 218
Vapor 3, 28, 32, 102–110, 116, 140–143, 150–151, 167, 170–173, 192–194, 203, 209, 250, 262, 265–266, 299–300, 306, 327, 362, 367, 369
Vapor pressure 151, 170, 172
Velocity 4, 86, 103, 107, 114, 118–120, 144, 147, 181, 254, 322, 338–340
Venn diagram 45
Ventilation 38–39, 106, 265–266, 354–355, 366–368, 370
Vermiculite 298
Viruses 8, 275, 355, 369, 373–374
Viscosity 120, 147, 231, 247, 267, 301
Visibility 342, 363
Volcanoes 364

W

Waste management 38, 98, 133, 217, 380–381
Wastewater 38, 96–97, 105, 118, 133, 208, 235, 274, 278–280, 325, 376–377, 380
Water systems 37–38, 325–326, 333, 371, 373–375
Weather 36, 198, 223, 249, 313, 315, 335–337, 339, 341–347
Weibull distribution 76–81, 87, 89, 97–98, 245, 313. *See also* Bathtub curve
Wildlife 25–26, 263, 365, 378–379

Y

Y2K 302
Yield 48, 64, 70, 73–74, 76–77, 79, 89, 116, 154, 185–186, 191, 231, 233–234, 239–240, 244, 255–256, 261, 280, 283, 292, 294–295, 301, 312–313, 330, 332, 357–358, 387–388

Z

Zeolite 113, 120, 230, 301
Zirconium 298–299